1983

ELEMENTS OF CALCULUS

with contemporary applications

ELEMENTS OF CALCULUS

with contemporary applications

Marcus M. McWaters
University of South Florida

James H. Reed
University of South Florida

Harcourt Brace Jovanovich, Inc.
New York   Chicago   San Francisco   Atlanta

Cover photo: Macrophoto of shrinking lacquer by Manfred P. Kage from Peter Arnold Photo Archives

ISBN: 0-15-522063-2

Library of Congress Catalog Card Number: 75-37065

Printed in the United States of America

To our children:
    James
    Reneé
    Sharon
    Stephen

Preface

This textbook has been designed to meet the needs of students whose major interests lie outside the physical sciences but who require a working knowledge of calculus. The content is suitable for a one-semester course but is easily adaptable to other time requirements. We assume that the student has studied basic high school algebra, though not necessarily recently. Thus the first two chapters consist of a leisurely and self-contained review of number systems, graphs, and functions (much of which can be skipped by the well-prepared student). Chapters 3–8 deal with limits, continuity, derivatives, and integrals of functions of a single variable. Chapter 9 covers functions of several variables.

We have devoted more than the usual number of pages to examples and exercises. The examples are worked in detail, and the exercises are sufficient in number, level, and variety to satisfy the most demanding student or instructor. As a rule, a model problem is worked in the text for each type of exercise. Since we have leaned toward explanations based on intuitive and geometric appeal rather than on rigorous proof, we have also included many illustrations and graphs.

When introducing the concepts of the calculus we have used examples cast in familiar, nontechnical settings. We believe that the special material of business, medicine, or social science has neither the universal appeal nor the appropriate simplicity for the introduction of the subtle ideas of the calculus. However, numerous applications in each of these areas (and such other areas as ecology, psychology, and biology) are presented with thorough explanations in nonintroductory examples and in the exercises.

Every attempt has been made to minimize the use of technical symbolism and language. Nonetheless, we have included the precise statements of those definitions and theorems warranted in a text written on this level. Where the proof of a theorem is considered accessible to the student, it is presented in the last example of the section in which the theorem first appears. Thus, the individual instructor may introduce, or skip, any given proof without interrupting the flow of the material critical to the section.

Self-tests are given at the end of each of the five parts of the book. Answers to all tests are given at the back of the book, and each answer is keyed to the relevant chapter and section so that the student may reread this material if necessary. The marginal notes should also help the student review. Many of the notes key important definitions and topics for easy reference. The remaining notes clarify fine points that could not be included in the main text discussion without impairing its continuity.

The background of the students and the material selected for emphasis or omission are key factors in determining an appropriate syllabus for a course. For a short course (about 37 sessions) for students with modest mathematical backgrounds, we have found the following sequence to be realistic.

Chapter 1	4 lectures
Chapter 2	4 lectures (Sections 2.1, 2.2, and 2.5)
Chapter 3	omit
Chapter 4	6 lectures
Chapter 5	5 lectures
Chapter 6	8 lectures
Chapter 7	5 lectures
Chapter 8	5 lectures
Chapter 9	omit

Full coverage of the sections omitted in the above syllabus would provide sufficient material for a two-term course. For instructors who wish to include trigonometric or logarithmic functions in the course, Appendices A and B can be covered any time after completion of Chapter 8.

We would like to express our appreciation to Diane Gossett and Sybil French for their patience and tireless efforts in typing the manuscript. We are also indebted to the reviewers, Douglas Crawford of the College of San Mateo, Larry Elbrink of Ohio State University, Theodore Laetsch of the University of Arizona, and Carla Oviatt of Montgomery College, Rockville Campus, Maryland, for their many helpful suggestions on both content and style. Particular thanks are due to the unusually talented staff of Harcourt Brace Jovanovich, Inc.; Marilyn Davis, Judy Burke, and Gail Lemkowitz were especially helpful.

Marcus M. McWaters

James H. Reed

Contents

PART I

**Functions
and
Graphs**

PART II
Limits and Continuity

PART III
The Derivative and its Applications

PART IV
The Integral and its Applications

PART V
Multivariable Calculus

Functions and Graphs

PART I

Elementary Concepts

1.1. Introduction

Before we begin our discussion of the basic concepts of the calculus, a brief review of some of the essential ideas of precalculus mathematics will be profitable. This is the intent of the present chapter.

1.2. Numbers

If we are asked to pick a number between 1 and 5, we might choose any of the numbers 1, 2, 3, 4, or 5; or, if we are suspicious of the word "between," we might limit ourselves to choosing 2, 3, or 4; but it is most unlikely that we would pick $3\frac{1}{2}$ or π, even though these are also numbers between 1 and 5. On the other hand, if we are asked to guess the length of a recently caught fish and are told that it was between 1 and 5 ft long, $3\frac{1}{2}$ ft is not so unnatural a guess; but a choice of π ft is still quite unlikely. However, if we remember our high school geometry and want to find the circumference of a circle of diameter 1 ft, we would know that the correct answer is π ft. So it appears that we naturally make distinctions between types of numbers, although we may never think about such distinctions and the grounds on which we make them may be somewhat obscure.

The numbers that we will use in this book can all be represented as lengths (more precisely, as *directed* lengths, that is, as lengths in some

Some intuitive distinctions between types of numbers

The number π is an important mathematical constant. Its approximate value is 3.14159.

The formula for the circumference c of a circle of diameter d is
$c = \pi d$.

The interpretation of numbers as lengths

Any convenient length may be chosen as a unit length; for example, we might choose the length •——• as 1 unit.

Choose any point on a line and label it 0: —————•—————
                                                        0

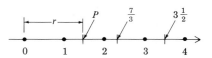

FIGURE 1.2-1   The choice of the unit length to be used on the line is arbitrary. It is determined by the choice of the two points on the line to be labeled 0 and 1.

The counting numbers are the numbers 1, 2, 3, 4, 5, 6, 7, ...

direction). At first we will restrict our attention to *positive numbers.* If we first decide on a *unit length* (say we decide to let 1 unit equal 1 ft), then the number $\frac{1}{2}$ is represented by a length of $\frac{1}{2}$ ft, 3 is represented by a length of 3 ft, and $\frac{5}{2}$ is represented by a length of $2\frac{1}{2}$ ft. A more systematic representation of positive numbers as lengths is obtained by picking a point on a horizontal line and labeling it 0. Then label each point P to the right of 0 with the number r which is the distance of the point P from 0 in terms of the chosen unit length (see Fig. 1.2-1). Thus the point 1 unit to the right of 0 is labeled 1, the point $3\frac{1}{2}$ units to the right of 0 is labeled $3\frac{1}{2}$ (or $\frac{7}{2}$), and so on.

If a scale were set up on a horizontal line as described above, and if a rabbit that jumps exactly 1 unit length at each jump were to start jumping at 0 and continue jumping indefinitely to the right, the points on which the rabbit would land would be labeled 1, 2, 3, 4, 5, 6, 7, . . . (the three dots mean "and so on"). These numbers are called the *counting numbers.* Since the rabbit can always jump 1 unit further to the right than it has already jumped, it is clear that there are infinitely many distinct counting numbers (see rabbit #2 in Fig. 1.2-2).

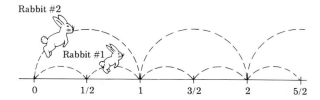

FIGURE 1.2-2

Suppose now that a second rabbit that always jumps exactly $\frac{1}{2}$ a unit length begins jumping at 0 and continues jumping indefinitely to the right (see rabbit #1 in Fig. 1.2-2). Then the points on which it would land would be numbered $\frac{1}{2}$, 1, $\frac{3}{2}$, 2, $\frac{5}{2}$, 3, $\frac{7}{2}$, 4, ... In order to locate the point on the line corresponding to the number $\frac{16}{7}$, we could imagine a rabbit that jumps $\frac{1}{7}$ of a unit length at each jump starting at 0 and jumping 16 times to the right (see Fig. 1.2-3). More generally, to locate

If we break the unit length into seven equal parts, then any of these parts represents a length of $\frac{1}{7}$.

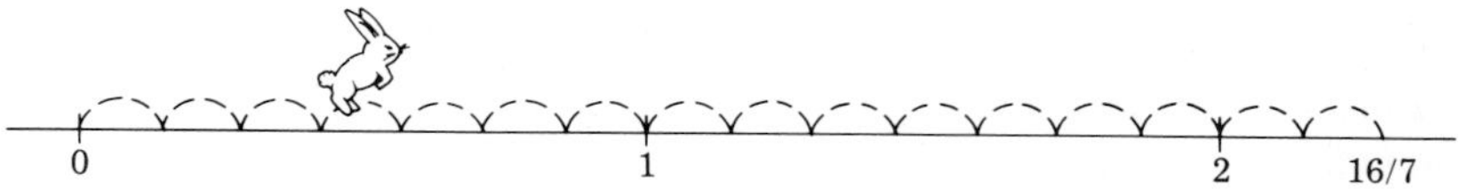

FIGURE 1.2-3

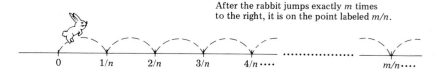

the point on the line corresponding to a fraction m/n, where m and n are both counting numbers, we imagine a rabbit that always jumps $1/n$ of a unit length starting at 0 and jumping m times to the right. The point on which it lands is then labeled m/n (see Fig. 1.2-4).

After the rabbit jumps exactly m times to the right, it is on the point labeled m/n.

FIGURE 1.2-4

Even after using all the numbers of the form m/n as labels for points on the line to the right of 0, it turns out that many points are still unlabeled. The numbers used to label the remaining points are called the *positive irrational numbers*, while those of the form m/n (as just described) are called the *positive rational numbers*. The positive rational numbers include the counting numbers, since any counting number m can be written in the form $m/1$ and consequently is of the form m/n with $n = 1$.

An example of an irrational number is $\sqrt{2}$, the positive number which when multiplied by itself yields 2. To demonstrate that there is a length which corresponds to $\sqrt{2}$, we need a little geometry. We construct a right triangle with legs of unit length and use the theorem of Pythagoras to conclude that the length of the hypotenuse is $\sqrt{1^2 + 1^2}$ $= \sqrt{2}$. We then measure off the length of the hypotenuse to the right of 0 on the line to find the point corresponding to $\sqrt{2}$ (see Fig. 1.2-5). To show that the number $\sqrt{2}$ is irrational (i.e., that it is not equal to the quotient of two counting numbers) requires additional effort, and we ask the student to accept this fact on faith for the time being.

The positive rational and irrational numbers

Any counting number m is also a rational number, since $m = m/1$. For example, $5 = 5/1$ and $113 = 113/1$.

$\sqrt{2}$ is an irrational number. It is the positive number with the property that $\sqrt{2} \cdot \sqrt{2} = 2$.

The theorem of Pythagoras states that given a right triangle with legs of length a and b, the length of the hypotenuse is $\sqrt{a^2 + b^2}$.

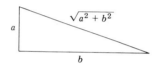

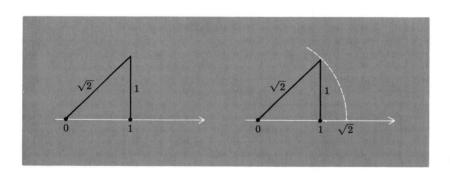

FIGURE 1.2-5

Let us suppose that all the points to the right of 0 on the line have now been labeled with appropriate positive numbers, so that each point now corresponds to the positive number (whether rational or irrational) that measures its distance from 0. The *negative numbers* are assigned to the points to the left of 0 as follows. If a point P is to the left of 0, and if x is the positive number located on the line to the right of 0 and the same distance from 0 as P, then label P with the number $-x$ (see Fig. 1.2-6). For example, the point 1 unit to the left of 0 is labeled -1, the point $2\frac{1}{2}$ units to the left of 0 is labeled $-2\frac{1}{2}$, the point $\sqrt{2}$ units to the left of 0 is labeled $-\sqrt{2}$, and so on.

Locating the points on a line that correspond to negative numbers

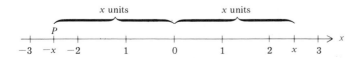

FIGURE 1.2-6

The real numbers are those numbers that correspond to the points on a line.

We have now associated a number with each point on the line. These numbers are called *real numbers*, and they are the numbers that we will be using throughout this book. Whenever we choose a unit length and a zero point and assign real numbers to the points on a line in terms of the unit length and the zero point, we say that we have *calibrated* the line.

The calibration of a line

If x is a real number, then x corresponds to a particular point P on the line, and the distance from P to 0 is called the *absolute value* of x, denoted $|x|$. For example, $|2| = 2$, $|-1| = 1$, $|-3\frac{1}{2}| = 3\frac{1}{2}$, and $|-\sqrt{2}| = \sqrt{2}$. In general, for any real number x, we see that $|x| = x$ if x is positive or 0, and $|x| = -x$ if x is negative. Thus $|x|$ is always positive or 0 (see Fig. 1.2-7).

The absolute value of a real number

FIGURE 1.2-7

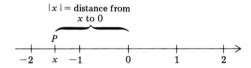

Given a positive number z, there are exactly two real numbers whose absolute value is z, namely, z and $-z$. For example, if $z = 3$, we have $|-3| = |3| = 3$.

Given two distinct real numbers x and y, we say that x is *less than* y if the point on a calibrated line corresponding to x lies to the left of the point on the line corresponding to y. If x is less than y, we write

Definition of $x < y$

$$x < y$$

(the symbol $<$ means "less than") or

$$y > x$$

(the symbol $>$ means "greater than"). If x is less than or equal to y, we write

Definition of $x \leqslant y$

$$x \leqslant y$$

or

$$y \geqslant x$$

For example, $2 > 0$, $-1 < 3$, $-1 \leqslant -1$, $2 \geqslant 2$, and $-4 < -2$ (see Fig. 1.2-8). It is often useful to know that if $x < y$ and $y < z$, then $x < z$.

FIGURE 1.2-8 Since -4 lies to the left of -2, we have $-4 < -2$. If x is to the left of y and y is to the left of z, then x is to the left of z; thus if $x < y$ and $y < z$, we can conclude that $x < z$.

The integers, Z, consist of the numbers $0, \pm 1, \pm 2, \pm 3, \ldots$.

It will be useful to specify certain kinds of real numbers. The counting numbers and their negatives, together with 0, are called the *integers*. We denote the collection of integers by the letter Z; thus Z consists of the numbers $0, \pm 1, \pm 2, \pm 3, \ldots$.

The positive rational numbers and their negatives, together with 0, are called the *rational numbers* (or fractions), and this collection of numbers will be denoted by Q. Thus Q consists of all numbers of the form m/n, where m and n are integers and $n \neq 0$.

The real numbers corresponding to points on a line which are not labeled with rational numbers are called *irrational numbers*, and this collection of numbers will be denoted by J.

To describe the irrational numbers more completely, we ask the student to accept the fact that every real number can be written in decimal notation. For example

$$\tfrac{1}{2} = 0.5$$

$$\tfrac{1}{4} = 0.25$$

$$\tfrac{1}{3} = 0.333 \ldots$$

$$\pi = 3.14159265 \ldots$$

and so on. Some real numbers have *repeating* decimal representations; by this we mean that there is a digit or block of digits that is repeated indefinitely in the decimal number. For example $\tfrac{1}{3}$ has the decimal representation $0.33333 \ldots$; the digit 3 is repeated indefinitely. When a digit or block of digits repeats itself in a decimal, we indicate this by placing a bar over the digit or block of digits that is repeated. Thus we would write $\tfrac{1}{3} = 0.\overline{3} = 0.333 \ldots$, and $\tfrac{5}{33} = 0.\overline{15} = 0.15151515 \ldots$. If a decimal has no digit or block of digits that repeats itself indefinitely, it is called a *nonrepeating* decimal. The decimal representation for π is nonrepeating; when we write $\pi = 3.14159265 \ldots$, the three dots indicate that there are more digits which follow the last digit given (5), but there is no block of digits that repeats itself indefinitely in the decimal representation for π.

The distinction between rational and irrational real numbers can now be made in terms of their decimal representations: a real number is irrational if it has a nonrepeating decimal expansion, and it is rational if it has a repeating decimal expansion. (Note that a finite decimal, such as 0.5, is repeating, since $0.5 = 0.5\overline{0}$ with the 0 repeated indefinitely.)

The rational numbers, Q, are those numbers that can be written as the quotient of two integers, m/n, where $n \neq 0$.

The irrational numbers, J

Some examples of irrational numbers are $\pm \sqrt{2}$, $\pm \sqrt{3}$, $\pm \pi$, $\pm (6\sqrt{2} + 1)$.

Decimal notation

The three dots at the end of a decimal number do not mean that the last digit given is repeated; they only indicate the existence of additional nonzero digits.

Repeating decimal representations

A bar above a digit or block of digits indicates that the digit or block of digits is repeated indefinitely. For example, $3.2\overline{14} = 3.21414141414$.

A decimal number with no repeating digit or block of digits is called a nonrepeating decimal.

A real number is an irrational number if it has a nonrepeating decimal representation.

A real number is a rational number if it has a repeating decimal representation.

A finite decimal is a repeating decimal.

EXAMPLE 1.2-1

Find the decimal representation of the number $\tfrac{3}{7}$.

SOLUTION

Since $\tfrac{3}{7}$ is a rational number, its decimal representation will be repeating. We simply divide 3 by 7 to find the decimal.

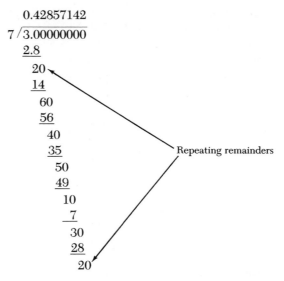

and we see that the remainders and consequently the quotient are beginning to repeat. Therefore, $\frac{3}{7} = 0.\overline{428571}$.

EXAMPLE 1.2-2

Find a rational number whose decimal representation is $3.\overline{12}$.

SOLUTION

Let $x = 3.\overline{12}$. Then

$$100x = 312.\overline{12}$$

(remember that $3.\overline{12} = 3.1212121\overline{2}$). Thus

$$100x - x = 312.\overline{12} - 3.\overline{12} = 309$$

and we have

$$99x = 309$$

Solving for x, we find

$$x = \tfrac{309}{99} \quad \text{or} \quad x = \tfrac{103}{33}$$

Multiplication of the decimal representation of a number by 10^n has the effect of shifting the decimal point n places to the right. Thus multiplication by 10 shifts the decimal point one place to the right; by 100, two places to the right; by 1000, three places, and so on.

In Example 1.2-2, notice that two digits are repeated in $x = 3.\overline{12}$, and we multiplied x by $10^2 = 100$ to produce a repeating decimal, $100x$, *which matched x exactly in each position to the right of the decimal point.* Thus subtracting x from $100x$ produced an *integer*, so division by 99 resulted in a *rational* expression for x. The trick in this type of problem is to produce two multiples of x which match identically to the right of the decimal point, subtract one multiple of x from the other to produce an integer, and then divide to obtain a rational expression for x.

EXAMPLE 1.2-3

Find a rational number whose decimal representation is $.01\overline{413}$.

SOLUTION

Let $x = .01\overline{413}$; then

$$10^5 x = 1413.\overline{413} \quad \text{and} \quad 10^2 x = 1.\overline{413}$$

Thus

$$10^5 x - 10^2 x = 1413.\overline{413} - 1.\overline{413} = 1413 - 1 = 1412$$

Then we have

$$100{,}000x - 100x = 1412$$

or

$$(100{,}000 - 100)x = 1412$$

or

$$99{,}900x = 1412$$

so

$$x = \frac{1412}{99{,}900}$$

EXAMPLE 1.2-4

Given that π is an irrational number, explain why 100π is also irrational.

> ## SOLUTION
>
> Since π is irrational, it has a nonrepeating decimal representation. But 100π has the same decimal representation with the decimal point shifted two places to the right. Thus the decimal representation for 100π is also nonrepeating, and hence 100π is irrational.

It may be well to point out here that the numbers that come out of calculators and those used in everyday business transactions are all rational numbers. An irrational number may appear in a formula or as the solution of an equation, but before being used in a business transaction or computation, it is converted to a rational number sufficiently close in size for the purposes at hand. For example, π might be approximated by $\frac{22}{7}$ or $\frac{314}{100}$, and $\sqrt{2}$ might be changed to $1.41 = \frac{141}{100}$; but these rational numbers are only convenient approximations of π and $\sqrt{2}$.

Exercises 1.2

In Exercises 1 through 10, calibrate a line using an appropriate unit of length and locate on the line the approximate position of the given numbers. Then find the smallest number and the largest number of each group by inspecting each diagram.

1. $0, 5, -3, -5$, and 6
2. $0, 2, -1, -2$, and 4
3. $0, 4, -3\frac{1}{2}, 2\frac{1}{4}$, and -2
4. $0, 6, -4\frac{1}{4}, 2\frac{1}{2}$, and -3
5. $0, 3, -2.5, 1.25$, and -3.25
6. $0, -4, -2.25, 1.75$, and 2.75
7. $0, 2, -2, -1.83$, and 0.61
8. $0, 2, -2, -0.02$, and 1.17
9. $0, 4, -4, \pi, -2\pi$, and $2\pi + 1$
10. $0, 3, -5, \pi - 1$, and $-(\pi + 1)$

In Exercises 11 through 16, locate the given numbers on a calibrated line and find a third number which lies strictly between the given numbers.

11. 2.1 and 2.01
12. -1.3 and -1.03
13. 0.02 and 0.17
14. -0.12 and -0.01
15. $\pi - 3$ and 0.12
16. -2π and -6.23

In Exercises 17 through 22, find the number which has the smallest absolute value.

17. -2, $5\frac{1}{2}$, 0, and 3
18. -1, $-3\frac{1}{2}$, 2, and 0.5
19. 2π, $-\pi$, 3, and -5.5
20. 2.7, -1.3, -0.5, and 3.1
21. -1, -1.5, -2, and -2.5
22. -0.5, -2.5, -0.15, and -0.1
23. Show that for any real number x, $(|x|)^2 = x^2$
24. If x and y are positive numbers and $x < y$, show (by means of their location on a line) that $|x| < |y|$.
25. If x and y are negative numbers and $x < y$, show (by means of their location on a line) that $|x| > |y|$.
26. Show that $|x - y| = |y - x|$, where x and y are any two real numbers. (*Hint*: $x - y = -(y - x)$.)

In Exercises 27 through 38, express the given rational number as a repeating decimal.

27. $\frac{1}{2}$
28. $\frac{1}{4}$
29. $\frac{3}{8}$
30. $\frac{5}{8}$
31. $\frac{3}{7}$
32. $\frac{5}{7}$
33. $\frac{7}{11}$
34. $\frac{7}{13}$
35. $1\frac{6}{7}$
36. $1\frac{4}{13}$
37. $-2\frac{1}{11}$
38. $-2\frac{5}{13}$

In Exercises 39 through 48, find the rational number whose decimal representation is the given repeating decimal.

39. $1.4\overline{9}$
40. $0.\overline{9}$
41. $27.\overline{5}$
42. $13.\overline{3}$
43. $17.3\overline{1}$
44. $52.\overline{76}$
45. $326.1\overline{73}$
46. $427.8\overline{29}$
47. $-18.13\overline{27}$
48. $-22.74\overline{37}$

In Exercises 49 through 56, determine whether the given number is rational or irrational.

49. $513.214587\overline{256}$
50. $-4798.217859\overline{7964}$
51. $\pi + 2.431\overline{0}$
52. $\pi - 47.2468\overline{0}$
53. 10π
54. $10\pi - 7.1\overline{0}$
55. 2π
56. $n\pi$, where n is any counting number.
57. Construct an irrational number that lies between 1.213 and 1.214.
58. Construct an irrational number that lies between -2.426 and -2.427.

59. Is the sum of two irrational numbers always irrational? Give examples.
60. Is the sum of two rational numbers always rational? Why or why not?
61. Show that the product of a nonzero rational number and an irrational number is always irrational.

1.3   Cartesian Planes

We may envision a *plane* as the surface of a flat sheet of glass which extends indefinitely in all directions. If we choose a point on the glass and paint a horizontal and a vertical line through it, we have a pair of perpendicular lines in the plane. Let us pick the point of intersection of the two lines as the 0 point on each line and calibrate each line with respect to some unit length (the two unit lengths used in calibrating the two lines do not have to be the same). We agree to locate positive numbers to the right of 0 on the horizontal line and above 0 on the vertical line (see Fig. 1.3-1). The horizontal line is usually called the *x axis*, the vertical line is usually called the *y axis*, and the point of intersection of the two lines is called the *origin*. A pair of calibrated lines constructed like this in a plane is called a *coordinate system* for the plane, and any plane in which such a coordinate system has been constructed is called a *Cartesian plane*. The coordinate system in a Cartesian plane divides the plane into four regions, called *quadrants*, which are numbered as shown in Fig. 1.3-2.

Definition of a coordinate system for a plane

A Cartesian plane is any plane with a coordinate system.

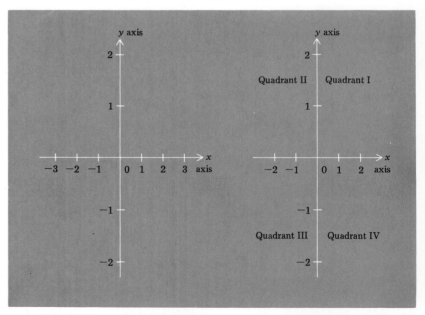

FIGURE 1.3-1 The Cartesian plane. Both lines extend indefinitely in each direction.

FIGURE 1.3-2

A coordinate system gives us a way to locate and label any point in a Cartesian plane. Let us first consider a point P_0 in the plane which does not lie on either axis (see Fig. 1.3-3). We draw two lines through P_0, one parallel to the x axis and the other parallel to the y axis. The vertical line intersects the x axis at a point which corresponds to some number x_0 on the (previously calibrated) x axis, and the horizontal line intersects the y axis at a point which corresponds to some number y_0 on the (previously calibrated) y axis. The numbers x_0 and y_0 can be thought of as instructions for finding the point P_0, starting from the origin. We first go to the point x_0 on the x axis. Then we move vertically $|y_0|$ units, upward if y_0 is positive and downward if y_0 is negative. This brings us to P_0. Thus the two numbers x_0 and y_0 locate the point P_0 in the plane, and we write $P_0 = (x_0, y_0)$, calling x_0 the *first coordinate* (or *x coordinate*, or *abscissa*) of P_0 and y_0 the *second coordinate* (or *y coordinate*, or *ordinate*) of P_0. If P_0 is on the x axis, then $y_0 = 0$ (the line through P_0 parallel to the x axis is the x axis, which intersects the y axis at $y_0 = 0$), and thus $P_0 = (x_0, 0)$. Similarly, if P_0 is on the y axis, then $x_0 = 0$ and $P_0 = (0, y_0)$. It is convenient, when no confusion is likely to result, to denote a point $(x_0, 0)$ on the x axis by the single number x_0; similarly, a point $(0, y_0)$ on the y axis may be denoted by y_0 (see Fig. 1.3-4).

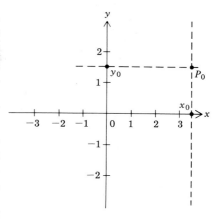

FIGURE 1.3-3

If x_0 is the first coordinate of P_0 and y_0 is the second coordinate of P_0, we write $P_0 = (x_0, y_0)$.

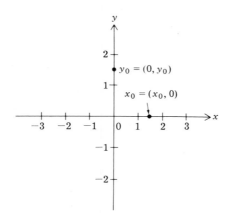

FIGURE 1.3-4

EXAMPLE 1.3-1

Locate the point $P_0 = (-2, 3)$ in a Cartesian plane.

SOLUTION

We construct a coordinate system and then, starting at the origin, move left 2 units to -2, and up $|3| = 3$ units (see Fig. 1.3-5) to arrive at P_0.

FIGURE 1.3-5

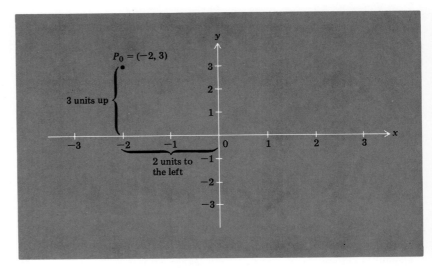

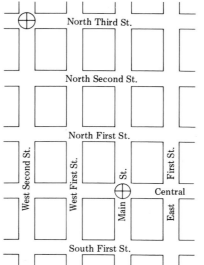

FIGURE 1.3-6

The idea of using coordinates to describe the position of a point in a Cartesian plane is similar to the way in which we might locate a particular corner in a city with rectangular street patterns, such as the corner of West Second Street and North Third Street (see Fig. 1.3-6). To reach this corner from Central and Main (which are like the x axis and y axis), we go 2 blocks west and 3 blocks north. Similarly, to say that a point has coordinates $(-2, 3)$ means that to reach it, we start at the origin and go 2 units in the direction of the negative side of the x axis and then 3 units in the direction of the positive side of the y axis (as in Example 1.3-1).

A convenient method of specifying the coordinates of several points in a Cartesian plane is to express the coordinates in tabular form. The following table gives the coordinates of five points, P_1, P_2, P_3, P_4, and P_5.

	P_1	P_2	P_3	P_4	P_5
x	1	-1	2	7	0
y	3	4	-5	0	-3

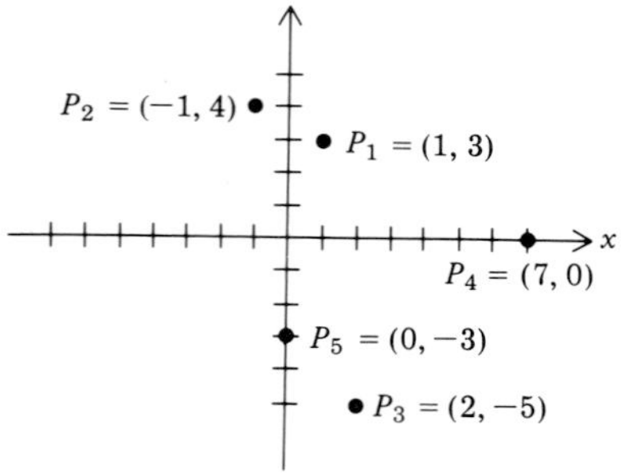

To read the coordinates of P_4 from this table, we look at the two numbers in the column directly below P_4. The number 7 is in the row of the table labeled x and is the x coordinate; similarly, the number 0 is the y coordinate of P_4. Thus, from the table we have $P_1 = (1, 3)$, $P_2 = (-1, 4)$, $P_3 = (2, -5)$, $P_4 = (7, 0)$, and $P_5 = (0, -3)$. The location of these points in a Cartesian plane is shown in Fig. 1.3-7.

FIGURE 1.3-7

EXAMPLE 1.3-2

A manufacturing company finds that its production cost per unit depends on the number of units produced per day as shown in the following table.

	P_1	P_2	P_3	P_4	P_5
Number of units produced	40	45	50	55	60
Cost per unit (dollars)	15	14	13	17	16

Locate the points P_1, P_2, P_3, P_4, and P_5 in a Cartesian plane, using a suitably calibrated coordinate system, with values on the x axis representing the number of units produced per day and values on the y axis representing the production cost per unit. Assuming that P_1 through P_5 represent the only production possibilities, how many units per day should the company produce in order to minimize the production cost per unit?

SOLUTION

We construct a coordinate system as shown in Fig. 1.3-8. Note that sections of the x and y axes near the origin are omitted (as is indicated by the broken lines), since the values on the x axis (number of units produced) that we are interested in are all between 40 and 60, and the important values on the y axis are between 13 and 17. Now, from the table, the coordinates of the points are $P_1 = (40, 15)$, $P_2 = (45, 14)$, $P_3 = (50, 13)$, $P_4 = (55, 17)$, and $P_5 = (60, 16)$. The location of these points on the coordinate system is as shown. The point $P_3 = (50, 13)$ is the "lowest" point, the one whose y coordinate (cost) is the smallest, so the company should produce 50 units per day (the x coordinate of P_3) to minimize the cost per unit.

FIGURE 1.3-8

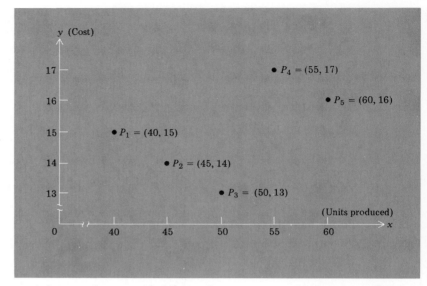

FIGURE 1.3-9 The distance between two points on a calibrated line
is defined to be $|x - y|$, where x and y
are the numbers corresponding to the
two points.

Since $|x_1 - x_2| = |x_2 - x_1|$, we
also have $|P_1P_2| = |x_2 - x_1|$.

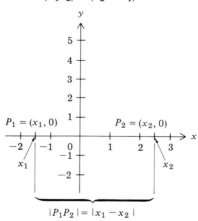

$$|P_1P_2| = |x_1 - x_2|$$

FIGURE 1.3-10

If the x and y axes of the coordinate system in a Cartesian plane are
calibrated with the same unit length, we can develop a formula that will
give us the (straight-line) distance between any two points P_1 and P_2 of
the plane in terms of their coordinates. We begin by defining the
distance between any two points x and y on a calibrated line to be
$|x - y|$ (see Fig. 1.3-9). Thus if our two points P_1 and P_2 in the plane
both lie on the x axis, so that $P_1 = (x_1, 0)$ and $P_2 = (x_2, 0)$ for some
numbers x_1 and x_2, then the distance between P_1 and P_2, which we
denote by $|P_1P_2|$, is given by

$$|P_1P_2| = |x_1 - x_2|$$

(see Fig. 1.3-10). For example, if $P_1 = (-2, 0)$ and $P_2 = (5, 0)$, then
$|P_1P_2| = |(-2) - 5| = |-7| = 7$ units (see Fig. 1.3-11). Similarly, if the
points P_1 and P_2 are both on the y axis, so that $P_1 = (0, y_1)$ and

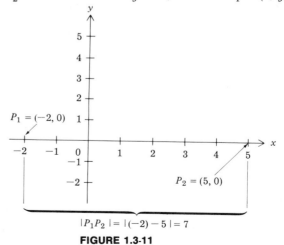

$$|P_1P_2| = |(-2) - 5| = 7$$

FIGURE 1.3-11

$P_2 = (0, y_2)$, then

$$|P_1 P_2| = |y_1 - y_2|$$

Now let $P_1 = (x_1, y_1)$ and $P_2 = (x_2, y_2)$ be two points in the plane, not lying on a common horizontal or vertical line. For simplicity we put P_1 and P_2 in the first quadrant, but the proof does not depend on this. To obtain the distance between P_1 and P_2, we construct a right triangle as shown in Fig. 1.3-12, by drawing a horizontal and a vertical line through the points P_1 and P_2 and then joining P_1 and P_2 with a straight line segment. Then the hypotenuse of this triangle has length $|P_1 P_2|$, and the lengths of the legs of the triangle are

$$d_1 = |x_1 - x_2|$$

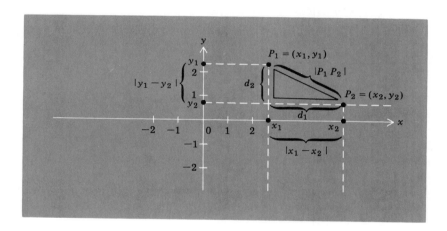

FIGURE 1.3-12

and

$$d_2 = |y_1 - y_2|$$

Using the Pythagorean theorem, we have

$$|P_1 P_2| = \sqrt{d_1^2 + d_2^2} = \sqrt{|x_1 - x_2|^2 + |y_1 - y_2|^2}$$

and since $|x_1 - x_2|^2 = (x_1 - x_2)^2$ and $|y_1 - y_2|^2 = (y_1 - y_2)^2$ (see Exercise 1.2-23), we see that

$$|P_1 P_2| = \sqrt{(x_1 - x_2)^2 + (y_1 - y_2)^2}$$

Although our derivation of this formula was dependent upon P_1 and P_2 not lying on a common horizontal or vertical line, the formula is valid for computing the distance between *any* two points in the plane.

If $P_1 = (x_1, y_1)$ and $P_2 = (x_2, y_2)$, then the distance between P_1 and P_2 is given by the formula

$$|P_1 P_2| = \sqrt{(x_1 - x_2)^2 + (y_1 - y_2)^2}$$

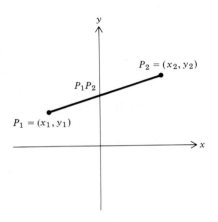

FIGURE 1.3-13   The line segment P_1P_2 determined by two points in a plane.

Definition of the slope of a line segment

If $x_1 = x_2$, the segment P_1P_2 is vertical and the slope does not exist. Vertical segments are the only segments for which slope is undefined.

If $y_1 = y_2$, the segment P_1P_2 is horizontal and the slope of the segment is 0.

EXAMPLE 1.3-3

Find the distance between the points $P_1 = (-1, 2)$ and $P_2 = (3, -4)$.

SOLUTION

We use the distance formula

$$|P_1P_2| = \sqrt{(x_1 - x_2)^2 + (y_1 - y_2)^2}$$

with $x_1 = -1$, $x_2 = 3$, $y_1 = 2$, and $y_2 = -4$. Thus we have

$$|P_1P_2| = \sqrt{(-1 - 3)^2 + (2 - (-4))^2} = \sqrt{16 + 36} = \sqrt{52}$$

units.

When we are given two distinct points $P_1 = (x_1, y_1)$ and $P_2 = (x_2, y_2)$ in a Cartesian plane, they determine a unique (straight) line segment, namely the line segment which begins at one of the points and ends at the other point (see Fig. 1.3-13). We denote this line segment by P_1P_2 (or P_2P_1, since it is the same line segment no matter which point we name first), and we define the *slope* of the line segment P_1P_2 to be the number

$$\frac{y_2 - y_1}{x_2 - x_1}$$

provided that $x_2 \neq x_1$ (if $x_1 = x_2$, then $x_2 - x_1 = 0$ and the number $(y_2 - y_1)/(x_2 - x_1)$ does not exist). Note that

$$\frac{y_2 - y_1}{x_2 - x_1} = \frac{y_1 - y_2}{x_1 - x_2}$$

so when the slope of the line segment joining two points is computed, it makes no difference which point is labeled P_1.

EXAMPLE 1.3-4

Compute the slope of the line segment joining the two points $(-1, 3)$ and $(4, -2)$.

SOLUTION

If we take P_1 to be $(-1, 3)$ and P_2 to be $(4, -2)$, then $x_1 = -1$, $x_2 = 4$, $y_1 = 3$, and $y_2 = -2$. Thus the slope is

$$\frac{y_2 - y_1}{x_2 - x_1} = \frac{-2 - 3}{4 - (-1)} = \frac{-5}{5} = -1$$

If we had taken $P_1 = (4, -2)$ and $P_2 = (-1, 3)$, then we would have had $x_1 = 4$, $x_2 = -1$, $y_1 = -2$, and $y_2 = 3$, and the slope again would have been

$$\frac{y_2 - y_1}{x_2 - x_1} = \frac{3 - (-2)}{-1 - 4} = \frac{5}{-5} = -1$$

(see Fig. 1.3-14).

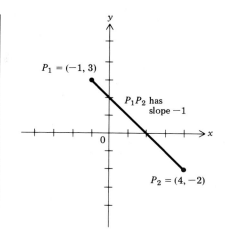

FIGURE 1.3-14

If $P_1 = (x_1, y_1)$ and $P_2 = (x_2, y_2)$ are points such that $x_1 = x_2$, then the line segment P_1P_2 is vertical, and, as mentioned above, the slope of P_1P_2 is undefined. Vertical segments are the only segments which do not have a slope. It can be shown that if the line segment P_1P_2 rises as we proceed from left to right, the slope is positive; if P_1P_2 falls as we go from left to right, the slope is negative; and horizontal line segments have slope 0. Thus the slope of a line segment is a measure of the "direction" of the segment in the plane (see Fig. 1.3-15).

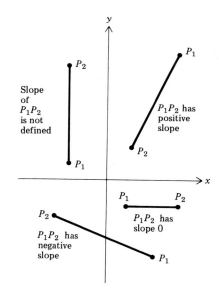

FIGURE 1.3-15

Exercises 1.3

In Exercises 1 through 6, construct a coordinate system and locate the given points in the Cartesian plane.

1. $P_1 = (1, 3)$, $P_2 = (-1, -3)$,
   $P_3 = (1, -3)$, and $P_4 = (-1, 3)$
2. $P_1 = (0, -2)$, $P_2 = (1, 0)$,
   $P_3 = (-1, 4)$, and $P_4 = (2, 3)$
3. $P_1 = (0.5, 2)$, $P_2 = (-1, 2.5)$,
   $P_3 = (4, -1.5)$, and $P_4 = (-1.5, 0.5)$
4. $P_1 = (-3, 2.5)$, $P_2 = (0, 1.5)$,
   $P_3 = (-0.5, -0.5)$, and $P_4 = (2.25, 1)$
5. $P_1 = (\frac{1}{4}, -1)$, $P_2 = (-3, \frac{1}{2})$,
   $P_3 = (0, 1.25)$, and $P_4 = (-2\frac{1}{2}, \frac{1}{2})$
6. $P_1 = (1, 2)$, $P_2 = (2\frac{1}{2}, 5)$,
   $P_3 = (-2, -4)$, and $P_4 = (-3\frac{1}{4}, -6\frac{1}{2})$

In Exercises 7 through 10, the coordinates of several points in a Cartesian plane are given in tabular form (as in Example 1.3-2). Construct a coordinate system and locate the points in the Cartesian plane.

7.

	P_1	P_2	P_3	P_4	P_5
x	0	-1	4	-3	2
y	2	3	2	0	1

8.

	P_1	P_2	P_3	P_4	P_5
x	2	-1	3	1	0
y	1	-2	2	0	-3

9.

	P_1	P_2	P_3	P_4	P_5
x	1.5	2.5	3	-1	-1.5
y	-1	2	1.5	-2	3

10.

	P_1	P_2	P_3	P_4	P_5
x	3.5	3	-1	0	2
y	1	2	0	1.5	-4

In Exercises 11 through 20, construct a coordinate system, calibrating each axis with the same unit length (this is the method usually employed), and locate the given points P_1 and P_2 in the Cartesian plane. Sketch the line segment P_1P_2, compute its slope (if it exists), and find the distance between the points P_1 and P_2.

11. $P_1 = (4, 6)$, $P_2 = (1, 2)$
12. $P_1 = (-4, 2)$, $P_2 = (1, 2)$
13. $P_1 = (-1, -1)$, $P_2 = (2, -3)$
14. $P_1 = (0, 3)$, $P_2 = (-3, 0)$
15. $P_1 = (2, -1)$, $P_2 = (2, 3)$
16. $P_1 = (1, -5)$, $P_2 = (3, -4)$
17. $P_1 = (3, -2)$, $P_2 = (-7, -6)$

18. $P_1 = (-4, 3)$, $P_2 = (2, -3)$
19. $P_1 = (-7, 2)$, $P_2 = (-7, -3)$
20. $P_1 = (2, -5)$, $P_2 = (-2, -5)$
21. (a) A triangle is said to be *isosceles* provided that two of its legs are of equal length. Show that the triangle whose vertices are the points $P_1 = (-2, 4)$, $P_2 = (-5, 1)$, and $P_3 = (-6, 5)$ is an isosceles triangle.

    (b) A triangle is said to be *equilateral* provided that all three of its legs are of equal length. Show that the triangle whose vertices are the points $P_1 = (-3, 0)$, $P_2 = (-1, 0)$, and $P_3 = (-2, \sqrt{3})$ is an equilateral triangle.

22. In a study of work and efficiency, assembly-line workers performing the same task of inserting cotter pins in connecting rod bolts were observed, and the number of units of work they performed per hour was measured for each hour of the working day. The results of the study were displayed on the coordinate system shown in Fig. 1.3-16.

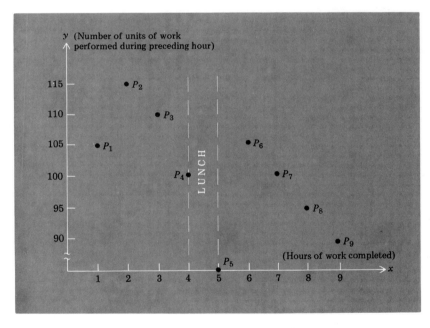

FIGURE 1.3-16

How many units of work were performed during the first hour of the day? During the first hour after lunch? What was the most productive hour of the day? What are the coordinates of the point P_6? Why does the point P_5 have y coordinate 0?

23. A hospital checks a patient's pulse rate at 15-min intervals. The data are displayed on the coordinate system in Fig. 1.3-17. When was the patient's pulse rate the highest? What was the pulse rate 45 min after admission? What are the coordinates of the point P_4? What was the pulse rate at the time of admission?

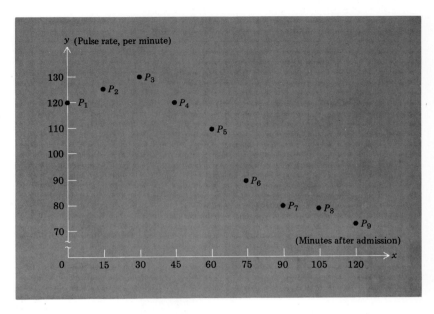

FIGURE 1.3-17

24. The total sales revenue generated by a product depends upon the selling price of the product. A company finds that for one of its products the yearly total sales revenue corresponds to the selling price of the product as shown in the following table.

	P_1	P_2	P_3	P_4	P_5
Selling price of product (dollars)	20	25	30	35	40
Total sales revenue (thousands of dollars)	36	39	42	43	38

Locate the points P_1, P_2, P_3, P_4, and P_5 in a Cartesian plane, using a suitably calibrated coordinate system, with values on the x axis representing the selling price of the product and values on the y axis representing revenue (in thousands of dollars). What price for the product will generate the most revenue? Which of the points P_1, P_2, P_3, P_4, and P_5 has this optimum price for its x coordinate? Why is the location of this point in the Cartesian plane "higher" than the other points? (See Example 1.3-2.)

25. In an experiment on learning, a rat is placed in a small box equipped with a device which presents the rat with a pellet of food whenever it depresses a lever. The following table gives the number of responses (depressions of the lever) measured against the total time the rat has been in the box.

In Exercise 25, the slope of the segment P_1P_2 is (by the definition of slope) equal to $(2 - 1)/(1 - 0.5) = 2$. The numerator is the number of responses that occurred during the second half-hour that the rat was in the box, and the denominator is the length of time during which these responses occurred, namely, 1/2 hr. Thus the slope of P_1P_2 is interpreted to mean that during the second half-hour, the rat was responding at the *rate* of $(2 - 1)$ responses per $(1 - 0.5)$ units of time (1 response per half-hour) or 2 responses per hour. Similarly, the slope of P_2P_3 will be the number of responses per half-hour during the third half-hour, etc. Thus when the slopes of these segments begin to increase, the rat has begun to learn.

	P_1	P_2	P_3	P_4	P_5
Time elapsed (hours)	0.5	1	1.5	2	2.5
Number of responses since placed in box	1	2	3	11	62

Locate the points P_1, P_2, P_3, P_4, and P_5 in a Cartesian plane, using a suitably calibrated coordinate system, with values on the x axis representing time elapsed in hours and values on the y axis representing the number of responses since entering the box. How long was the rat in the box before showing signs of "learning" how to get food? *Hint*: compute the slopes of the segments P_1P_2, P_2P_3, P_3P_4, and P_4P_5. When did the slopes of the segments begin to increase? (See marginal note.)

26. A company that manufactures tape players purchases small electric motors for use in the manufacture of its product. The cost per unit for the motors depends on the quantity of motors purchased as shown in the following table.

	P_1	P_2	P_3	P_4	P_5
Quantity purchased (thousands)	1	2	3	4	5
Cost per unit (dollars)	12	11	10	10	10

Locate the points P_1, P_2, P_3, P_4, and P_5 in a Cartesian plane, using a suitably calibrated coordinate system, with values on the x axis representing the number of motors purchased (in thousands) and values on the y axis representing the cost per unit. What is the smallest number of motors the company can buy and still pay the most favorable price per unit? Compute the slopes of the line segments P_1P_2, P_2P_3, P_3P_4, and P_4P_5. The first of these segments to have slope 0 is P_3P_4. What is the significance of this in terms of the cost per unit? (See marginal note.)

In Exercise 26, the slope of the segment P_2P_3 is (by definition) $(10 - 11)/(3 - 2) = -1$. The numerator is the change in the cost per unit, and the denominator is the change in the quantity of motors purchased. Thus the slope of P_2P_3 measures the *rate* at which the cost per unit is changing as the quantity purchased changes from 2000 to 3000. Therefore, since the slope of P_2P_3 is -1, the cost per unit is decreasing (as seen from the negative sign) at the rate of $1 per 1000 units purchased.

27. A hospital records a patient's temperature every 12 hr. The following table summarizes the patient's temperature over a 60-hr period.

	P_1	P_2	P_3	P_4	P_5	P_6
Time elapsed (hours) since admission	0	12	24	36	48	60
Temperature (°F)	103	104	105	104	101	98.6

In Exercise 27, since the slope of the segment

$$P_1P_2 = \frac{104 - 103}{12 - 0} = \frac{1}{12},$$

the segment P_1P_2 is rising (since the slope is positive), and thus the patient's temperature is increasing at the rate of 1° per 12 hr during the first 12-hr period. The first segment which has negative slope will indicate that the patient's temperature is falling and that he or she is beginning to recover.

Locate the points P_1, P_2, P_3, P_4, P_5, and P_6 in a Cartesian plane, using a suitably calibrated coordinate system, with values on the x axis representing time since admission (in 12-hr intervals) and values on the y axis representing the patient's temperature. Compute the slopes of the segments P_1P_2, P_2P_3, P_3P_4, P_4P_5, and P_5P_6. When did the patient begin to recover? (*Hint*: When do the slopes of the segments P_1P_2, P_2P_3, etc., change from positive to negative?) (See marginal note.)

1.4.   Functions

It is commonplace to observe that the values of certain quantities depend on other quantities. Such observations lead quite naturally to the idea of a *function*, as we shall soon see. We have all encountered tables which relate the value of one quantity to that of another, as in the following examples.

Tables provide a convenient way of displaying relationships between quantities.

EXAMPLE 1.4-1

We might display the correspondence between the years 1940 and 1945 and the number of traffic fatalities on Labor Day of each year as shown in Table 1.4-1.

TABLE 1.4-1

Year	1940	1941	1942	1943	1944	1945
Number of fatalities	615	537	712	749	802	611

EXAMPLE 1.4-2

The average number of pedestrians crossing a given intersection between 8:00 AM and 5:00 PM depends upon the day of the week as shown in Table 1.4-2 (the first day of the week is Sunday).

TABLE 1.4-2

Day of week	1	2	3	4	5	6	7
Number of pedestrians	142	817	845	845	814	875	416

EXAMPLE 1.4-3

The Paris Paint Company projects that the number of gallons of its premium quality paint that it will sell during the coming year depends upon the price per gallon as shown in Table 1.4-3.

TABLE 1.4-3

Price per gallon (dollars)	2.67	2.68	2.69	2.70	2.71
Number of gallons sold (thousands)	250	247	240	212	200

EXAMPLE 1.4-4

Table 1.4-4 displays the readings on a thermometer as the thermometer is brought closer to a flame.

TABLE 1.4-4

Distance from flame (inches)	3	2	1.75	1.50	1.25	1
Temperature (°F)	90	91	93	97	102	108

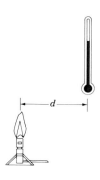

d = distance of thermometer from flame.

Each of these tables gives a correspondence between the numbers in the top row and the numbers in the bottom row. From Table 1.4-1 we see that the number (of traffic fatalities) corresponding to the number 1942 (year) is 712. From Table 1.4-2, the number corresponding to 1 (Sunday) is 142 (pedestrians). Similarly, in Table 1.4-3 we see that 200,000 corresponds to 2.71, and in Table 1.4-4 the number 97 corresponds to 1.50. Mathematicians would say that the quantity named in the second row of each table *is a function of* the quantity named in the first row. Thus, based on Table 1.4-3, the number of gallons of paint sold *is a function of* the price per gallon of the paint. We may interpret the phrase *is a function of* to mean *depends upon*. Thus the income tax that a person pays is a function of income, and the rental fee for a car charged by an auto rental agency is a function of the make of the car.

The phrase *is a function of* means *depends upon.*

Explicit rules, or formulas, are often used to specify the dependence of one quantity upon another.

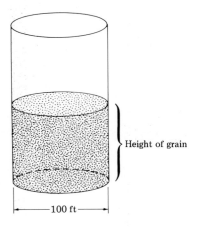

Height of grain

100 ft

FIGURE 1.4-1

Any sample values may be used for the remaining entries in the top row of the table.

Tables are not the only means at our disposal to indicate that the value of one quantity is a function of (depends upon) the value of another. Sometimes a rule or formula is used. For example, the quantity of grain (in cubic feet) contained in a cylindrical storage bin 100 ft in diameter can be determined once the height (in feet) of the grain is known, by the rule

$$\text{volume of grain} = \pi(50)^2 \cdot \text{height of grain}$$

(see Fig. 1.4-1). Thus if the height of the grain is 1 ft, then the volume of grain is 2500π cubic ft, or approximatey 7853 cubic ft (using 3.141 as an approximation of π). The volume of grain *is a function of* the height of the grain.

Here is another example of the use of a rule or formula to describe the dependence of one quantity on the value of another. The cost of manufacturing a product usually depends upon the number of units produced. A particular firm may find that the total cost C (in dollars) of producing x units is given by the rule $C = \frac{1}{3}x + 50$. Thus if 3000 units are manufactured, then $C = \frac{1}{3}(3000) + 50 = \$1050$, and we see that C *is a function of* (depends upon) x.

Each of the above formulas serves the same purpose as the tables presented earlier. In fact, the formulas may be written in tabular form as shown in Tables 1.4-5 and 1.4-6. Note that these tables differ slightly from Tables 1.4-1 through 1.4-4, in that one of the entries in the top row of each of these tables is a *letter* rather than a number. This is because the choices for the entries in the top row of each of these tables are unlimited, that is, any positive number or 0 could be taken as the height of the grain (Table 1.4-5), and any counting number could be taken as the number of items produced (Table 1.4-6). Thus we cannot enter *all* possible values in the top rows of these tables; instead, we use a letter as the first entry and then use some sample values of that letter for the

TABLE 1.4-5

Height of grain h (feet), any real number $h \geqslant 0$	h	1	1.5	7
Volume of grain (cubic feet)	$\pi(50)^2 h$	$\pi(50)^2$	$\pi(50)^2(1.5)$	$\pi(50)^2(7)$

TABLE 1.4-6

Number x of items produced, x a positive integer	x	1000	2500	3000
Total cost	$\frac{1}{3}x + 50$	$\frac{1}{3}(1000) + 50$	$\frac{1}{3}(2500) + 50$	$\frac{1}{3}(3000) + 50$

remaining entries in the top row of each table. Letters used in this way to represent any one selection from a (possibly) unlimited supply of values are called *variables*.

A close examination of the six tables presented thus far will show that each number from a given set of numbers (the entries for the top row) is coupled with a second number (the corresponding number in the bottom row). This is precisely what we adopt as the definition of a function for the remainder of this text. A *function* is a rule which assigns to each number in a given set of numbers, called the *domain* of the function, a unique second number. Tables such as Tables 1.4-1 through 1.4-4 are to be considered functions where the entries in the top row constitute the domain. Tables such as 1.4-5 and 1.4-6 are considered functions whose domain is described in the top left column. For each of these tables the rule is to assign to each number in the top row the number appearing directly below it. The domain of the function given by Table 1.4-2 is the set of numbers 1, 2, 3, 4, 5, 6, and 7; the domain of the function given by Table 1.4-5 consists of all real numbers $h \geqslant 0$.

If we have occasion to refer to a particular function very often, the tabular representation of the function is cumbersome. It is much more convenient to refer to a function like the one given in Table 1.4-6 by writing

$$C = f(x), \quad \text{where} \quad f(x) = \tfrac{1}{3}x + 50$$

Here, the symbol $f(x)$ is read "f of x." Thus, "$C = f(x)$" tells us that C (the cost of production) is a function of x (the number of units sold), and the *equation* $f(x) = \tfrac{1}{3}x + 50$ tells us *how* the cost $C = f(x)$ depends upon x. Care must be taken to remember that the symbol $f(x)$ does not mean f multiplied by x; it stands for *the value of f at x*. In the example just cited, the value of $C = f(x)$ when $x = 3000$ would be denoted by $f(3000)$ (read "f of 3000") and computed as

$$f(3000) = \tfrac{1}{3}(3000) + 50$$

Similarly, the value of $C = f(x)$ when $x = 1200$ is

$$f(1200) = \tfrac{1}{3}(1200) + 50$$

In fact, whenever we want the value of the function $C = f(x)$ at a particular positive integer a, we simply replace x by a in the equation $f(x) = \tfrac{1}{3}x + 50$ to get

$$C = f(a) = \tfrac{1}{3}a + 50$$

All functions could be set up in tabular form in much the same way as we have done until now, but the notation just introduced will be well worth the initial inconvenience of becoming accustomed to it. One additional convention should be noted: when a function is given by an equation and the domain is not specified, we intend the domain to consist of all real numbers for which the computations can be carried out (that this, for which there is a real number that is the answer to the computations). For example, if the function $f(x) = x^2$ is given, it is understood that the domain consists of all real numbers, since any real number can be squared. On the other hand, if the function $f(x) = \sqrt{x}$ is given, the domain consists only of nonnegative real numbers, since it is impossible to find a real number that is the square root of any negative number.

Determining the domain of a function given by an equation

Two important facts to keep in mind when determining the domain of a function given by an equation are that (1) division by 0 is impossible (i.e., undefined), and (2) it is impossible to find a real number that is the square root (or any even root) of any negative number. Thus the domain of the function

$$f(x) = \frac{3x}{2x - 1}$$

consists of all real numbers x for which $2x - 1 \neq 0$, i.e., all real numbers except $\frac{1}{2}$. The domain of the function

$$f(x) = \sqrt{x - 3}$$

consists of all real numbers x for which $x - 3$ is greater than or equal to 0, i.e., all real numbers greater than or equal to 3.

Writing a function f in tabular form

When writing a function f in tabular form, it is useful (1) to indicate the domain of the function in the top row of the left (first) column, (2) to use the top and bottom row of the second column to exhibit the rule (or equation) for determining $f(x)$, where x represents any number chosen from the domain, and (3) to use the remaining columns of the table to display the values $f(x)$ of the function f for some particular numbers x *arbitrarily* chosen from the domain. Tabular forms for the functions $f(x) = 3x/(2x - 1)$ and $f(x) = \sqrt{x - 3}$ are shown in Tables 1.4-7 and 1.4-8.

TABLE 1.4-7

All real numbers x, except $x = \frac{1}{2}$	x	0	1	$\frac{5}{3}$	-1
$f(x)$	$\dfrac{3x}{2x - 1}$	0	3	$\frac{15}{7}$	1

TABLE 1.4-8

All real numbers $x \geqslant 3$	x	3	$\frac{13}{4}$	7
$f(x)$	$\sqrt{x - 3}$	0	$\frac{1}{2}$	2

It is worth remarking that g, h, and other letters are commonly used to denote functions as well as f, and that y, z, and other letters are frequently used in place of x. The function $f(x) = x^2$ could be written equally well as $f(y) = y^2$, $f(z) = z^2$, $g(x) = x^2$, $g(y) = y^2$, $h(z) = z^2$, and so on. All these expressions have the same meaning.

EXAMPLE 1.4-5

Table 1.4-9 defines a function h. Fill in the missing entries in the table.

TABLE 1.4-9

	y	2	-1	10	0
$h(y)$	$\dfrac{2}{y-1}$				

SOLUTION

Since division by 0 is impossible, we see that y cannot be equal to 1; hence the domain of the function h consists of all real numbers y such that $y \neq 1$. This information is entered in the top row of the first column. From the table we see that

$$h(y) = \frac{2}{y-1}$$

hence when $y = 2$ we have $h(2) = 2/(2-1) = \frac{2}{1} = 2$, and this number is entered in the bottom row of the table under 2. Similarly, $h(-1) = 2/(-1-1) = -1$; $h(10) = 2/(10-1) = \frac{2}{9}$, and $h(0) = 2/(0-1) = -2$. These numbers are entered in the table under the entries -1, 10, and 0. The completed table is Table 1.4-10.

TABLE 1.4-10

All real numbers y, $y \neq 1$	y	2	-1	10	0
$h(y)$	$\dfrac{2}{y-1}$	2	-1	$\frac{2}{9}$	-2

EXAMPLE 1.4-6

Find an equation that defines a function f whose values $f(x)$ for selected values of x appear in Table 1.4-11. Compute the domain of f, as determined by the equation. Insert this information in the appropriate spaces of the table.

TABLE 1.4-11

	x	0	-2	3	4	5
$f(x)$		2	0	5	6	7

SOLUTION

We see that each number in the bottom row of Table 1.4-11 is the number directly above it increased by 2. Thus an equation of a function f which has the desired value at each value of x is $f(x) = x + 2$. This expression is inserted in the table in the bottom row of the second column. The domain of f consists of all real numbers x, and this information is inserted in the top row of the first column. The completed table is Table 1.4-12.

The function f found in Example 1.4-6 is only one of many equally acceptable functions. Another is $f(x) = (x^2 - 4)/(x - 2)$; but this is certainly more complicated.

TABLE 1.4-12

All real numbers x	x	0	-2	3	4	5
$f(x)$	$x + 2$	2	0	5	6	7

EXAMPLE 1.4-7

An appliance store estimates that the number of units N of its economy model blender that it can sell per week is given by the equation

$$N(x) = -\tfrac{1}{2}x + 50$$

A function N which expresses the demand for a certain commodity in terms of the selling price x (and perhaps some additional variables, such as television advertising time, special promotions, etc.) is called a *demand function*.

where x is the selling price per unit (in dollars). The equation is considered valid for values of x between \$20 and \$40 inclusive. Express the function N in tabular form, and include in the table the values of $N(x)$ when x assumes the values \$20, \$26, \$34, and \$38. How many blenders will be sold per week if they are priced at \$29.50 each?

SOLUTION

Since it is given that $N(x) = -\frac{1}{2}x + 50$ is valid for x between $20 and $40, the domain consists of all real numbers between 20 and 40 inclusive (in fact, only those values of x between $20 and $40 that can be interpreted in standard coinage are meaningful, but we can ignore this for our purposes). From the equation $N(x) = -\frac{1}{2}x + 50$, we have $N(20) = -\frac{1}{2}(20) + 50$ $= 40$. Similarly, $N(26) = -\frac{1}{2}(26) + 50 = 37$, $N(34)$ $= -\frac{1}{2}(34) + 50 = 33$, and $N(38) = -\frac{1}{2}(38) + 50 = 31$. The representation of N in tabular form is shown in Table 1.4-13. If $x = \$29.50$ or $29\frac{1}{2}$, then $N(29\frac{1}{2}) = -\frac{1}{2}(29\frac{1}{2}) + 50 \doteq 35\frac{1}{4}$. Thus $35\frac{1}{4}$ blenders will be sold per week if they are priced at $29.50 each. Since it is impossible to sell $35\frac{1}{4}$ blenders, we interpret this to mean that approximately 35 blenders would be sold each week, or more accurately, that $4(35\frac{1}{4}) = 141$ blenders would be sold every four weeks.

TABLE 1.4-13

All real numbers x between 20 and 40	x	20	26	34	38
$N(x)$	$-\frac{1}{2}x + 50$	40	37	33	31

EXAMPLE 1.4-8

A farmer has 400 ft of fencing. He decides to make a rectangular garden and enclose it with the fencing to protect it. Assuming that he uses all the fencing, find the equation which expresses the area A of the garden in terms of the length x of one side of the rectangle. Then write the function A in tabular form, and include in the table the values of $A(x)$ when x is 80, 90, 100, and 110 ft. Which of these values of x should be used in order to have a garden of maximum area?

FIGURE 1.4-2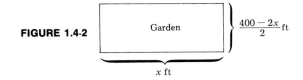

SOLUTION

Consider the diagram in Fig. 1.4-2. Clearly, if the length of the garden is x ft, then a total of $2x$ ft of fencing will be used for two opposite sides, leaving $400 - 2x$ ft of fencing to be used for the remaining two opposite sides. Thus if the garden is x ft long, then it will be $(400 - 2x)/2$ ft wide; and the area $A(x)$ (length times width) will be equal to

$$x \cdot \left(\frac{400 - 2x}{2} \right)$$

Thus the equation which expresses A as a function of x is

$$A(x) = x \cdot \left(\frac{400 - 2x}{2} \right)$$

and when $x = 80$, we have $A(80) = 80(400 - 160)/2 = 9600$ square ft. Similarly, $A(90) = 90(400 - 180)/2 = 9900$ square ft; $A(100) = 100(400 - 200)/2 = 10{,}000$ square ft; and $A(110) = 110(400 - 220)/2 = 9900$ square ft. The domain of A consists of the possible values of x which might be used as the length of one side of the garden, or all real numbers x which are greater than 0 and less than 400. Thus the representation of $A(x) = x(400 - 2x)/2$ in tabular form is as shown in Table 1.4-14. We see from the table that of the specific values of x given, the one which makes the area of the garden the largest is $x = 100$ ft (note that when $x = 100$ ft the garden is square).

TABLE 1.4-14

All real x strictly between 0 and 400	x	80	90	100	110
$A(x)$	$x\left(\dfrac{400 - 2x}{2} \right)$	9600	9900	10000	9900

Exercises 1.4

In each of Exercises 1 through 15, a function is defined in tabular form.
Fill in the missing entries in each table.

1.

All real numbers x	x	0	-1	5	-3	2
$f(x)$	$2x$	0				

2.

All real numbers x	x	2	-2	5	0	1
$f(x)$	$3x - 1$	5				

3.

	x	-1	3	2	0	-3
$f(x)$	$2x + 5$					

4.

	x	-2	4	6	0	-8
$f(x)$	$\frac{1}{2}x + 10$					

5.

	x	0	-5	5	-10	15
$f(x)$	$\frac{1}{5}x - 5$					

6.

	x	1	0	3	-2	-1
$h(x)$	x^2	1				

7.

	x	-1	3	0	2	-7
$p(x)$	$x^2 + 4$	5				

8.

	x	0	-2	1	3	-4
$s(x)$	$x^2 + x - 1$					

9.

All real numbers x which are $\geqslant 0$	x	4	0	9	25	100
$t(x)$	$\sqrt{x}$					

10.

	x	5	10	1	17	26
$n(x)$	$\sqrt{x-1} + 2$					

11.

All real numbers x except $x = -1$	x	0	-4	2	5	-7
$r(x)$	$\dfrac{x}{x+1}$					

12.

	t	1	97	33	6	13
$f(t)$	$\dfrac{t}{\sqrt{t+3}}$					

13.

	s	0	3	-3	10	-5
$r(s)$	$\dfrac{s}{s^2 + 2}$					

14.

	z	0	-3	7	-5	-11
$s(z)$	$\sqrt{z^2}$					

15.

	x	0	25	9	4	16
$q(x)$	$(\sqrt{x})^2$					

In each of Exercises 16 through 25, find an equation which defines a function f whose values $f(x)$ for selected values of x appear in the given table. Compute the domain of f, as determined by the equation. Insert this information in the appropriate spaces of the table.

16.

	x	0	-1	1	2	-3
$f(x)$		0	-1	1	2	-3

17.

	x	0	1	2	3	-4
$f(x)$		0	2	4	6	-8

18.

	x	0	-2	2	3	-4
$f(x)$		1	-1	3	4	-3

19.

	x	0	2	4	-4	1
$f(x)$		-2	0	2	-6	-1

20.

	x	1	4	6	-3	-10
$f(x)$		1	$\frac{1}{4}$	$\frac{1}{6}$	$-\frac{1}{3}$	$-\frac{1}{10}$

21.

	x	0	3	2	-2	-4
$f(x)$		0	9	4	4	16

22.

	x	0	4	9	16	25
$f(x)$		0	2	3	4	5

23.

	x	1	2	3	-1	-2
$f(x)$		3	5	7	-1	-3

24.

	x	1	2	3	-1	-2
$f(x)$		1	3	5	-3	-5

25.

	x	1	2	3	-2	-5
$f(x)$		1	$\frac{1}{4}$	$\frac{1}{9}$	$\frac{1}{4}$	$\frac{1}{25}$

Note in Exercises 26 through 35 that the domains of the functions are determined by the context of each problem (see Examples 1.4-7 and 1.4-8).

26. A piece of industrial machinery which cost $2500 new is depreciated for tax purposes according to the function $v(t) = 2500 - 400t$, where v gives the taxable value of the machinery and t is the age of the machinery in years. The equation defining v as a function of t is considered valid for values of t between 0 and 6 inclusive. Express the function v in tabular form, and include in the table the values of $v(t)$ when t assumes the values 1, 2, 3, and 5. What is the taxable value of the machinery when it is $2\frac{1}{2}$ years old?

27. A company estimates that its total sales revenue S, in dollars per week, is given by the equation $S(x) = 3000 + 100x$, where x is the number of 30-sec television advertisements per week. The equation is considered valid for values of x between 0 and 100 inclusive. Express the function S in tabular form and include in the table the values of $S(x)$ when x assumes the values 10, 50, 75, and 100. What will the estimated sales revenue be per week if the company decides not to advertise on television at all?

28. A small lake is contaminated with coliform bacteria due to abnormally high amounts of rainfall which have caused septic tanks in nearby dwellings to malfunction. The lake is treated with bactericidal agents, and ecologists estimate that the equation $N(t) = 100(10 - t)^2$ will give the number of viable bacteria N per milliliter of water in terms of the time t elapsed after treatment (where t is measured in days). The equation is considered valid for

values of t between 0 and 9 inclusive. Express the function N in tabular form and include in the table the values of $N(t)$ when t assumes the values 0, 4, 6, and 8. If a coliform count of 100 per milliliter or less is considered safe for swimming, how many days after treatment must elapse before the lake is safe?

29. A large mass of polluted air from a northern industrial city is observed spreading south according to the equation $d(t) = 9t^2 + 10t$, where d is the distance from the center of the city in miles and t is time measured in days. Predictions are such that the equation is considered valid for values of t between 2 and 10 days inclusive. Express the function d in tabular form and include in the table the values of $d(t)$ when t assumes the values 2, 4, 5, 7, and 9 days. If Day $t = 1$ is Tuesday, when will the polluted air mass reach a resort village located 656 mi south of the city?

30. The Ace Electric Tool Company estimates that the demand function for its economy model $\frac{1}{4}$-in. electric hand drill is $N(x) = 35 - 2x$, where N is the number of thousands of drills that can be sold per year and x is the selling price of the drill in dollars. The equation is considered valid for values of x between \$9 and \$13 inclusive. Express the function N in tabular form and include in the table the values of $N(x)$ when x assumes the values 9, 10, 11, 12, and 13. Which selling price (in whole dollars) will produce the largest gross profit for the company? (Gross profit is the number of items sold times the selling price per item.)

31. A large metropolitan hospital is investigating the relationship between weight and heart disease in women. By observing the female cardiac patients admitted to the hospital, it is found that the average number N of patients weighing x lbs admitted per month is approximated closely by the equation $N(x) = 40 - \frac{1}{25}(x - 140)^2$. The equation is considered valid for integer values of x between 110 and 160 lbs, inclusive. Express the function N in tabular form and include in the table the values of $N(x)$ when x assumes the values 110, 120, 130, 140, 150, and 160. For which of these values of x is the value of $N(x)$ the largest? How many patients weighing 145 lbs are admitted per month?

32. A manufacturing company estimates that the cost of producing a certain item will be \$5 per unit for labor and \$3 per unit for material. Find the equation which expresses the cost C of manufacturing the items in terms of the number x of units produced. Then write the function C in tabular form, and include in the table the values of $C(x)$ when x is 100, 500, 1000, and 10,000 units. How much will it cost the company to manufacture 350 units?

33. As contracted, a sales representative receives a weekly salary of \$125 plus a commission of one-tenth of the weekly sales made. Find the equation that expresses the sales representative's weekly salary S in terms of the amount x of the weekly sales (x measured in dollars).

Then write the function S in tabular form, and include in the table the values of $S(x)$ when x is 300, 450, 600, and 750. During a particularly good week, the sales representative sold $1240 worth of merchandise. What was the sales representative's salary for that week?

34. The checking accounts at a certain bank are subject to a charge of $1 per month plus $.10 for each check written. Find the equation that expresses the monthly charge C for the account in terms of the number x of checks written. Then write the function C in tabular form, and include in the table the values of $C(x)$ when x is 5, 10, 20, and 25. What is the charge on an account for a month in which 17 checks are written?

35. Scientists studying air pollution in a particular city find that the average value of the pollution index at 10:00 AM is 40 parts per million (ppm) and that the index increases steadily by 10 units during each succeeding hour until 4:00 PM. Find the equation that expresses the pollution index P in terms of the number of hours t elapsed since 10:00 AM ($t = 0$ at 10:00 AM, $t = 4$ at 2:00 PM, etc.). Then write the function P in tabular form and include in the table the values of $P(t)$ when t is 1, 3, 4, and 5. What is the pollution index at 2:30 PM?

Some Basic Techniques

2.1. Graphs of Functions

Now that we have discussed functions and how they arise in various contexts, we are ready to introduce the geometric representation of a function, its *graph*. By the *graph* of a function f we mean all those points in a Cartesian plane whose coordinates are of the form $(x, f(x))$ for some x in the domain of f. For functions whose domain consists of a finite collection of numbers, the graph is a finite set of points in a Cartesian plane. Such functions can be expressed by a table in which all the numbers in the domain are included in the top row. The graph then consists of all those points in a Cartesian plane that have their first coordinate equal to some number in the top row of the table and their second coordinate equal to the number directly below.

Definition of the graph of a function

If the domain of a function consists of a finite collection of numbers, the graph of the function is a finite set of points in a Cartesian plane.

EXAMPLE 2.1-1

Sketch the graph of the function given by the following table.

Sketching the graph of a function whose domain consists of a finite number of values

$x = -2, 0, 1, 2, 3$	x	-2	0	1	2	3
$f(x)$	$2x + 1$	-3	1	3	5	7

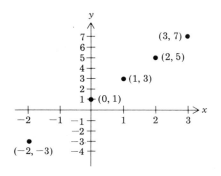

FIGURE 2.1-1

SOLUTION

The domain of the function is specified by the table as the numbers -2, 0, 1, 2, and 3. The point with coordinates $(-2, -3)$ is on the graph of the function, since -3 appears directly below -2 in the table. Similarly, the points with coordinates $(0, 1)$, $(1, 3)$, $(2, 5)$, and $(3, 7)$ are on the graph. To sketch the graph, we construct a Cartesian coordinate system and locate these five points in the resulting Cartesian plane, as shown in Fig. 2.1-1. The five points constitute the graph of the function.

EXAMPLE 2.1-2

Sketch the graph of the function given by Table 1.4-2.

SOLUTION

The domain of this function is finite and consists of the numbers 1, 2, 3, 4, 5, 6, and 7. Referring to Table 1.4-2, we see that the point with coordinates $(1, 142)$ is on the graph of the function, since 142 appears directly below 1 in the table. Similarly, the points with coordinates $(2, 817)$, $(3, 845)$, $(4, 845)$, $(5, 814)$, $(6, 875)$, and $(7, 416)$ are on the graph. The graph is shown in Fig. 2.1-2.

FIGURE 2.1-2

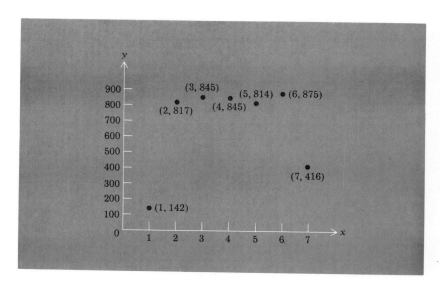

To sketch the graph of a function whose domain consists of more than a finite number of values, we locate a few sample points on the graph and then employ some intelligent guesswork. Skill in this art is gained by experience. Let us consider some examples.

───────────────   **EXAMPLE 2.1-3**   ───────────────

Sketch the graph of the function $f(x) = x$.

Sketching the graph of a function whose domain includes an infinite number of values

SOLUTION

The domain of this function consists of all real numbers and thus is not finite. We begin by writing the function in tabular form, selecting some sample values from the domain of the function for inclusion in the table. The more sample values we pick, the more accurate our graph is likely to be, but of course it is not practical to use a very large number of values. Here we use six sample values:

All real numbers x	x	0	1	-1	2	-2	3
$f(x)$	x	0	1	-1	2	-2	3

From this table, using the sample values of x that appear in the top row, we see that the points whose coordinates are $(0, 0)$, $(1, 1)$, $(-1, -1)$, $(2, 2)$, $(-2, -2)$, and $(3, 3)$ are on the graph of the function. We construct a Cartesian coordinate system and locate these six points in the resulting Cartesian plane (see Fig. 2.1-3). Notice that all six points appear to lie on a straight line. Using a straight edge, draw a line through these six points. Now pick any point P on this line, and notice that if we denote the coordinates of P by (x_0, y_0), then it appears that $y_0 = x_0 = f(x_0)$ (see Fig. 2.1-3). So the coordinates of P apparently have the form $(x_0, f(x_0))$, and thus P should be included in the graph of $f(x) = x$. So it seems reasonable to accept all the points on the line shown in Fig. 2.1-3 as the graph of the function $f(x) = x$.

Recall that the graph of a function f is the collection of all points whose coordinates are of the form $(x, f(x))$ for some x in the domain of f.

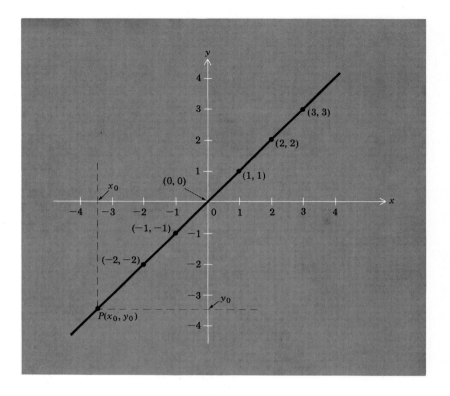

FIGURE 2.1-3

EXAMPLE 2.1-4

Sketch the graph of the function $g(x) = x^2$.

SOLUTION

As in the previous example, we begin by selecting some sample values from the domain of the function and writing the function in tabular form.

All real numbers x	x	0	1	-1	2	-2	3	-3
$g(x)$	x^2	0	1	1	4	4	9	9

From this table we see that the points $(0, 0)$, $(1, 1)$, $(-1, 1)$, $(2, 4)$, $(-2, 4)$, $(3, 9)$, and $(-3, 9)$ all are on the graph of g. We locate these points in a Cartesian plane and connect them

The more sample values chosen from the domain of the function, the more accurate the graph is likely to be.

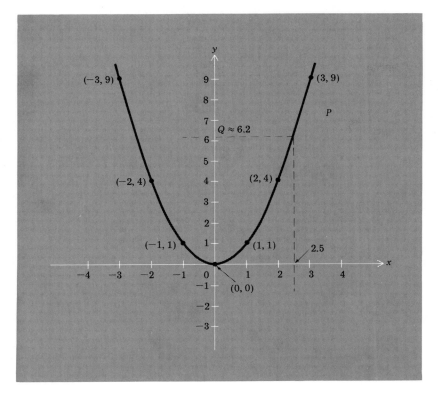

When sketching the graph of a function, the letters used to label the coordinate axes depend upon the letters used to define the function. Traditionally, when given a function such as $g(x) = x^2$, many mathematicians think of it as "the function $y = g(x)$ where $g(x) = x^2$," so the graph of the function consists of the points (x, y), where $y = g(x)$. Thus the horizontal and vertical axes are usually labeled with the letters x and y, respectively. However, if we are given the function "$z = h(w)$, where $h(w) = w^2$," then the horizontal and vertical axes would be labeled, respectively, the w axis and the z axis. Otherwise, the appearance of the graph of $z = h(w) = w^2$ is the same as the graph shown in Fig. 2.1-4.

FIGURE 2.1-4

with a *smooth* curve (see Fig. 2.1-4). It is our hope that this smooth curve is a reasonable approximation of the graph of $g(x) = x^2$. To bolster our confidence that this is indeed the case, we use the curve to estimate values of $g(x)$ corresponding to values of x which do not appear in the table. For example, we might choose $x = 2.5$; we then locate 2.5 on the x axis and go vertically upward until we intersect the curve at the point P (see Fig. 2.1-4), then we move horizontally from P until we intersect the y axis at the point Q. The number corresponding to Q on the y axis is now estimated; in this instance (see Fig. 2.1-4) we might guess that Q corresponds to 6.2. This means that the coordinates of P are approximately $(2.5, 6.2)$. If the curve in Fig. 2.1-4 is the graph of $g(x) = x^2$, the coordinates of P (since P is on the curve) should be $(2.5, g(2.5))$, so $g(2.5)$ should be approximately 6.2. Actually, $g(2.5) = (2.5)^2 = 6.25$, so our estimate that $g(2.5) = 6.2$, based on the assumption that the curve is actually the graph of g, is not far wrong. Additional estimates afford equally good results, and this convinces us that the curve in Fig. 2.1-4 is an acceptable graph of the function $g(x) = x^2$.

EXAMPLE 2.1-5

Sketch the graph of the function $h(x) = \sqrt{x}$.

SOLUTION

The domain of this function consists of the positive real numbers and 0. Choosing some sample values from the domain and writing the function in tabular form, we have

All positive real numbers and 0	x	0	1	4	9	16
$h(x)$	$\sqrt{x}$	0	1	2	3	4

From this table we see that the points (0, 0), (1, 1), (4, 2), (9, 3), and (16, 4) are on the graph of h. We locate these points in a Cartesian plane and connect them with a smooth curve (see Fig. 2.1-5). Remember that the domain of h consists only of real numbers x which are $\geqslant 0$, so no part of the curve (which is our estimate of the graph) will lie to the left of the y axis. As in Example 2.1-4, before accepting this curve as the graph of our function, we use the curve to estimate some values $h(x)$ for values of x not in the table. For $x = 7$, we might estimate from the curve in Fig. 2.1-5 (using the method described in Example 2.1-4) that $h(x) = h(7) = 2.6$; since

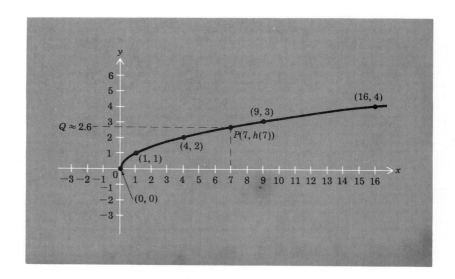

FIGURE 2.1-5

$h(7) = \sqrt{7}$, we conclude that $\sqrt{7}$ should be equal to 2.6. Squaring 2.6 yields 6.76, so our estimate that $h(7) = \sqrt{7} = 2.6$ is reasonable. Additional estimates further convince us that the curve shown in Fig. 2.1-5 is the graph of $h(x) = \sqrt{x}$.

To estimate the value of $h(7)$ from the curve shown in Fig. 2.1-5, locate 7 on the x axis, go vertically upward to P on the curve, then horizontally to Q on the y axis, and estimate the number corresponding to Q.

In each of the previous three examples, our method for sketching the graph was to locate a finite number of points on the graph in a Cartesian plane and then connect these points with a smooth curve. This procedure is not entirely foolproof. In spite of its shortcomings, however, it is still the best technique presently available to us for constructing graphs of functions whose domains consist of more than a finite number of values. The number of sample values chosen from the domain of the function remains the option of the person doing the graphing; you must decide when you have enough values to determine the shape of the graph. The next example illustrates the opportunity for error.

EXAMPLE 2.1-6

Sketch the graph of the function $f(x) = 4x^3 - x$.

SOLUTION

As before, we select some sample values from the domain of the function and write the function in tabular form. If we choose sample values 0, 1, -1, 2, and -2, we get the following table.

FIGURE 2.1-6

All real numbers x	x	0	1	-1	2	-2
$f(x)$	$4x^3 - x$	0	3	-3	30	-30

From this table, if we plot the points $(0, 0)$, $(1, 3)$, $(-1, -3)$, $(2, 30)$, and $(-2, -30)$ and connect them with a smooth curve, we arrive at the graph shown in Fig. 2.1-6(a). However, let us choose additional sample values of x between 1 and -1 and produce the following supplemental table.

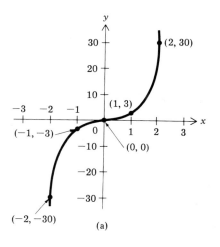

(a)

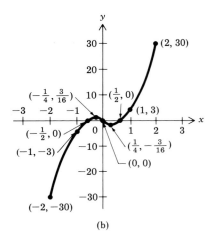

(b)

FIGURE 2.1-6

All real numbers x	x	1	-1	$\frac{1}{2}$	$-\frac{1}{2}$	$\frac{1}{4}$	$-\frac{1}{4}$
$f(x)$	$4x^3 - x$	3	-3	0	0	$-\frac{3}{16}$	$\frac{3}{16}$

When we plot the points on the graph from this table in addition to the points from the first table and connect all the points with a smooth curve, we obtain the graph shown in Fig. 2.1-6(b). Our graph is now much more accurate; we have used more sample values from the domain of the function. A comparison of the two curves in Fig. 2.1-6 reveals significant differences. How do we know when our construction of the graph of a function is accurate? For the moment, checking the curve by using it to estimate the value of the function at several values in the domain (as done in Examples 2.1-4 and 2.1-5) is our best method of insuring accuracy. If the curve does not check well, more points on the graph should be plotted and the curve redrawn. We shall see later that one of the successes of the calculus is in determining useful information about the graph of a function which will improve our ability to construct accurate graphs.

EXAMPLE 2.1-7

The transit authority of a large city finds that the daily gross income I from the operation of a particular commuter train is closely approximated by the equation $I(x) = 600x - 50x^3$ where x is the fare charged per person. The equation is considered valid for values of x between \$1 and \$3 inclusive. Write the function I in tabular form and sketch its graph. From the graph, estimate the fare that should be charged so that the authority will receive the maximum income from the operation of the train.

SOLUTION

The domain of this function consists of all real numbers between 1 and 3 inclusive (actually, only values of x which can be interpreted in standard coinage are meaningful, but we can ignore this for our purposes). Choosing sample values of 1, 1.5, 2, 2.5, and 3 from the domain, we construct the following table.

All real numbers x between 1 and 3	x	1	1.5	2	2.5	3
$I(x)$	$600x - 50x^3$	550	731.25	800	718.75	450

Thus, from this table, the points $(1, 550)$, $(1.5, 731.25)$, $(2, 800)$, $(2.5, 718.75)$, and $(3, 450)$ are on the graph of the function. We plot these points and connect them with a smooth curve to get the graph shown in Fig. 2.1-7. Note that the entire graph lies over the part of the x axis that is between 1 and 3, since these values of x constitute the domain of the function. Since the highest point on the graph appears to be directly above $x = 2$, this means that the largest value of $I(x)$ occurs when x is 2, so a fare of \$2 will yield the maximum income.

Our estimate that a fare of \$2 will yield the maximum income is entirely dependent on the accuracy of the graph, and should be accepted with some reservation until checked by other methods. In Chapter 6 we will meet some methods of checking this information.

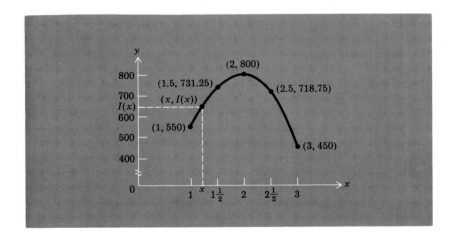

FIGURE 2.1-7

It may have occurred to the reader that even when the graph of a function is just a finite set of points, these points could still be connected by a smooth curve and some useful interpretation of the resulting curve made. This procedure is often utilized, but caution must be exercised in making use of the "new" information generated (see Exercises 2.1-32 and 2.1-33).

Extension of a finite graph and subsequent interpretation of "new" information

FIGURE 2.1-8

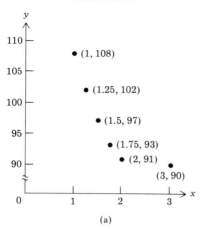

(a)

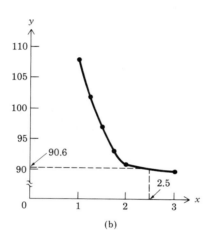

(b)

Locate 2.5 on the x axis in Fig. 2.1-8(b), go upward to the curve, and then go horizontally to the left to the y axis at 90.6.

To use the curve in Fig. 2.1-8(b) to estimate the distance at which the temperature reaches 100°F, locate 100 on the y axis and go horizontally to the curve; then go downward to the x axis. The point located on the x axis in this manner is approximately 1.35.

EXAMPLE 2.1-8

Sketch the graph of the function given by Table 1.4-4. Connect the points of the graph by a smooth curve and discuss the interpretation of this curve.

SOLUTION

The function presented in Table 1.4-4 has a finite domain consisting of the numbers 3, 2, 1.75, 1.50, 1.25, and 1. Thus the graph is a finite set of points in a Cartesian plane, namely the points (3, 90), (2, 91), (1.75, 93), (1.50, 97), (1.25, 102), and (1, 108). The graph is shown in Fig. 2.1-8(a), and the curve obtained by connecting the points of the graph by a smooth curve is shown in Fig. 2.1-8(b). If the curve in Fig. 2.1-8(b) actually reflects the manner in which the temperature depends upon the distance from the flame (see Example 1.4-4), then we might expect that at 2.5 in. from the flame, the temperature would be 90.6°F. No measurement has been made to determine this, however, so before we place much confidence in this kind of newly acquired information, it would be well to see how some sample projections actually correspond to direct measurement. According to the curve in Fig. 2.1-8(b) the temperature reaches 100°F at a distance of approximately 1.35 in. Again a direct check of this information would be beneficial. If, after several projected values are checked by direct experiment, we are satisfied with the results, we can feel reasonably confident about using information generated by the simple process of drawing a smooth curve through the points of a graph. Direct experience and common sense are irreplaceable in this kind of enterprise. Some experience will be obtained through the exercises, but we must rely on the reader for the common sense.

EXAMPLE 2.1-9

The monthly production costs of the Nifty Noodle Company for each of six months of the past year are given in the following table (Month 1 is January, 3 is March, etc.).

Month	1	3	5	7	9	11
Production cost (thousands of dollars)	5	5	4.5	4	5	5.5

Construct the graph of this function, and join the points of the graph with a smooth curve. Use the curve to estimate the production costs in February and June. From the curve, for which month of the year is the estimated production cost the greatest?

Note that the domain of the function given in the table in Example 2.1-9 is finite and consists of the numbers 1, 3, 5, 7, 9, and 11. Thus the graph will consist of six points in a Cartesian plane.

SOLUTION

The graph consists of the six points in a Cartesian plane whose coordinates are (1, 5), (3, 5), (5, 4.5), (7, 4), (9, 5), and (11, 5.5). The graph is shown in Fig. 2.1-9(a). In Fig. 2.1-9(b) we have

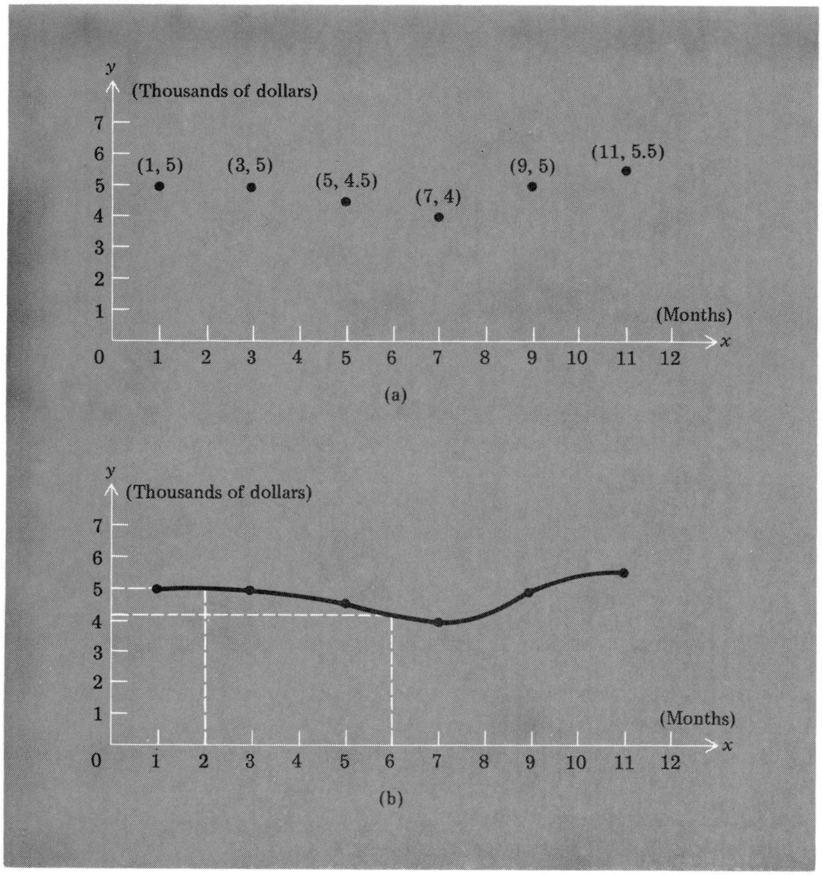

The smooth curve joining the six points is *not* the graph; the graph consists of *only* the six points in the plane given by the table.

FIGURE 2.1-9

joined the six points of the graph with a smooth curve, and from this curve we see that the estimated production cost for Month 2 (February) is 5 (thousands of dollars); for Month 6 (June), $4250; notice that no estimate can be made for Month 12 (December), since the graph could go up or down after November. Since the highest point on the curve occurs above Month 11, the estimated production cost is greatest for the month of November. It is to be noted that since these estimates depend entirely on the smooth curve used to join the points of the graph, we might (through slight variations in drawing the curve) equally well have estimated the production cost for June to be $4300 or $4200 or, for that matter, any amount between these two figures. Also, unless we have some assurance that the production costs closely follow, from month to month, the trend dictated by the information given in the table, we have no reason to believe that our estimates are even close to the actual costs. In practice, additional information would be used to substantiate these estimates before placing much confidence in their accuracy.

Exercises 2.1

In Exercises 1 through 10, complete the given table and carefully sketch the graph of the function.

1.

$x = 1, 2, 3, 4, 5$	x	1	2	3	4	5
$f(x)$	$3x$					

2.

$x = -2, -1, 0, 1, 2$	x	-2	-1	0	1	2
$f(x)$	$3x - 1$					

3.

$x = -4, -2, 0, 2, 4$	x	-4	-2	0	2	4
$f(x)$	$x^2 + x$					

4.

$x = -\frac{1}{2}, 0, 2, 3, 4$	x	$-\frac{1}{2}$	0	2	3	4
$f(x)$	$4x + 2$					

5.

All real numbers x	x	-2	0	1	2	3
$f(x)$	$3x + 2$					

6.

All real numbers x	x	-2	0	2	1	-1
$f(x)$	$-3x$					

7.

All real numbers x	x	-4	-2	0	2	4
$f(x)$	$-\frac{1}{2}x + 5$					

8.

All real numbers x	x	-3	-2	0	2	3
$f(x)$	$-x^2$					

9.

All real numbers x	x	-3	-1	0	1	3
$f(x)$	$|x|$					

10.

All real numbers x greater than or equal to 3	x	3	4	7	12	19
$f(x)$	$\sqrt{x - 3}$					

In Exercises 11 through 20, find the domain of the given function, and select at least five sample values from the domain. Express the function in tabular form, including in the table the values of the function at the numbers chosen from the domain. Then sketch the graph of the function.

11. $f(x) = 2x - 1$
12. $f(x) = -2x - 1$
13. $f(x) = x^2 + 3$
14. $f(x) = -x^2 + 3$
15. $f(x) = |x| + 2$
16. $f(x) = |x + 2|$
17. $t(x) = \sqrt{(x + 2)^2}$
18. $s(x) = x^2 - 2x + 1$
19. $t(v) = \sqrt{v^2} + 2$
20. $w(z) = -(z + 2)^2$

21.   A sales representative's weekly salary S (in dollars) is determined by the equation $S(x) = 150 + 0.2x$, where x (in dollars) is the amount of weekly sales. The equation is to be used for values of x between \$100 and \$400 inclusive. Write the function S in tabular form (taking the domain to be all real numbers between 100 and 400 inclusive) and sketch its graph. Does the graph have a familiar shape? Use the graph to estimate the representative's salary for a week in which \$160 worth of merchandise is sold. Check your estimate by evaluating the expression for $S(x)$ at $x = 160$.

22.   A company finds that its monthly profit P (total sales less production costs) from the manufacture and sale of one of its products is closely approximated by the equation $P(x) = 200x - x^2$, where x is the number of units produced per month. The equation is considered valid for values of x between 80 and 120. Write the function P in tabular form (take the domain to be all real numbers between 80 and 120 inclusive), including in the table the values of $P(x)$ when x is 80, 90, 100, 110, and 120. Sketch the graph of the function P. From the graph, what value of x yields the maximum monthly profit?

23.   Suppose the retail price P of a certain product is given by the equation $P(x) = 8 + (x - 10)^2$, where x is the number of units produced (in thousands), and suppose the equation is valid for values of x between 2 and 10 inclusive. Write the function in tabular form (take the domain to be all real numbers between 2 and 10 inclusive), including in the table the values of $P(x)$ when x is 2, 4, 6, 8, and 10. Sketch the graph of the function. From the graph, how many units should be produced to achieve a minimum retail selling price?

24.   The population P (in thousands) of a certain bacteria culture is given by the equation $P(t) = 4t^2 - 2t$ where t is measured in

hours. The equation is valid for values of t between 4 and 10 inclusive. Write the function P in tabular form and sketch the graph of the function. From the graph, estimate the value of P when t is 6.5 hr. Check your estimate by evaluating the expression for $P(t)$ at $t = 6.5$.

25. The monthly production costs of the Ace Electric Motor Company for each of six months of the past year are given in the following table (Month 1 is January, 3 is March, etc.).

Month	1	3	5	7	9	11
Production cost (thousands of dollars)	4	4.5	5	5.5	6	7

The graph of the function given in Exercise 25 consists of six points in a Cartesian plane, since there are only six values in the domain of the function, namely, 1, 3, 5, 7, 9, and 11. Note that the curve joining the six points is *not* the graph; the graph is just the six points in the plane given by the table.

Construct the graph of this function, and join the points of the graph with a smooth curve. Use the curve to estimate the production costs in February and October.

26. An ecologist, investigating the effects of air pollution on plant life, compiles the following data in measuring the percent of diseased ponderosa pine trees as a function of the distance from a large industrial city.

Distance (miles)	50	100	150	200	250	300	350	400
Percent of diseased trees	39	35	31	25	10	8	6	4

Again, the domain of the function given by the table in Exercise 26 is finite, consisting of the numbers 50, 100, 150, 200, 250, 300, 350, and 400.

Construct the graph of this function, and join the points of the graph with a smooth curve. Use the curve to estimate the percent of diseased trees at a location 125 mi from the city. Suppose that under normal conditions fewer than 6% of the trees are diseased. Then, from the curve, how far from the city does the pollution affect the trees?

27. A firm that manufactures small transistor radios estimates that the number of radios it can sell (the demand) per month is a function of the selling price per radio according to the following table.

Price per radio (dollars)	6	7	8	9	10	11
Demand per month (hundreds)	6.5	6	5.5	4.5	4	2

Construct the graph of this function. Join the six points of the graph with a smooth curve, and use the curve to estimate the number of radios the company would sell per month if they were priced at $6.50 each. If they were priced at $8.50 each?

28. The *total sales* per month of a product is defined as the number of units sold times the selling price per unit. The following table gives the total sales per month as a function of the selling price per unit for the firm described in Exercise 27.

Price per radio (dollars)	6	7	8	9	10	11
Total sales per month (hundreds of dollars)	39	42	44	40.5	40	22

Construct the graph of this function, and join the six points of the graph with a smooth curve. Use the curve to estimate the total sales per month if the radios are sold for $7.50 each.

29. Medical researchers, investigating growth characteristics in human males, obtain the following data about the average weight of human males as a function of age.

Age (years)	0 (birth)	3	6	9	12	15	18
Weight (pounds)	7	35	50	65	85	130	145

Construct the graph of this function, and join the seven points of the graph with a smooth curve. Use the curve to estimate (1) the weight of an average $4\frac{1}{2}$-yr-old boy and (2) the weight of an average 8-yr-old boy.

30. The town council of Pleasantville is investigating the possible need for a traffic light at an intersection which has become increasingly congested due to the recent opening of a nearby shopping center. A traffic count is conducted at the intersection, and the following data are collected during the six months from January through June, giving the number of vehicles passing through the intersection per month as a function of the month (Month 1 is January, Month 2 is February, etc.).

Month	1	2	3	4	5	6
Number of vehicles (thousands)	11	13.4	15.8	17.5	19	20.5

Construct the graph of this function and join the points of the graph with a smooth curve. Use the curve to estimate the traffic count in July. The council agrees to authorize installation of the signal when the monthly traffic count reaches 23,000 vehicles. Use the curve to estimate the month when this figure will be reached.

31. Researchers investigating the increase in wages and prices in the United States for the seven months from January through July accumulate the information given in the following two tables. The first table gives the percent increase in the average hourly wage as a function of the month; the second table gives the percent increase in the consumer price index as a function of the month (Month 1 is January, 2 is February, etc.):

Month	1	2	3	4	5	6	7
Percent increase in hourly wage	0	0.2	0.5	0.9	1.4	1.7	2

Month	1	2	3	4	5	6	7
Percent increase in consumer price index	0.2	0.6	1.2	2.1	3	3.5	4.1

Construct the graphs of both these functions on the same Cartesian coordinate system. Join the points of each graph with a smooth curve. Use these two curves to compare the rates of increase in wages and prices: are prices increasing (approximately) (1) as fast as wages (2) twice as fast as wages or (3) three times as fast as wages? Use the wage increase curve to estimate the percent increase in hourly wages one might expect in August.

32. The following table gives the number of turkeys consumed on Thanksgiving Day in the United States as a function of the year of the Thanksgiving holiday.

Year of the Thanksgiving holiday	1965	1966	1967	1968	1969	1970
Number of Thanksgiving turkeys consumed (millions)	20	18	22	21	22	23

Construct the graph of this function, and join the points of the graph with a smooth curve. Use the curve to estimate the number of turkeys consumed on Thanksgiving in 1971. Now locate the point on the horizontal axis of the coordinate system which (approximately) corresponds to May 1, 1967, and use the curve to estimate the number of turkeys consumed on May Day of 1967. Does this seem to be a reasonable estimate? Where does the difficulty lie? (*Hint*: Can any day, such as May 1, 1967, be interpreted as a meaningful value for the domain of the function?)

33. Rainfall measurements taken in a small western city yielded the following data on the amount of rainfall (in millimeters) that occurs each month as a function of the month (Month 1 is January, 2 is February, etc.).

Month	1	3	5	7	9	11
Amount of rainfall (millimeters)	240	140	40	5	10	90

Construct the graph of this function, and join the points of the graph with a smooth curve. Use the curve to estimate the amount of rainfall the city receives in February. From the curve, during which month is the estimated rainfall the least? Now locate the point on the horizontal axis which corresponds (approximately) to February 15, and use the curve to estimate the amount of rainfall the city receives on February 15. Does this seem to be a reasonable estimate? Where does the difficulty lie? (*Hint*: can any day, such as February 15, be interpreted as a meaningful value for the domain of the function?)

2.2.   The Algebra of Functions

The following two tables give the average number of gallons of gasoline that the Zesto Gas Station pumps as a function of the day of the week (Day 1 is Sunday, 2 is Monday, etc.). Table 2.2-1 displays regular gasoline sales (and defines a function f) while Table 2.2-2 shows premium gasoline sales (and defines a function g).

TABLE 2.2-1.   Regular gasoline.

Day of week	x	1	2	3	4	5	6	7
Number of gallons pumped	$f(x)$	0	1300	1250	1230	1200	1300	1450

TABLE 2.2-2.   Premium gasoline.

Day of week	x	1	2	3	4	5	6	7
Number of gallons pumped	$g(x)$	0	800	745	718	732	760	850

The 0 entry in the second row, first column of each of these tables reflects the fact that the station is closed on Sunday. To determine the average *total* number of gallons of gasoline (regular and premium) that the station pumps per day, as a function of the day of the week, we add the number of gallons of regular gasoline pumped on a given day (a number from the bottom row of Table 2.2-1) to the number of gallons of premium gasoline pumped that same day (the corresponding number from the bottom row of Table 2.2-2). The resulting function h is given in Table 2.2-3.

To compute the entry below Day 6, for example, in Table 2.2-3, we add the entries below Day 6 from Tables 2.2-1 and Table 2.2-2 and get 1300 (gallons of regular gasoline) + 760 (gallons of premium gasoline) = 2060, the total number of gallons of gasoline sold on Day 6.

TABLE 2.2-3. Total gasoline sales.

Day of week	x	1	2	3	4	5	6	7
Number of gallons pumped	$h(x)$	0	2100	1995	1948	1932	2060	2300

Since f, g, and h denote, respectively, the functions given in Tables 2.2-1, 2.2-2, and 2.2-3, we see that for $x = 3$ (Tuesday), the quantity $f(3)$ is the average number of gallons of regular gasoline pumped on a Tuesday, $g(3)$ is the average number of gallons of premium gasoline pumped on a Tuesday, and $h(3)$ (which is equal to $f(3) + g(3)$) is the average total number of gallons of gasoline pumped on a Tuesday. Actually, if x is any particular day of the week, then we have the rule $h(x) = f(x) + g(x)$. Thus h is a function whose value at x can be computed by adding the value of the function f at x, $f(x)$, to the value of the function g at x, $g(x)$. This idea of "adding" two functions to produce a new function, called the *sum* of the two functions, is important in the study of calculus. Consequently, we define the sum of *any* two functions f and g to be the function h given by the rule

Definition of the sum of two functions f and g

$$h(x) = f(x) + g(x)$$

for each x common to the domains of f and g. To remind us that h is the sum of f and g we frequently write $(f + g)(x)$ rather than $h(x)$, so that

$$(f + g)(x) = f(x) + g(x)$$

When defining $h(x)$ by the equation $h(x) = f(x) + g(x)$, we require that x be in the domain of each of the functions f and g so that the computations on the right-hand side of the equation can be carried out. If x is not in the domain of f, there is no $f(x)$ to be added to $g(x)$; similarly, if x is not in the domain of g, there is no $g(x)$ to be added to $f(x)$.

EXAMPLE 2.2-1

Given the functions $f(x) = x^2 + 1$ and $g(x) = -\sqrt{x}$, find an equation that defines the function $f + g$. Determine the domain of the function $f + g$, and write the function in tabular form.

SOLUTION

Since for each x common to the domains of f and g, $(f + g)(x)$ is defined to be $f(x) + g(x)$, we have

$$(f + g)(x) = f(x) + g(x) = x^2 + 1 - \sqrt{x}$$

This is an equation which defines the function $f + g$. Now, the domain of f consists of all real numbers x, and the domain of g consists of all real numbers that are greater than or equal to 0. Thus the real numbers common to the domains of f and g are those real numbers which are greater than or equal to 0, and these numbers constitute the domain of $f + g$. Selecting sample values 0, 9, 16, 25, and 36 from the domain of $f + g$, we write the function in tabular form as follows.

The domain of the function $g(x)$ $= -\sqrt{x}$ consists only of real numbers that are greater than or equal to 0 because it is not possible to take the square root of a negative real number.

All real numbers $x \geqslant 0$	x	0	9	16	25	36
$(f + g)(x)$	$x^2 + 1 - \sqrt{x}$	1	79	253	621	1291

Let us now return to the Zesto Gas Station. Suppose the station's profit on a gallon of gasoline is $.07. Then by multiplying each entry of the bottom row of Table 2.2-3 by 0.07 we can calculate the daily profit (in dollars) as a function p of the day of the week. Since the function defined by Table 2.2-3 is h, we see that the value of p for Day x is computed by multiplying 0.07 by $h(x)$, i.e.,

$$p(x) = (0.07)h(x)$$

Since the entries in the bottom row of Table 2.2-3 give the total number of gallons of gasoline sold on a given day, multiplying these entries by the profit per gallon ($.07) yields the total profit for the day. For example, to compute the entry below Day 4 in Table 2.2-4, we multiply the entry below Day 4 from Table 2.2-3 by 0.07 and get 0.07 (profit per gallon) × 1948 (total number of gallons sold on Day 4) = $136.36, the total profit for Day 4 (Wednesday).

The function p is given in Table 2.2-4. This procedure of multiplying each value of a given function by a fixed real number to produce a new function occurs frequently in mathematics, and warrants the following definition. For a given fixed real number c and a given function h, we define a new function p, called the *product* of h by c, by the rule $p(x) = c \cdot h(x)$. The product function p has the same domain as the function h.

Definition of the product of a function h and a fixed number c

If $p(x) = c \cdot h(x)$, then for any value of x for which $h(x)$ is defined, $p(x)$ is also defined. Thus the domain of $p(x)$ is the same as the domain of $h(x)$.

TABLE 2.2-4

Day of week	x	1	2	3	4	5	6	7
Daily profit (dollars)	$p(x)$	0	147.00	139.65	136.36	135.24	144.20	161.00

EXAMPLE 2.2-2

Let h be the function $h(x) = \sqrt{x - 1}$, and let c be the real number 5. Let p be the product of h by c. Find the domain of p and write p in tabular form.

SOLUTION

The function h is defined only for real numbers x for which $x - 1 \geqslant 0$, since we cannot extract the square root of a negative number. Thus the domain of h is all real numbers x which are greater than or equal to 1, and this is also the domain of the product function $p(x) = ch(x) = 5\sqrt{x - 1}$. Selecting sample values 1, 5, 10, 17, and 26 from the domain of $p(x)$, we write p in tabular form as follows.

Instead of $c \cdot h(x)$, we write simply $ch(x)$.

All real numbers $x \geqslant 0$	x	1	5	10	17	26
$p(x)$	$5\sqrt{x - 1}$	0	10	15	20	25

Thus far in this section we have defined the sum $f + g$ of two functions f and g and the product ch of a function h by a fixed number c. We now define three other methods of combining two given functions f and g to produce a new function.

(1) The *difference* function, $f - g$, is defined by the rule $(f - g)(x) = f(x) - g(x)$. The domain of $f - g$ consists of those values of x common to the domains of f and g.

(2) The *product* function fg, is defined by the rule $(fg)(x) = f(x) \cdot g(x)$. The domain of fg is again those values of x common to the domains of f and g.

(3) The *quotient* function, f/g, is defined by the rule $(f/g)(x) = f(x)/g(x)$. The domain of f/g consists of those values of x common to the domains of f and g for which $g(x) \neq 0$.

The product function fg, bears resemblance to the product cg, of the function g, by a fixed number c. Indeed, the definition of fg includes as a special case the product cg, for if f is the "constant" function defined by $f(x) = c$ for all numbers x, then for each x in the domain of g we have $(fg)(x) = f(x)g(x) = cg(x)$.

The equations that define the values of the functions $f - g$, fg, and f/g at x all require that $f(x)$ and $g(x)$ both be defined. Thus, only values of x that are in the domains of *both* f and g will be in the domains of these new functions. In addition, since the value of f/g at x is $f(x)/g(x)$, and since division by 0 is not possible, we must exclude from the domain of f/g those values of x for which $g(x) = 0$.

The domain of f/g does not include -2 or 2, since $g(2) = g(-2) = 0$.

EXAMPLE 2.2-3

Let f and g be the functions $f(x) = \sqrt{x + 3}$ and $g(x) = x^2 - 4$. Write equations which define each of the functions $f - g$, fg, and f/g. Find the domains of each of these functions and express each function in tabular form.

SOLUTION

The domain of $f(x) = \sqrt{x + 3}$ consists of those real numbers x for which $x + 3 \geqslant 0$, i.e., $x \geqslant -3$. The domain of $g(x) = x^2 - 4$ is all real numbers x. Note that $g(2) = g(-2) = 0$. Thus, $f - g$ and fg both have the same domain, all real numbers $x \geqslant -3$, and the domain of f/g consists of all these numbers except -2 and 2. Equations defining $f - g$, fg, and f/g are

$$(f - g)(x) = f(x) - g(x) = \sqrt{x + 3} - (x^2 - 4)$$

$$(fg)(x) = f(x)\,g(x) = (\sqrt{x + 3}\,)(x^2 - 4)$$

$$\left(\frac{f}{g}\right)(x) = \frac{f(x)}{g(x)} = \frac{\sqrt{x + 3}}{x^2 - 4}\;;\; x \neq \pm 2$$

These functions may be written in tabular form as follows.

All real numbers $x \geqslant -3$	x	-3	1	6	13	22
$(f - g)(x)$	$\sqrt{x + 3}$ $- (x^2 - 4)$	-5	5	-29	-161	-475

All real numbers $x \geqslant -3$	x	-3	1	6	13	22
$(fg)(x)$	$\sqrt{x + 3}\,(x^2 - 4)$	0	-6	96	660	2400

All real numbers $x \geqslant -3$ except $-2, 2$	x	-3	1	6	13	22
$(f/g)(x)$	$\sqrt{x+3}\,/(x^2-4)$	0	$-\frac{2}{3}$	$\frac{3}{32}$	$\frac{4}{165}$	$\frac{1}{96}$

EXAMPLE 2.2-4 *(Optional.)*

The following illustration, a problem in pollution control, shows how differences, products, and quotients of functions might arise in a specific situation. Throughout the discussion the time, t, is measured in minutes, with the beginning of the 8-hr work day corresponding to $t = 0$ and the end of the work day corresponding to $t = 480$.

The Bilford city council requests that industries dumping pollutants into a nearby river supply complete information concerning their daily discharge of polluted water into the river and the percent of pollutants in the discharge. Company A pipes water and pollutants into a mixing complex from which a polluted discharge flows into the river (see Fig. 2.2-1). The function determining the rate of flow of water into the complex is $f(t) = 500,000 + 3t$ cubic in. per minute. The company is unable to test directly for the percent of pollutants in the total discharge into the river but has determined that the total discharge (water plus pollutants) flows into the river at a rate of $h(t) = -t^2 + 843t + 502,460$ cubic in. per minute. Thus the function g giving the rate of flow of pure pollutants into the river at time t is the *difference* function $h - f$, i.e.,

$$g(t) = (\text{rate of total discharge}) - (\text{rate of water discharge})$$

$$= h(t) - f(t) = -t^2 + 840t + 2460$$

cubic in. per minute. The percent p of pollutants in the total discharge at time t is then obtained by multiplying the *quotient* function $g(t)/h(t)$ by 100, i.e., $p(t) = 100(g(t)/h(t))$. Company B can directly analyze samples of its total discharge into the river and determine that the percent of pollutants in the discharge at time t is given by the function $q(t) = [30 + (0.1)t]\%$. The total discharge at time t from Company B is also measured and found to be given by

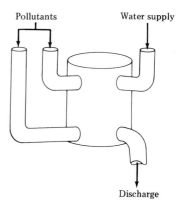

FIGURE 2.2-1 Company A mixing complex.

The percent of pollutants in the discharge at time t is given by 100 times the quotient (pollutant discharge)/(total discharge).

$r(t) = -2t^2 + 1600t + 50{,}000$ cubic in. per minute. Consequently, the rate of discharge s of pure pollutants at time t into the river is the *product*

Recall that to convert a percent into a numerical value, we must divide by 100. For example, the numerical value of 34% is $34/100 = 0.34$.

$$\left(\frac{\text{percent of pollutants}}{100} \right) \times (\text{total discharge})$$

i.e.,

$$s(t) = \frac{1}{100}\, q(t) r(t)$$

$$= \frac{1}{100}\, (30 + 0.1t)(-2t^2 + 1600t + 50{,}000)$$

cubic in. per minute. Thus, for example, after 6 hr (360 min) of operation, the rate of discharge of pollutants from Company B is $s(360) = 242{,}088$ cubic in. per minute and from Company A is $g(360) = 275{,}260$ cubic in. per minute.

Exercises 2.2

In Exercises 1 through 6, recall that when a function is defined entirely by a table (rather than by an equation), the domain of the function consists of the entries in the top row of the table.

In each of Exercises 1 through 6, two tables are given which define functions f and g. Construct the tables which define $f + g$, $f - g$, fg, and f/g (pay particular attention to the domains of each of these four functions as you construct the tables).

1.

x	1	2	3	4	5	6
$f(x)$	-10	2	4	-6	5	1

x	1	2	3	4	5	6
$g(x)$	-3	1	-4	-2	3	2

2.

x	2	4	6	8	10	12
$f(x)$	-1	0	3	4.5	-2	1

x	2	4	6	8	10	12
$g(x)$	3	2	0	5	7	-9

3.

x	1	3	5	7	9	11
$f(x)$	1	1	1	1	1	1

x	1	3	5	7	9	11
$g(x)$	2	-3	4	-5	7	1

4.

x	-2	-1	1	2	5	9
$f(x)$	-7	-3	5	-1	0	2

x	-2	-1	0	1	2	3
$g(x)$	4	3	2	1	-1	-2

5.

x	-5	-3	-1	1	3	5
$f(x)$	0	0	0	0	0	0

x	-5	-2	-1	1	2	5
$g(x)$	0	1	2	3	4	5

6.

x	2	-1	3	5	-2	1
$f(x)$	5	2	7	5	3	5

x	-3	2	4	-4	-5	-2
$g(x)$	1	0	0	0	0	1

In Exercises 7 through 22, for the given functions f and g, find equations which define each of the functions $f + g$, $f - g$, fg, and f/g. Determine the domains of each of these four functions and write the functions in tabular form.

7. $f(x) = x + 2$
   $g(x) = x^2 + 1$

8. $f(x) = 3x - 1$
   $g(x) = x$

9. $f(x) = \sqrt{x - 2}$
   $g(x) = 7$

   $f(x) = \sqrt{x + 3}$
   $g(x) = x - 1$

11. $f(x) = x - 5$
    $g(x) = x - 5$

12. $f(x) = 3x + 2$
    $g(x) = x^2 - 4$

13. $f(x) = x^2 + 4$
    $g(x) = x^2$

14. $f(x) = x^2 + 1$
    $g(x) = \dfrac{1}{x}$

15. $f(x) = x^2 + 5x + 1$
    $g(x) = \sqrt{x - 3}$

16. $f(x) = \sqrt{x + 2}$
    $g(x) = \dfrac{1}{x}$

17. $f(x) = x - 3$
    $g(x) = x^2 - 9$

18. $f(x) = |x|$
    $g(x) = x$

19. $f(x) = |x + 1|$
    $g(x) = |x| + 1$

20. $f(x) = |x| + 1$
    $g(x) = |x + 1|$

21. $f(x) = \sqrt{x}$
    $g(x) = \sqrt{x}$

22. $f(x) = 1 + \sqrt{x}$
    $g(x) = 1 - \sqrt{x}$

23. Let $f(x) = x^2$ and $g(x) = 2$ be given functions. Write the functions $f + g$ and $f - g$ in tabular form and sketch the graphs of f, $f + g$, and $f - g$ on the same coordinate system.

24. Given the functions f and g in Exercise 23, write the functions fg and f/g in tabular form and sketch the graphs of f, fg, and f/g on the same coordinate system.

25. Let $f(x) = x^3$. Write the functions f, $2f$, and $(-1)f$ in tabular form and sketch the graphs of these three functions on the same coordinate system.

26. Let $f(x) = x^2 - 1$ and $g(x) = x - 1$. Write f/g in tabular form and sketch its graph (give close attention to the domain of f/g).

27. Let $f(x) = |x|$ and $g(x) = x$. Write the functions $f + g$ and $f - g$ in tabular form and sketch the graphs of f, g, $f + g$, and $f - g$ on the same coordinate system.

28. Given the functions f and g in Exercise 27, write the functions fg and f/g in tabular form and sketch their graphs.

29. The Gro-Fast Seed Company sells its bahia grass seed for $.79 a pound. Find an equation which defines the selling price S (in dollars) of x lbs of seed as a function of x. If an order of seed is

delivered, there is a delivery charge T of \$1.50 plus \$.01 a pound. Find an equation which defines the delivery charge T (in dollars) of an order of x lbs of seed as a function of x. Find an equation which defines the function $S + T$. What do the values of $S + T$ represent? How much will an order of 100 lbs of seed cost if it is delivered?

30. The Mow-Rite Company manufactures and sells power lawn-mowers. The demand function D (number of mowers per year that can be sold) for its super-deluxe model mower is defined by the equation $D(x) = 1500 - 3x$, where x is the selling price per mower (in dollars) and $75 \leqslant x \leqslant 100$. The profit P that the company realizes from the sale of one of the mowers is also a function of the selling price x and is given by $P(x) = x - 65$ for $75 \leqslant x \leqslant 100$. Find an equation which defines the product function DP as a function of x. What do the values of the function DP represent? What total profit will the company incur from the sale of the mowers if they are priced at \$79 each?

31. A manufacturing company finds that to produce x units of its product per day it incurs (1) a fixed cost E of \$1200 per day in employees' wages, (2) a unit production cost U of \$1.20 per day per article, and (3) a maintenance cost M of $x^2/100,000$ dollars per day. Thus, as functions of x, the daily costs of the company are given by $E(x) = 1200$, $U(x) = 1.20x$, and $M(x) = x^2/100,000$. Find an equation which defines the function $E + U + M$. What do the values of the function $E + U + M$ represent? What is the total production cost on a day when 300 units are produced?

32. The Riverside Travel Club arranges a round trip charter flight to Paris for a total cost C (in dollars) given by the equation $C(x) = 50,000 + 420x$, where x is the number of people in excess of 100 making the trip. Assuming that more than 100 people make the trip, express the total number of travelers N as a function of x. Find an equation that defines the quotient function C/N. What do the values of the function C/N represent? What is the charge per person if 125 people make the trip?

33. A farmer finds that the number N of bushels of apples she can produce per acre is a function of the number x of trees planted per acre as given by the equation $N(x) = 10x - 0.2x^2$ for $x \geqslant 25$. She also finds, since the quality of the apples diminishes if the trees are too crowded, that the price P per bushel (in dollars) for which she can sell the apples is a function of x and is given by

$$P(x) = 12.20 - \frac{x}{5}$$

for $x \geqslant 25$. Find an equation that defines the product function NP. What do the values of the function NP represent? How much income will the farmer receive per acre from the apples if she plants 30 trees per acre?

34. Perma-Plastics, Inc. manufactures shock-resistant plastic cases for electronic hand calculators, and each case is sold for $15. Let R be the total revenue received from the sale of x cases; find an equation which defines R as a function of x. The company estimates that its production cost C (in dollars) in producing x cases is given by the equation $C(x) = x^2/200 + 2x + 150$. Find an equation which defines the difference function $R - C$ as a function of x. The function $R - C$ (total revenue minus total production cost) is called the *profit* function. What is the company's profit if it produces 2000 cases?

35. The city of Pleasantdale authorized the construction of a new steel mill. The air pollution index P_1 (parts per million) in Pleasantdale before the opening of the new plant was a function of the time t (in hours) given by $P_1(t) = 3t + 10$, with $t = 0$ corresponding to 8:00 AM. The function was considered valid for $0 \leqslant t \leqslant 10$. After the opening of the new plant, it was found that the pollution index was approximated by the function $P_2(t) = t^2/10 + 5t + 15$ for $0 \leqslant t \leqslant 10$. Find an equation which defines the difference function $P_2 - P_1$. What do the values of the function $P_2 - P_1$ represent? How much pollution at 11:00 AM is due to the new plant?

36. At the beginning of a biological experiment, a culture of bacteria is observed to have a count of 30,000. The culture is treated with a bactericidal agent, and the number N of viable bacteria remaining after t hr is found to be given by the function $N(t) = 30,000 - 625t$ for $0 \leqslant t \leqslant 10$. It is known that if the culture is not treated, the number S of bacteria after t hr is given by the equation $S(t) = 10,000(3 + 2\sqrt{t} + t^2)$ for $0 \leqslant t \leqslant 10$. Find an equation that defines the function $S - N$. The values of the function $S - N$ represent the number of bacteria whose growth was prevented by the agent after t hr. Now find an equation that defines the function $100(S - N)/S$ as a function of t. Use this function to compute the percent of bacteria whose growth was prevented by the agent after 4 hr.

2.3.   Linear Functions, Linear Graphs, and Graphing Techniques

Definition of a linear function

No vertical line can intersect the graph of a function at more than one point (see Fig. 2.3-1). Thus no vertical line can be the graph of a (linear) function, since the line intersects itself in more than one point.

In Example 2.1-3 we graphed the function $f(x) = x$ and discovered that its graph was the straight line containing the two points $(0, 0)$ and $(1, 1)$. Any function whose graph is a straight line is called a *linear function*. Thus every nonvertical straight line is the graph of some linear function.

For any linear function f, there are always fixed real numbers a and b such that the value of f at any x is given by the rule

$$f(x) = ax + b$$

Every linear function f can be given by the rule $f(x) = ax + b$, where a and b are fixed real numbers.

The domain of any linear function f, since f is given by $f(x) = ax + b$, consists of all real numbers.

Although this can be proved, we will not do so here; the following examples will show how to find the values of a and b.

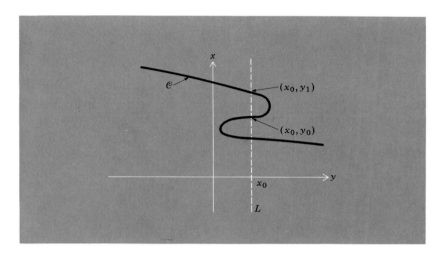

FIGURE 2.3-1 If a curve $\mathcal{C}$ is the graph of a function g, then each point on $\mathcal{C}$ has coordinates of the form $(x, g(x))$. Thus if a vertical line L intersects $\mathcal{C}$ at two points (x_0, y_0) and (x_0, y_1), then $g(x_0) = y_0$ and $g(x_0) = y_1$ and $y_0 \neq y_1$. But then g cannot be a function, since it does not have a unique value at x_0. Thus no vertical line can intersect the graph of a function at more than one point.

EXAMPLE 2.3-1

Find the real numbers a and b such that the graph of the linear function

$$g(x) = ax + b$$

is the straight line containing the points $(1, 5)$ and $(3, 1)$.

SOLUTION

Any point on the graph of g has coordinates of the form $(x, g(x))$ for some value of x. Thus, since $(1, 5)$ is on the graph of $g(x) = ax + b$, we have

$$5 = g(1) = a(1) + b$$

(see Fig. 2.3-2). But $(3, 1)$ is also on the graph of g, so

$$1 = g(3) = a(3) + b$$

An alternative method of finding a and b from the two equations $5 = a + b$ and $1 = 3a + b$ is to subtract the first equation from the second, yielding $1 - 5 = (3a + b) - (a + b)$ or $-4 = 2a$; hence $a = -2$. Now substitute $a = -2$ into either of the two equations to get $b = 7$.

Thus we have

$$5 = a + b \quad \text{and} \quad 1 = 3a + b$$

Since $5 = a + b$, we have

$$b = 5 - a$$

Replacing b by this value in the equation $1 = 3a + b$, we obtain

$$1 = 3a + 5 - a \quad \text{or} \quad 2a = -4$$

So $a = -2$, and using this value in the equation $5 = a + b$, we get $b = 7$. Thus g is given by the rule

$$g(x) = -2x + 7$$

As a simple check for errors, we substitute the values $x = 1$ and $x = 3$ into the equation $g(x) = -2x + 7$ to see whether we obtain the correct values $g(1) = 5$ and $g(3) = 7$. Since $g(1) = -2(1) + 7 = 5$ and $g(3) = -2(3) + 7 = 1$, our defining rule for g is correct.

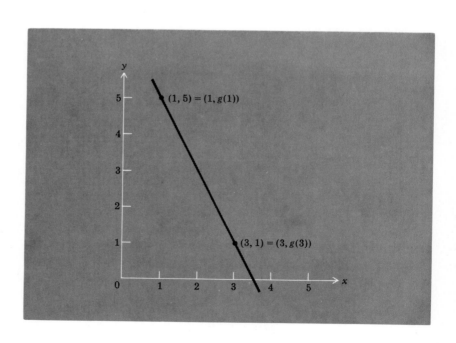

FIGURE 2.3-2

EXAMPLE 2.3-2

The Southern Shirt Company finds that its cost C to produce 2000 of its premium quality shirts is \$13,750 and the cost to produce 2500 of the shirts is \$15,250. Assuming that the production cost C is a linear function of the number x of shirts produced, find the equation which defines $C(x)$. What is the production cost for 1500 shirts?

SOLUTION

Since C is assumed to be a linear function of x, we know that $C(x) = ax + b$ for some numbers a and b. Our job is to find a and b. From the information given, we have

$$13{,}750 = C(2000) = a(2000) + b$$

and

$$15{,}250 = C(2500) = a(2500) + b$$

Thus from $13{,}750 = 2000a + b$ we see that

$$b = 13{,}750 - 2000a$$

Replacing b by this value in the equation $15{,}250 = 2500a + b$, we get

$$15{,}250 = 2500a + 13{,}750 - 2000a \quad \text{or} \quad 500a = 1500$$

So $a = 3$, and using this value in the equation $13{,}750 = 2000a + b$, we find that $b = 7750$. Thus

$$C(x) = ax + b = 3x + 7750$$

is the equation defining $C(x)$. We can check this rule for $C(x)$ by substituting the values $x = 2000$ and $x = 2500$ into $C(x) = 3x + 7750$ to see whether we obtain the correct values $C(2000) = 13{,}750$ and $C(2500) = 15{,}250$. When $x = 1500$, we have $C(1500) = 3(1500) + 7750 = \$12{,}250$, and this is the cost of producing 1500 shirts.

Every linear function C is defined by an equation of the form $C(x) = ax + b$ for some numbers a and b.

Again, an alternative method of finding a and b from the two equations $13{,}750 = 2000a + b$ and $15{,}250 = 2500a + b$ is to subtract the first equation from the second, yielding $15{,}250 - 13{,}750 = (2500a + b) - (2000a + b)$ or $1500 = 500a$; hence $a = 3$. Now substitute $a = 3$ into either of the two equations to get $b = 7750$.

To graph any particular linear function $f(x) = ax + b$, we simply compute the value of $f(x)$ at two convenient values of x, plot the two corresponding points $(x, f(x))$ on the graph, and draw the straight line between these two points.

Graphing linear functions

EXAMPLE 2.3-3

Sketch the graph of the function $f(x) = 3x - 4$.

SOLUTION

Since $f(x)$ is given by an equation of the form $f(x) = ax + b$ (with $a = 3$ and $b = -4$), $f(x)$ is a linear function and its graph will be a straight line. The domain of f consists of all real numbers. We choose two convenient values from the domain of f, say 0 and 1, and evaluate $f(0) = 3 \cdot 0 - 4 = -4$ and $f(1) = 3 \cdot 1 - 4 = -1$. Since the graph of f consists of all points in a Cartesian plane of the form $(x, f(x))$, the two points $(0, f(0)) = (0, -4)$ and $(1, f(1)) = (1, -1)$ lie on the graph of f. We plot these two points in a Cartesian plane and connect them with a straight line; the line is then the graph of $f(x) = 3x - 4$ (see Fig. 2.3-3).

The two values chosen from the domain of f are arbitrary; we have chosen 0 and 1, but any two other real numbers could be used.

FIGURE 2.3-3 The graph of $f(x) = 3x - 4$.

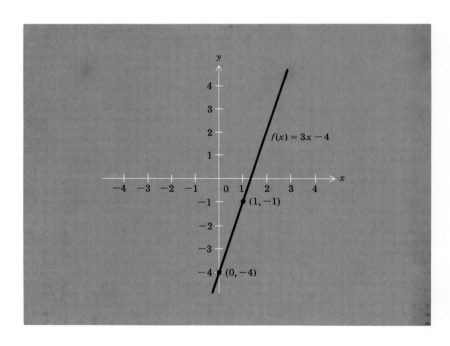

EXAMPLE 2.3-4

Sketch the graph of the function $g(x) = 2$.

SOLUTION

Since g is given by an equation of the form $g(x) = ax + b$ (with $a = 0$ and $b = 2$), g is a linear function. So we choose two values from the domain of g, say -3 and 2, and find that $g(-3) = 2$ and $g(2) = 2$. Thus $(-3, 2)$ and $(2, 2)$ lie on the graph of g, and the horizontal straight line drawn through these two points is the graph of g (see Fig. 2.3-4). There is nothing special in this example about the number 2; if c is any fixed real number, then the graph of $g(x) = c$ is the horizontal straight line through the point $(0, c)$ (see Fig. 2.3-5). Such functions are called *constant* functions.

The domain of $g(x) = 2$ consists of all real numbers x.

If c is a fixed real number, then the function $g(x) = c$ is called a *constant* function. A constant function is also a linear function; $g(x) = c$ is of the form $g(x) = ax + b$ with $a = 0$ and $b = c$.

In Example 2.3-3 we discovered that $(0, -4)$ and $(1, -1)$ were points on the graph of $f(x) = 3x - 4$. If we compute the slope of the line segment joining $(0, -4)$ and $(1, -1)$, we get

$$\frac{-4 - (-1)}{0 - 1} = \frac{-3}{-1} = 3$$

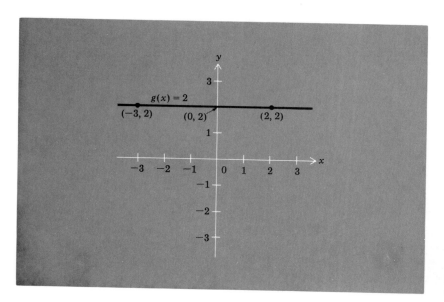

FIGURE 2.3-4 The graph of $g(x) = 2$.

FIGURE 2.3-5   The graph of $g(x)$ = c.

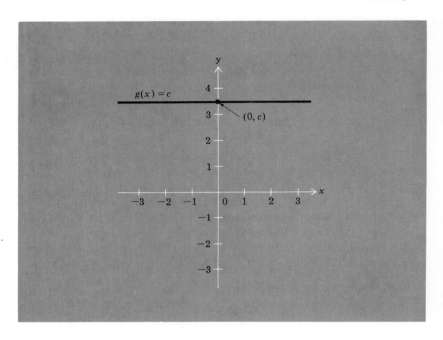

$Fig.$ 2.3-5

Now, since

$$f(-1) = 3(-1) - 4 = -7$$

and

$$f(3) = 3 \cdot (3) - 4 = 5$$

the points $(-1, -7)$ and $(3, 5)$ also lie on the line that is the graph of $f(x) = 3x - 4$, and the slope of the line segment joining these two points is

$$\frac{-7 - (5)}{-1 - (3)} = \frac{-12}{-4} = 3$$

Recall that if $P_1 = (x_0, y_0)$ and $P_2 = (x_1, y_1)$ are two points in a Cartesian plane, then the slope of the line segment joining P_1 and P_2 is the number $\dfrac{y_1 - y_0}{x_1 - x_0}$ (see Section 1.3).

again. Repeated calculations of this kind leads us to suspect that any two line segments on a given line have the same slope and that furthermore, if the given line is the graph of the linear function $f(x) = ax + b$, then the slope of any line segment on the line is the number a. This is indeed

Definition of the slope of a line

the case, and thus we define *the slope of a line* to be the slope of any line segment on the line. The expression $f(x) = ax + b$ is called the *slope–intercept* form of the linear function f. Here $f(0) = a(0) + b$, so $(0, b)$ is a point on the line and also on the y axis. This point is called the *y intercept* of the line. Since we discovered previously that horizontal

A horizontal line has slope 0. A vertical line has no slope.

line segments have slope 0 and vertical segments have no slope, the same is true for the slopes of horizontal and vertical lines.

EXAMPLE 2.3-5

Find the slope of the line that is the graph of the linear function $h(x) = (-\frac{1}{2})x + 7$.

SOLUTION

The function h is given by the equation $h(x) = ax + b$, where $a = -\frac{1}{2}$ and $b = 7$. Thus the slope of the line is just the number a, which is $-\frac{1}{2}$ (see Fig. 2.3-7).

If the slope of a line is positive, the line rises as we move from left to right; if the slope is negative, the line falls as we move from left to right (see Figs. 2.3-6 and 1.3-14).

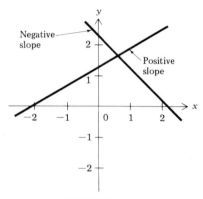

FIGURE 2.3-6

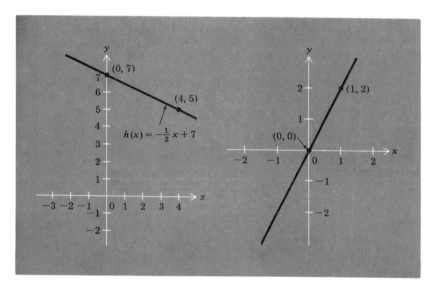

FIGURE 2.3-7   The line $h(x)$ $= (-\frac{1}{2})x + 7$ has slope $-\frac{1}{2}$.

FIGURE 2.3-8

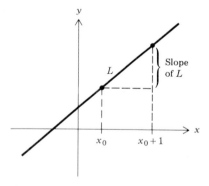

EXAMPLE 2.3-6

Find the slope and the equation of the line shown in Fig. 2.3-8.

SOLUTION

Since the points $(0, 0)$ and $(1, 2)$ lie on the line, the slope of the line is the same as the slope of the line segment joining $(0, 0)$ and $(1, 2)$, which is

$$\frac{2 - 0}{1 - 0} = \frac{2}{1} = 2$$

Since the y intercept of this line is $(0, 0)$, we know that $b = 0$; we also know that $a = 2$, so this is the graph of the linear function

$$f(x) = 2x + 0 = 2x$$

Note that the slope of this line (which is 2) gives the vertical change corresponding to a unit change on the x axis.

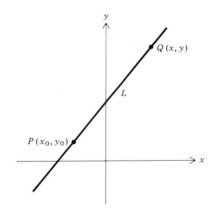

If a point $Q = (x, y)$ lies on the graph of a function f, then the coordinates of Q are of the form $(x, f(x))$, so $y = f(x)$. Because of this, the function f is frequently presented as the equation $y = f(x)$. For example, the equation $y = 3x - 1$ represents the function f defined by $f(x) = 3x - 1$, and $y = \sqrt{x}$ represents the function h defined by $h(x) = \sqrt{x}$.

To illustrate the use of this notation, let us find the equation $y = f(x)$ of a line L through a particular given point $P = (x_0, y_0)$ and having a given slope m. If $Q = (x, y)$ is any other point at all lying on L, then the slope of the segment PQ must be the slope of L, i.e., we must have

$$\frac{y - y_0}{x - x_0} = m$$

(see Fig. 2.3-9). Thus

$$y - y_0 = m(x - x_0)$$

FIGURE 2.3-9   If L has slope m, then $m = \dfrac{y - y_0}{x - x_0}$.

Since this equation holds true for *any* point (x, y) on L, it defines a linear function f whose graph is L. Another way of expressing this equation is

$$y = y_0 + m(x - x_0)$$

From this we see that f is defined by

$$f(x) = y_0 + m(x - x_0)$$

The point-slope formula for the equation of a line: The equation of the line L through $P = (x_0, y_0)$ with slope m is $y - y_0 = m(x - x_0)$.

The equation $y - y_0 = m(x - x_0)$ is called the *point-slope* formula for the equation of a line.

EXAMPLE 2.3-7

Find the equation of the line L through $P = (-1, 3)$ with slope $m = -4$.

SOLUTION

The equation of the line L is given by $y - y_0 = m(x - x_0)$ with $x_0 = -1$, $y_0 = 3$, and $m = -4$. Thus the equation is $y - 3 = -4(x - (-1))$, or $y = -4x - 1$. This equation tells us that the graph of the linear function $f(x) = -4x - 1$ is the line L (see Fig. 2.3-10). We can partially check this by substituting the value $x = -1$ in this expression to get $f(-1) = -4(-1) - 1 = 3$, as asserted.

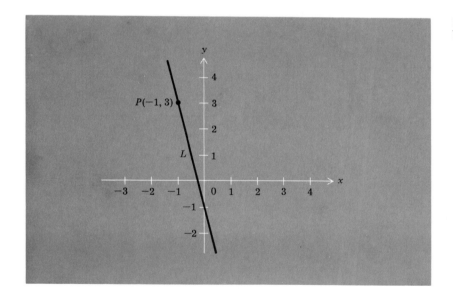

FIGURE 2.3-10 The line L defined by the equation $y = -4x - 1$.

Suppose now that a line L is known to pass through *two* given points in the plane, say $P = (x_0, y_0)$ and $Q = (x_1, y_1)$. Then the slope of L is

$$\frac{y_1 - y_0}{x_1 - x_0}$$

The two-point formula for the
equation of a line: The equation of
the line L through $P = (x_0, y_0)$
and $Q = (x_1, y_1)$ is
$$y - y_0 = \left(\frac{y_1 - y_0}{x_1 - x_0} \right)(x - x_0).$$

and using the point-slope formula, the equation of L is

$$y - y_0 = \frac{y_1 - y_0}{x_1 - x_0}(x - x_0)$$

This equation is called the *two-point* formula. From this equation, we see that the linear function f whose graph is the line L through $P = (x_0, y_0)$ and $Q = (x_1, y_1)$ is defined by

$$f(x) = y_0 + \frac{y_1 - y_0}{x_1 - x_0}(x - x_0)$$

EXAMPLE 2.3-8

Find the equation of the line L through the points $P = (1, -2)$ and $Q = (-3, 4)$.

SOLUTION

The equation of the line L is given by

$$y - y_0 = \frac{y_1 - y_0}{x_1 - x_0}(x - x_0)$$

with $x_0 = 1$, $y_0 = -2$, $x_1 = -3$, and $y_1 = 4$. Thus the equation is

$$y - (-2) = \frac{4 - (-2)}{-3 - 1}(x - 1)$$

or

$$y = -\tfrac{3}{2}x - \tfrac{1}{2}$$

This means that the graph of the linear function f defined by $f(x) = -\tfrac{3}{2}x - \tfrac{1}{2}$ is the line L (see Fig. 2.3-11). We can check this equation by substituting the values $x = -3$ and $x = 1$ into it to verify that $f(-3) = 4$ and $f(1) = -2$.

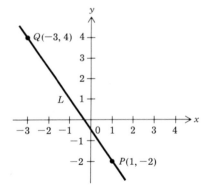

FIGURE 2.3-11   The line L defined by the equation $y = -\tfrac{3}{2}x - \tfrac{1}{2}$.

Our discussion so far in this section has certainly simplified the task of sketching the graph of any function g of the form $g(x) = ax + b$; we need only locate two points on the graph of g and then draw a line through these two points. Are there ways of shortening the graphing procedure for other functions? The remainder of this section is devoted to some special techniques of graphing which are frequently helpful in this respect.

Let f be a given function, and suppose that h is the function defined by $h(x) = (-1)f(x)$ for each x in the domain of f. Then h is usually denoted by $-f$, i.e.,

$$-f(x) = (-1)f(x)$$

Now, for each number x in the domain of f, the values $f(x)$ and $-f(x)$ are at the same (vertical) distance from the x axis but on opposite sides of the axis (see Fig. 2.3-12). Thus the graph of $-f$ is the inverted graph of f located the same distance from, but on the opposite side of, the x axis (see, for example, the graphs in Fig. 2.3-13). Thus if the graph of f is known, the graph of the function $-f$ is easily drawn.

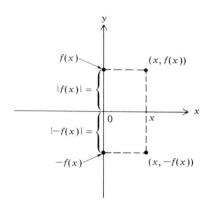

FIGURE 2.3-12 The numbers $f(x)$ and $-f(x)$ on the y axis are the same distance from the origin, i.e., $|f(x)| = |-f(x)|$.

FIGURE 2.3-13

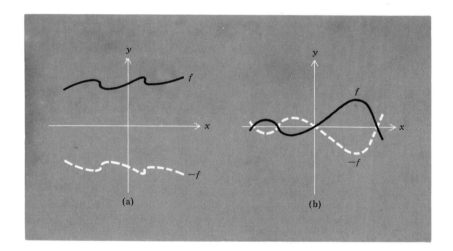

EXAMPLE 2.3-9

Sketch the graph of the function $g(x) = -x^2$.

SOLUTION

The graph of the function $f(x) = x^2$ was drawn in Example 2.1-4 and is shown in Fig. 2.3-14. We invert this graph and locate it the same distance from, but on the opposite side of, the x axis to get the graph of $g(x) = -f(x) = -x^2$, as shown in Fig. 2.3-15.

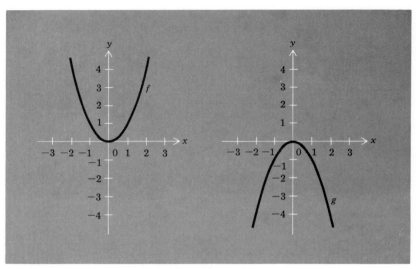

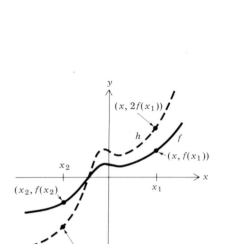

FIGURE 2.3-16

FIGURE 2.3-14   The graph of $f(x)$ = x^2.

FIGURE 2.3-15   The graph of $g(x)$ = $-x^2$.

Suppose now that for a given function f, the function h is defined by $h(x) = 2f(x)$. Then for each point $(x, f(x))$ on the graph of f, the point $(x, 2f(x))$ is on the graph of h, so at each point x the graph of h is twice as far above (or below) the x axis as the graph of f (see Fig. 2.3-16). Thus the graph of h can be obtained by "stretching" the graph of f vertically by a factor of 2. More generally, for any real number c, the graph of

FIGURE 2.3-17   **FIGURE 2.3-18**

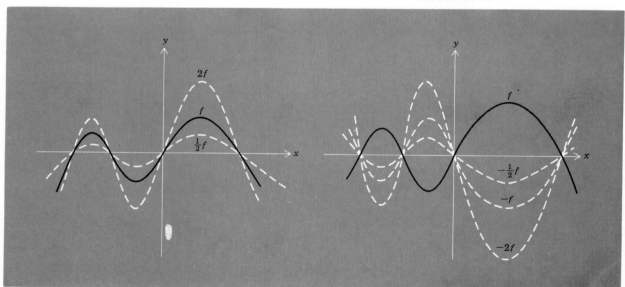

$h(x) = cf(x)$ can be obtained by treating $|c|$ as a vertical "stretching" or "shrinking" factor of the graph of f or of $-f$. If c is positive, then $|c|$ stretches the graph if $|c| > 1$ and shrinks it if $|c| < 1$; similarly, if c is negative, then $|c|$ stretches the graph of $-f$ if $|c| > 1$ and shrinks it if $|c| < 1$. Figures 2.3-17 and 2.3-18 illustrate some of the possibilities.

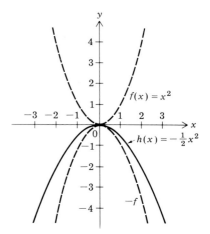

FIGURE 2.3-19   The graph of $h(x) = (-1/2)x^2$.

EXAMPLE 2.3-10

Sketch the graph of the function $h(x) = -\frac{1}{2}x^2$.

SOLUTION

The function h is given by $h(x) = cf(x)$, where $f(x) = x^2$ and $c = -\frac{1}{2}$. Since c is negative and $|c| = \frac{1}{2} < 1$, the graph of h is obtained by shrinking the graph of $-f$ (shown in Fig. 2.3-15) by a factor of $|c| = \frac{1}{2}$ (see Fig. 2.3-19).

Given two functions f and g, whose graphs are known, there is a simple method for graphing the function $f + g$. To illustrate the method, let us suppose that both f and g are positive-valued functions, with graphs as depicted in Fig. 2.3-20. Imagine, for the moment, a vertical wire through each point x on the x axis. Now, for each x, lengths of drinking straws with colored wooden beads at the top are cut to

Recall that the function $f + g$ is defined by the rule $(f + g)(x) = f(x) + g(x)$ for each x common to the domains of f and g.

A function f is positive-valued if for each x in the domain of f, $f(x) > 0$.

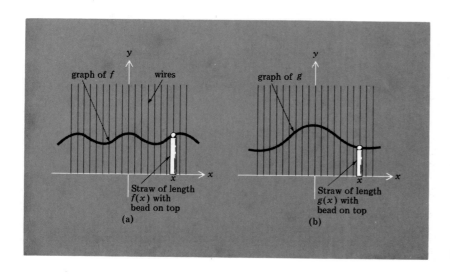

FIGURE 2.3-20

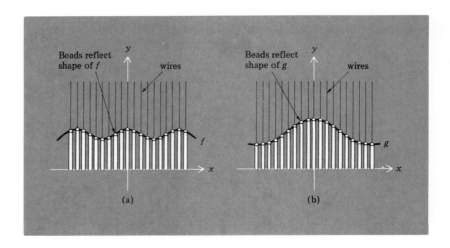

FIGURE 2.3-21

correspond to the values of $f(x)$ and $g(x)$ and slid onto the wires (see Fig. 2.3-20). After such straws are slid onto all the wires, the beads on the top of the straws reflect the shape of f and g, as shown in Fig. 2.3-21. If the straws and beads in Fig. 2.3-21(a) are removed from the wires and slid onto the corresponding wires in Fig. 2.3-21(b), then the top beads indicate the shape of the graph of $f + g$, since at each x, the value of $f + g$ is $f(x) + g(x)$ (see Fig. 2.3-22).

FIGURE 2.3-22

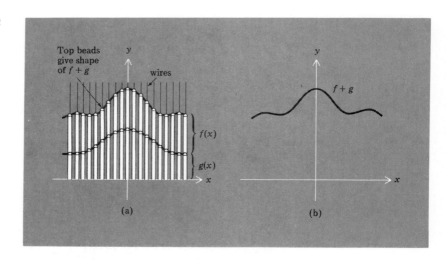

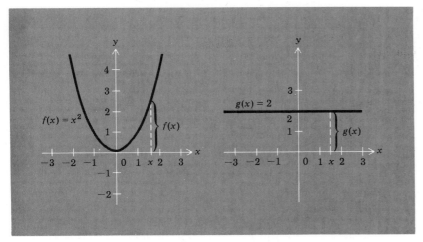

FIGURE 2.3.-23 The graph of $f(x)$ = x^2.

FIGURE 2.3-24 The graph of $g(x)$ = 2.

EXAMPLE 2.3-11

Sketch the graph of $h(x) = x^2 + 2$.

SOLUTION

The function h is given by $h(x) = f(x) + g(x)$, where $f(x)$ = x^2 and $g(x) = 2$. The graphs of f and g are shown in Figs. 2.3-23 and 2.3-24. Now for each x, the value of $h(x)$ is $x^2 + 2$, i.e., the height of the graph of h above x is the height of f above x plus the height of g above x (as shown by the straws and beads). So we select several values of x, locate the height of the graph of h above x, (namely, $f(x) + g(x)$), and sketch the graph of h as shown in Fig. 2.3-25. Note that the graph of $h(x) = x^2 + 2$ appears to be the graph of $f(x) = x^2$ shifted upward two units. This is indeed the case, and in general, it can be shown that for any function f and any number c, the graph of $h(x) = f(x) + c$ is the graph of f shifted upward c units if c is positive and downward $|c|$ units if c is negative (see Fig. 2.3-26).

For any function f and real number c, the graph of $h(x)$ = $f(x) + c$ is the graph of f shifted upward c units if c is positive and shifted downward $|c|$ units if c is negative.

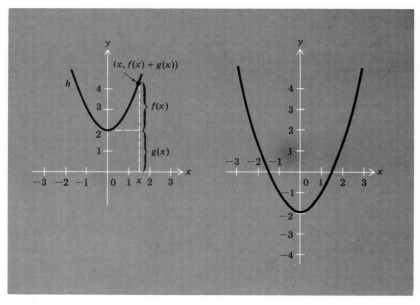

FIGURE 2.3-25   The graph of $h(x)$ = $x^2 + 2$.

FIGURE 2.3-26   The graph of $h(x)$ = $x^2 - 2$ is the graph of $f(x) = x^2$ shifted downward 2 units.

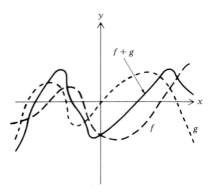

FIGURE 2.3-27

It is to be emphasized that our illustration of the shape of the graph of $f + g$ in terms of straws and beads is valid *only* for *positive-valued* functions f and g. However, when graphing the sum of *any* two functions (positive-valued or not), the following observations are helpful (see Fig. 2.3-27).

(1)   The graph of $f + g$ will coincide with that of f at any point x for which $g(x) = 0$ and with that of g at any point x for which $f(x) = 0$.

(2)   The graph of $f + g$ will be above both graphs at any point x for which $f(x)$ and $g(x)$ are both positive and below both graphs if $f(x)$ and $g(x)$ are both negative.

(3)   The graph of $f + g$ will be between the graphs of f and g at any point x for which one of the values $f(x)$ or $g(x)$ is positive and the other is negative.

EXAMPLE 2.3-12

Sketch the graph of the function $f + g$ where $f(x) = -x^2 + 1$ and $g(x) = x - 1$.

SOLUTION

We first sketch the graphs of f and g on the same coordinate system, as shown in Fig. 2.3-28; the graph of f is that of $h(x) = -x^2$ shifted upward one unit, and the graph of g is the straight line through the points $(0, -1)$ and $(1, 0)$. Then, from these graphs and the above discussion, we see that the graph of $f + g$ will coincide with that of f at $x = 1$ (since $g(1) = 0$) and with that of g at $x = 1$ and $x = -1$ (since $f(1) = f(-1) = 0$). Furthermore, the graph of $f + g$ will be below both graphs when $x < -1$ (since both $f(x)$ and $g(x)$ are negative for $x < -1$) and between the two graphs when $x > -1$ (since one of the values $f(x)$ or $g(x)$ is positive and the other negative for any value of $x > -1$, except $x = 1$). Now, after locating a few points on the graph of $f + g$ (locate some points of the form $(x, f(x) + g(x))$ for some arbitrarily chosen values of x), we use this information to sketch the graph of $f + g$ as shown in Fig. 2.3-29. Note that we have also redrawn the graphs of f and g in dotted lines in Fig. 2.3-29 to aid us in properly positioning the graph of $f + g$.

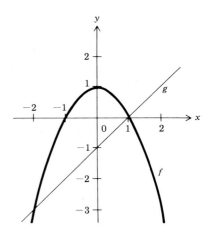

FIGURE 2.3-28 The graphs of $f(x)$ $= -x^2 + 1$ and $g(x) = x - 1$.

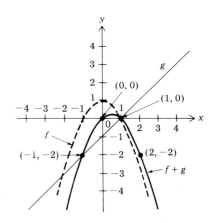

FIGURE 2.3-29

Exercises 2.3

In Exercises 1 through 9, find real numbers a and b such that the graph of the linear function $f(x) = ax + b$ is the straight line containing the two given points. Sketch the line and compute its slope. Check your answers as in Example 2.3-1.

1. $(0, 0), (1, 1)$ 2. $(-1, 2), (0, 0)$
3. $(-2, 2), (2, -2)$ 4. $(1, 3), (2, 4)$
5. $(-1, 4), (0, -2)$ 6. $(-2, 4), (0, -3)$
7. $(-3, \frac{3}{2}), (1, \frac{3}{2})$ 8. $(-3, -\frac{1}{2}), (1, \frac{3}{2})$
9. $(-\frac{1}{3}, \frac{1}{4}), (\frac{1}{4}, -\frac{1}{3})$

In Exercises 10 through 19, sketch the graph of the given linear function and give its slope.

10. $f(x) = 2x + 4$ 11. $f(x) = -2x + 4$
12. $f(x) = 2x - 4$ 13. $f(x) = -2x - 4$
14. $f(x) = 5x + 3$ 15. $f(x) = -3x + \frac{1}{2}$
16. $f(x) = -\frac{3}{2}x$ 17. $f(x) = \frac{7}{2}$
18. $f(x) = -\frac{1}{2}x + \frac{3}{2}$ 19. $f(x) = \frac{4}{5}x - \frac{1}{2}$

In Exercises 20 through 29, use the point-slope formula to find the equation of the line through the given point and with the given slope; then sketch the line. Use the method of Example 2.3-7 to make a partial check of your answers.

20. $P = (1, 1), m = 3$

21. $P = (4, 2), m = 2$

22. $P = (3, -2), m = 6$

23. $P = (-4, 0), m = 4$

24. $P = (-7, 0), m = 3$

25. $P = (6, -4), m = -7$

26. $P = (-\frac{2}{3}, -3), m = -2$

27. $P = (-5, \frac{3}{4}), m = -6$

28. $P = (\frac{1}{2}, 0), m = \frac{1}{2}$

29. $P = (-\frac{5}{3}, \frac{1}{2}), m = -\frac{3}{4}$

In Exercises 30 through 39, use the two-point formula to find the equation of the line through the given points; then sketch the line. Check your answers as in Example 2.3-8.

30. $P = (3, 4), Q = (4, 6)$

31. $P = (10, 7), Q = (5, 12)$

32. $P = (-1, 4), Q = (1, -2)$

33. $P = (3, -2), Q = (4, -4)$

34. $P = (-3, -5), Q = (-2, -1)$

35. $P = (-7, -3), Q = (-5, -1)$

36. $P = (-2, \frac{1}{2}), Q = (-4, 2)$

37. $P = (7, -3), Q = (-\frac{3}{4}, -2)$

38. $P = (\frac{1}{5}, -3), Q = (2, -\frac{1}{2})$

39. $P = (4, -\frac{1}{3}), Q = (\frac{1}{2}, 7)$

40. Show that the graph of the linear function $f(x) = ax + b$ (where a and b are fixed real numbers) intersects the y axis at the point $(0, b)$. (As mentioned before, the number b is called the y intercept of the line.)

41. Show that, for fixed real numbers a and b, $a \neq 0$, the graph of the linear function $f(x) = ax + b$ intersects the x axis at the point $(-b/a, 0)$. (The number $-b/a$ is called the x intercept of the line.)

In Exercises 42 through 50, use the techniques of graphing described in this section (when possible) to sketch the graphs of the indicated functions.

42. Let $f(x) = 2x + 1$ and $g(x) = 3$. On a common coordinate system, sketch the graphs of the functions f, g, $f + g$, fg, and $-f$.

43. Let $f(x) = -4x - 2$ and $g(x) = \frac{1}{2}$. On a common coordinate system, sketch the graphs of the functions f, g, $f - g$, fg, and $-f$.

44. Let $f(x) = x^2$ and $g(x) = x$. On a common coordinate system, sketch the graphs of the functions f, g, $f + g$, $-g$, and $f - g$.

45. Let $f(x) = x^2$ and $g(x) = \frac{1}{4}$. On a common coordinate system, sketch the graphs of the functions f, g, fg, $fg + g$, and $f + g$.

46. Let $f(x) = x^3$ and $g(x) = x^2$. On a common coordinate system, sketch the graphs of the functions f, g, $f + g$, $f + 1$, and $(\frac{1}{4})f$.

47. Let $f(x) = |x|$ and $g(x) = -x + 2$. On a common coordinate system, sketch the graphs of the functions f, g, $f + 2$, $(f + 2) + g$, and $(f + 2) - g$.

48. Let $f(x) = \sqrt{x - 1}$ and $g(x) = x$. On a common coordinate system, sketch the graphs of the functions f, g, $f + g$, $3(f + g)$, and $\frac{1}{3}(f + g)$.

49. Let $f(x) = x^4$ and $g(x) = x^2$. On a common coordinate system, sketch the graphs of the functions f, g, $g - 2$, $f + (g - 2)$, and $-[f + (g - 2)]$.

50. Let $f(x) = x^5$ and $g(x) = |x|$. On a common coordinate system, sketch the graphs of the functions f, g, $-f$, $-f + g$, and $-(-f + g)$.

51. Ace Auto Electric Company finds that its cost C to produce 150 of its premium quality automobile batteries is $2800 and the cost to produce 200 of the batteries is $3400. Assuming that the production cost C is a linear function of the number x of batteries produced, find an equation which defines $C(x)$. (Check your answer.) What is the cost to produce 300 batteries?

52. A northern industrial city finds that on an average workday, the pollution index P at 11:00 AM is 45 ppm and at 4:00 PM the index is 70 ppm. Assuming that the pollution index is a linear function of the time t (with t measured in hours and $t = 0$ corresponding to 8:00 AM) for $0 \leqslant t \leqslant 10$, find an equation which defines $P(t)$. (Check your answer.) What is the pollution index at 6:00 PM?

53. A group of students arrange a round-trip chartered flight to London. The total cost C of the flight is $22,000 if 60 students make the trip and $27,250 if 75 students go. It is known that the cost of the flight is a linear function of the number x of students making the trip provided $x \geqslant 50$. Assuming that more than 50 students go, find an equation which defines the cost C of chartering the flight as a function of the number x of students making the trip. (Check your answer.) How much will it cost to charter the flight if 86 students go?

54. The Quick-Hot Company manufactures and sells home water heaters. For its premium quality 50-gal heaters, the company finds that it can sell 1250 units per year if they are priced at $125 each, and 1100 units if they are priced at $140 each. Assuming that the demand D (number of heaters per year that can be sold) is a linear function of the selling price x if $110 \leqslant x \leqslant 160$, find an equation which defines $D(x)$. (Check your answer.) How many heaters per year can the company sell if they are priced at $150 each?

55. Economy Calculators, Inc., manufactures and sells electronic calculators. For its economy-model hand calculator, the company finds that it can sell 700 units per month if they are priced at $65

each and 610 units per month if they are priced at $80 each. Assuming that the demand D (number of calculators that can be sold per month) is a linear function of the selling price x, $55 \leqslant x \leqslant 90$, find an equation which defines $D(x)$. (Check your answer.) How many calculators per month can the company sell if they are priced at $70 each?

56. A paper mill discharges waste sulfuric acid into a river. The rate R at which the acid flows into the river at 10:00 AM is 60 cubic ft per hour. At 2:00 PM the acid is discharged into the river at 76 cubic ft per hour. Assuming that the rate R of discharge into the river is a linear function of the time t (with t measured in hours and $t = 0$ corresponding to 8:00 AM) for $2 \leqslant t \leqslant 8$, find an equation which defines $R(t)$. (Check your answer.) At 12:30 PM, what is the rate of discharge of acid into the river?

57. A factory starts operation at 8:00 AM. The rate of emission of smoke from the factory at 11:00 AM is 120 cubic ft per minute. At 3:00 PM the rate of emission is 144 cubic ft per minute. Assuming that the rate R of emission of smoke is a linear function of the time t (with t measured in hours and $t = 0$ corresponding to 8:00 AM) for $0 \leqslant t \leqslant 10$, find an equation which defines $R(t)$. (Check your answer.) At 5:00 PM, what is the rate of emission of smoke from the factory?

58. It is found experimentally that the increase I in pulse rate due to an injection of a certain drug is a linear function of the amount x of the drug injected for $0.1 \leqslant x \leqslant 0.5$ cc (cubic centimeters). If an injection of 0.2 cc of the drug causes the pulse rate to increase by an amount $I = 10$ beats per minute and an injection of 0.4 cc results in an increase of $I = 30$ beats per minute, find an equation which defines $I(x)$. (Check your answer.) How much will the pulse rate be increased if 0.35 cc of the drug is injected?

2.4.　Power Functions and Polynomials

Definition of a power function: $f(x) = x^n$ for some positive integer n.

Functions such as $g(x) = x^2$, $h(x) = x^3$, $p(x) = x^4$, $q(x) = x^5$, ... are called *power functions*. That is, a power function f is given by the rule $f(x) = x^n$ for some fixed positive integer n. The graphs of the power functions for $n = 2$, 3, 4, and 5 (obtained by plotting points and connecting them with a smooth curve) are shown in Fig. 2.4-1. Notice that the graphs of $f(x) = x^n$ for $n = 2$ and $n = 4$ are quite similar, and so are the graphs of $f(x) = x^n$ for $n = 3$ and $n = 5$. As a matter of fact, the graph of $f(x) = x^n$, where n is *any* even positive integer, is similar to the graph of $f(x) = x^2$, while the graph of $f(x) = x^n$ resembles that of $f(x) = x^3$ if n is odd (see Fig. 2.4-2).

A positive integer n is even if it is of the form $n = 2k$, for some integer k. If n can be expressed as $n = 2k + 1$ for some integer k, then n is odd.

By using the graphs of the power functions and the graphing techniques of the previous section, little effort is required to sketch the graphs

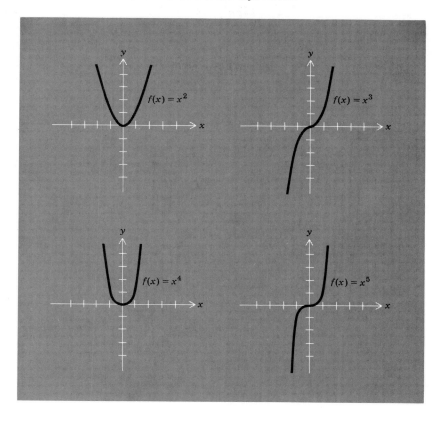

FIGURE 2.4-1

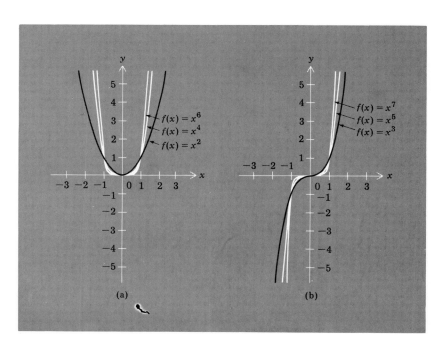

If n is an even positive integer, the graph of $f(x) = x^n$ resembles that of $f(x) = x^2$. If n is odd, the graph of $f(x) = x^n$ resembles that of $f(x) = x^3$.

FIGURE 2.4-2   Graphs of $f(x) = x^n$ for (a) n even and (b) n odd.

In Fig. 2.4-3, recall that the graph of $g(x) = 3x^3$ is obtained by vertically stretching the graph of the power function $f(x) = x^3$ by a factor of 3. Then the graph of $h(x) = 3x^3 + 2$ is that of $g(x) = 3x^3$ shifted vertically upward 2 units.

Definition of polynomial functions

Given a polynomial function $p(x) = a_n x^n + a_{n-1} x^{n-1} + \cdots + a_1 x + a_0, a_n \neq 0$, the *degree* of $p(x)$ is n and the real number a_n is the *leading coefficient*. For any positive integer $k \leq n$, the real number a_k is called the *coefficient* of x^k.

of functions such as $g(x) = 3x^3$, $h(x) = 3x^3 + 2$, and $q(x) = x^6 + 3x^3 + 2$ (see Fig. 2.4-3). In fact, we could (perhaps with considerable effort) graph any function p which can be defined by an equation of the form

$$p(x) = a_n x^n + a_{n-1} x^{n-1} + \cdots + a_1 x + a_0$$

where n is some positive integer or 0 and $a_n, a_{n-1}, \ldots, a_0$ are fixed real numbers with $a_n \neq 0$. Such functions are called *polynomials*; the integer n is called the *degree* of the polynomial, and the real number a_n is called the *leading coefficient*.

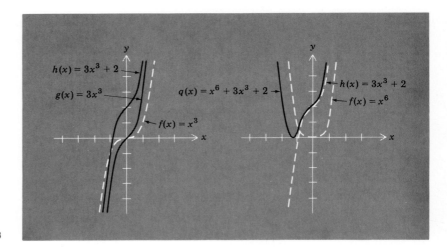

FIGURE 2.4-3

EXAMPLE 2.4-1

Find the degree and the leading coefficient of each of the polynomial functions $f(x) = 4x^5 + 5x^7 + \sqrt{2}$, $g(x) = \sqrt{2}\, x - 2$, and $h(x) = 5.7$.

SOLUTION

We first rewrite the function f so that we have descending powers of x. Then

$$f(x) = 5x^7 + 4x^5 + \sqrt{2}$$

so $f(x)$ is of the form

$$a_n x^n + a_{n-1} x^{n-1} + \cdots + a_1 x + a_0$$

where $n = 7$, $a_7 = 5$, $a_6 = 0$, $a_5 = 4$, $a_3 = 0$, $a_2 = 0$, $a_1 = 0$, and $a_0 = \sqrt{2}$. Thus the degree of f is 7 and the leading coefficient is 5. For the function g, we see that $g(x)$ is a polynomial of degree $n = 1$ and leading coefficient $\sqrt{2}$. Finally, $h(x)$ is of the form

$$a_n x^n + a_{n-1} x^{n-1} + \cdots + a_1 x + a_0$$

where $a_0 = 5.7$ and $a_k = 0$ for all $k > 0$. Thus n, the degree, is 0, and the leading coefficient a_0 is 5.7. Note that $g(x) = \sqrt{2}\, x - 2$ is a linear function and that $h(x) = 5.7$ is a constant function. Any linear function q, since it is of the form $q(x) = ax + b$, is a polynomial; it is of degree 1 if $a \neq 0$ and of degree 0 if $a = 0$. Any constant function (other than the 0 function) is a polynomial of degree 0.

A constant function (other than the 0 function) is a polynomial of degree 0, since $p(x) = c$ can be written $p(x) = cx^0$ $(x^0 = 1)$.

Recall that every linear function q is described by an equation of the form $q(x) = ax + b$ for some real numbers a and b.

Linear functions and constant functions are polynomials.

EXAMPLE 2.4-2

State which of the functions $f(x) = 3\sqrt{x} + 2$, $g(x) = \sqrt{2}\, x + \frac{1}{2}$, $h(x) = (x^2 - 9)/(x - 3)$, and $q(x) = x^7/100 + 4x$ are polynomials; find the degree and leading coefficient of each function that is a polynomial.

SOLUTION

The functions g and q are defined by equations of the form $a_n x^n + a_{n-1} x^{n-1} + \cdots + a_1 x + a_0$ and are thus polynomial functions; the functions f and h are not polynomials. The degree of g is 1, and its leading coefficient is $\sqrt{2}$; the degree of q is 7, and its leading coefficient is $\frac{1}{100}$.

Note that the function $h(x) = (x^2 - 9)/(x - 3)$ is not defined at $x = 3$ (since the denominator is 0). However, for $x \neq 3$, we have $h(x) = (x^2 - 9)/(x - 3) = (x - 3)(x + 3)/(x - 3) = x + 3$, so the values of h and the values of the polynomial function $p(x) = x + 3$ coincide for all $x \neq 3$.

Polynomial functions of degree 2, i.e., functions f which can be described by equations of the form $f(x) = a_2 x^2 + a_1 x + a_0$ $(a_2 \neq 0)$, are called *quadratic* functions. Since only three coefficients are used, it is customary to avoid the use of subscripts and present quadratic functions with the notation

$$f(x) = ax^2 + bx + c \qquad (a \neq 0)$$

A quadratic function is a polynomial function of degree 2. Its graph is called a *parabola*.

The graph of a quadratic function is called a *parabola* and is similar in shape to the graph of the (quadratic) power function $f(x) = x^2$ (see Fig. 2.4-4). It can be shown that the graph of

$$f(x) = ax^2 + bx + c \quad (a \neq 0)$$

opens upward (as does that of $f(x) = x^2$) if a is positive and opens downward (as does that of $f(x) = -x^2$) if a is negative (see Fig. 2.4-5). Furthermore, the *vertex* of the parabola, which is its low point if $a > 0$ and its high point if $a < 0$, occurs at the point $(p, f(p))$, where

$$p = \frac{-b}{2a}$$

(see Fig. 2.4-5).

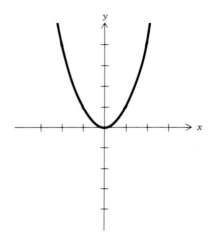

FIGURE 2.4-4   The graph of $f(x)$ $= x^2$.

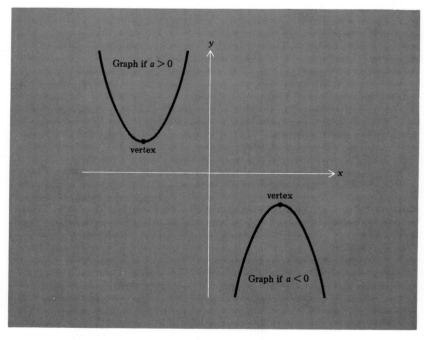

FIGURE 2.4-5   The graph of $f(x)$ $= ax^2 + bx + c$ opens upward if $a > 0$ and downward if $a < 0$. The low point (if $a > 0$) or high point (if $a < 0$) of the graph is called the *vertex* and has coordinates $(p, f(p))$, where $p = -b/2a$.

EXAMPLE 2.4-3

Sketch the graph of the quadratic function $f(x) = -x^2 - 2x$ and find the coordinates of the vertex.

SOLUTION

We see that $f(x) = -x^2 - 2x$ is of the form $f(x) = ax^2 + bx + c$ where $a = -1 < 0$. Thus the parabola opens downward. Since $b = -2$, we have

$$p = \frac{-b}{2a} = \frac{-(-2)}{2(-1)} = -1$$

so

$$f(p) = f(-1) = -(-1)^2 - 2(-1) = 1$$

and the vertex is the point $(p, f(p)) = (-1, 1)$. The graph is shown in Fig. 2.4-6. Notice that since $f(0) = f(-2) = 0$, the points $(0, 0)$ and $(-2, 0)$ are on the graph, so the graph intersects the x axis at $x = 0$ and $x = -2$.

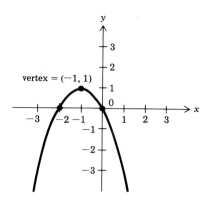

FIGURE 2.4-6

The graph of a quadratic function $f(x) = ax^2 + bx + c$ may or may not intersect the x axis (see Figs. 2.4-5 and 2.4-6).

If $(x, f(x))$ is a point of the graph which is also on the x axis, then the second coordinate, $f(x)$, must equal 0. Thus to determine the points where the graph of f crosses the x axis we must solve the equation $f(x) = 0$, or $ax^2 + bx + c = 0$. Such an equation is called a *quadratic equation* and can be solved by the method of "completing the square" (see Example 2.4-8). The formulas obtained by this method for the solutions to this equation are

$$x = \frac{-b + \sqrt{b^2 - 4ac}}{2a} \quad \text{and} \quad x = \frac{-b - \sqrt{b^2 - 4ac}}{2a}$$

If $b^2 - 4ac < 0$, there are no real numbers which solve the equation and the parabola does not intersect the x axis. If $b^2 - 4ac = 0$, the two formulas for the solutions give the same value for x and the parabola touches the x axis at one point. If $b^2 - 4ac > 0$, the parabola crosses the x axis at two distinct points.

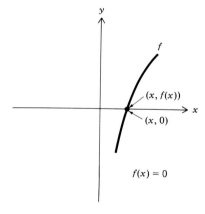

Solution of a quadratic equation

EXAMPLE 2.4-4

Sketch the graph of the quadratic function $f(x) = x^2 + 2x + 1$. Find the vertex of the parabola and find the points (if any) where the parabola intersects the x axis.

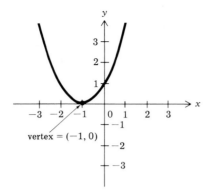

FIGURE 2.4-7   The graph of $f(x)$ = $x^2 + 2x + 1$.

Since $4 - 4(1)(1) = 0$, the parabola touches the x axis at only one point.

Since $(-2)^2 - 4(-1)(0) = 4 > 0$, the parabola intersects the x axis at two distinct points.

SOLUTION

The graph of $f(x) = x^2 + 2x + 1$ opens upward, since the leading coefficient a is 1 and $1 > 0$. Also, $p = -b/2a = -2/2 = -1$, so $f(p) = f(-1) = 0$, thus the vertex is $(p, f(p)) = (-1, 0)$. The parabola intersects the x axis at

$$x = \frac{-b + \sqrt{b^2 - 4ac}}{2a} = \frac{-2 + \sqrt{4 - 4}}{2} = -1$$

and at

$$x = \frac{-b - \sqrt{b^2 - 4ac}}{2a} = \frac{-2 - \sqrt{4 - 4}}{2} = -1$$

Thus the graph intersects the x axis in just one point, namely $x = -1$. The parabola is shown in Fig. 2.4-7.

EXAMPLE 2.4-5

Verify that the graph of the quadratic function given in Example 2.4-3 intersects the x axis at $x = 0$ and $x = -2$.

SOLUTION

The quadratic function given in Example 2.4-3 is $f(x) = -x^2 - 2x$. Thus $a = -1$, $b = -2$, and $c = 0$, so the parabola intersects the x axis at

$$x = \frac{-b + \sqrt{b^2 - 4ac}}{2a} = \frac{2 + \sqrt{4 - 0}}{2(-1)} = -2$$

and at

$$x = \frac{-b - \sqrt{b^2 - 4ac}}{2a} = \frac{2 - \sqrt{4 - 0}}{2(-1)} = 0$$

The parabola is shown in Fig. 2.4-6. Of course, if the quadratic expression factors easily, the solutions of the quadratic equation are quickly read off from the factors. Here

$$-x^2 - 2x = -x(x + 2)$$

so $-x^2 - 2x = 0$ if and only if $-x(x+2) = 0$, i.e., when $x = 0$ or $x + 2 = 0$; thus $x = 0$ or $x = -2$.

EXAMPLE 2.4-6

Sketch the graph of the quadratic function $g(x) = -x^2 - 1$. Find the vertex of the parabola and the points where the parabola intersects the x axis.

SOLUTION

The graph of $g(x) = -x^2 - 1$ opens downward, since $a = -1 < 0$. Since $b = 0$ and $a = -1$, we have $p = -b/2a = 0/-2 = 0$ and $f(p) = f(0) = -1$; so the vertex is the point $(p, f(p)) = (0, -1)$. The parabola intersects the x axis at

$$x = \frac{-b + \sqrt{b^2 - 4ac}}{2a} = \frac{0 + \sqrt{0 - 4}}{2(-1)}$$

But we cannot compute $\sqrt{-4}$; that is, x cannot be computed (as a real number), since $b^2 - 4ac = -4 < 0$. Similarly,

$$\frac{-b - \sqrt{b^2 - 4ac}}{2a}$$

cannot be computed, so the parabola does not intersect the x axis. The graph is shown in Fig. 2.4-8.

From $g(x) = -x^2 - 1$ we get $(0)^2 - (4)(-1)(-1) = -4 < 0$; thus the parabola does not intersect the x axis.

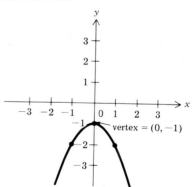

FIGURE 2.4-8   The graph of $g(x) = -x^2 - 1$.

EXAMPLE 2.4-7

A company estimates that its daily profit P (in dollars) from the manufacture and sale of x units of its product is given by the equation $P(x) = -x^2 + 60x - 50$. Sketch the graph of the function P and, by finding the vertex of the parabola, find the number of units that the company should manufacture and sell per day in order to maximize its profit. What is the maximum possible daily profit?

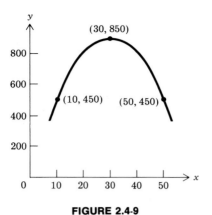

FIGURE 2.4-9

SOLUTION

The function $P(x) = -x^2 + 60x - 50$ is a quadratic function with $a = -1 < 0$, so the graph (a parabola) opens downward. Since $a = -1$ and $b = 60$, we see that $-b/2a = -60/2(-1) = 30$, and the vertex of the parabola is the point $(30, P(30)) = (30, 850)$. The graph of the function P is shown in Fig. 2.4-9. Since the highest point on the graph represents maximum profit and the vertex $(30, 850)$ is the highest point, we see that when the company manufactures and sells $x = 30$ units per day, its daily profit P is greatest and is $P(30) = \$850$.

EXAMPLE 2.4-8

Use the method of "completing the square" to solve the general quadratic equation $ax^2 + bx + c = 0$ $(a \neq 0)$.

SOLUTION

First verify that if d is any real number at all, then

$$(x + d)^2 = x^2 + 2dx + d^2$$

Now, given an equation of the form $x^2 + px = q$, we can "complete the square," that is, transform the equation in such a way that its left-hand side is of the form $x^2 + 2dx + d^2$. We do this by adding the square of one-half the coefficient of x (namely, $(p/2)^2$) to both sides of our equation to obtain

$$x^2 + px + \left(\frac{p}{2} \right)^2 = q + \left(\frac{p}{2} \right)^2$$

This new equation can be rewritten

$$\left(x + \frac{p}{2} \right)^2 = q + \left(\frac{p}{2} \right)^2$$

because (taking $d = p/2$ in the first equation)

$$x^2 + px + \left(\frac{p}{2} \right)^2 = \left(x + \frac{p}{2} \right)^2$$

We now start with the general quadratic equation and rewrite this equation in a variety of different forms, complete the square, and determine the solutions.

For $a \neq 0$, $ax^2 + bx + c = 0$ becomes

$$x^2 + \frac{bx}{a} + \frac{c}{a} = 0$$

on dividing by a; subtracting c/a from both sides gives

$$x^2 + \frac{bx}{a} = \frac{-c}{a}$$

Completing the square gives

$$x^2 + \frac{bx}{a} + \left(\frac{b}{2a} \right)^2 = -\frac{c}{a} + \left(\frac{b}{2a} \right)^2$$

or

$$\left(x + \frac{b}{2a} \right)^2 = \left(\frac{b}{2a} \right)^2 - \frac{c}{a}$$

Adding the fractions on the right-hand side yields

$$\left(x + \frac{b}{2a} \right)^2 = \frac{b^2 - 4ac}{4a^2}$$

Thus, if there are solutions, $b^2 - 4ac$ must be nonnegative and

$$\left(x + \frac{b}{2a} \right)^2 = \left(\frac{\sqrt{b^2 - 4ac}}{2a} \right)^2$$

Taking the square root of both sides gives

$$x + \frac{b}{2a} = \frac{\pm \sqrt{b^2 - 4ac}}{2a}$$

or

$$x = \frac{-b \pm \sqrt{b^2 - 4ac}}{2a}$$

The solution

$$x = \frac{-b \pm \sqrt{b^2 - 4ac}}{2a}$$

is called the *quadratic formula*.

Exercises 2.4

In each of Exercises 1 through 9, two power functions f and g are given, and a polynomial function p is defined in terms of f and g. Write an equation which defines $p(x)$ for each x, and state the degree and leading coefficient of p. Then, on a common coordinate system, sketch the graphs of f and g, and use these graphs to sketch the graph of p.

1. $f(x) = x^3$, $g(x) = x^2$, $p = f + g$

2. $f(x) = x^4$, $g(x) = x^3$, $p = f + g + 1$

3. $f(x) = x^3$, $g(x) = x$, $p = f + g - 2$

4. $f(x) = x^5$, $g(x) = x^2$, $p = f + g - 1$

5. $f(x) = x^7$, $g(x) = x^5$, $p = 2f + 2g - 1$

6. $f(x) = x^2$, $g(x) = x^5$, $p = 2f - 2g + 1$

7. $f(x) = x^4$, $g(x) = x^3$, $p = 4f + 4g - 2$

8. $f(x) = x^3$, $g(x) = x^4$, $p = 3f - 3g - \frac{1}{4}$

9. $f(x) = x$, $g(x) = x^6$, $p = 5f + 5g + \frac{1}{2}$

In Exercises 10 through 15, state which of the given functions are polynomial functions; find the degree and leading coefficient of each function that is a polynomial.

10. $f(x) = 6x^2 - 7x^5 + \frac{1}{2}$, $g(x) = x - 1$,
    $h(x) = \sqrt{x - 1}$, $q(x) = \frac{1}{2}x^5 + 2$

11. $f(x) = 3\sqrt{x} - x^2 + 1$, $g(x) = \sqrt{2}\,x + 2$,
    $h(x) = 7.5$, $q(x) = \frac{3}{2}x$

12. $f(x) = 2 + 3x^4 - 6x^5$, $g(x) = 3x^2 - \pi$,
    $h(x) = \sqrt{x^2 + 2x + 1}$, $q(x) = \dfrac{\sqrt{3}\,x - 1}{x}$

13. $f(x) = \dfrac{x^2 - 2}{x - 1}$, $g(x) = \dfrac{3 - 2x}{4}$,
    $h(x) = 5x^2 - x^7 + \sqrt{3}$, $q(x) = |x^2 - 1|$

14. $f(x) = 1.6x^3 - 0.7x^5 + \sqrt{2}$, $g(x) = \dfrac{x - 1}{x - 1}$,
    $h(x) = \pi x^4 - \sqrt{x}\,x^3 + 2$, $q(x) = \pi$

15. $f(x) = 3.7x^{16} - 7x^{19} + 3x$, $g(x) = \dfrac{x^2 - 4}{x - 2}$,
    $h(x) = |x - 1| + 3$, $q(x) = \sqrt{x^2} - 1$

16. Show by direct computation that

$$ax^2 + bx + c = a(x - r_1)(x - r_2)$$

where

$$r_1 = \frac{-b + \sqrt{b^2 - 4ac}}{2a} \quad \text{and} \quad r_2 = \frac{-b - \sqrt{b^2 - 4ac}}{2a}$$

In Exercises 17 through 25, sketch the graph of the given quadratic function. Find the vertex of the parabola and the points (if any) where the parabola intersects the x axis.

17. $f(x) = x^2 - 3x - 4$

18. $f(x) = -x^2 + 3x + 4$

19. $f(x) = -x^2 + 2x - 1$

20. $f(x) = \frac{1}{4}x^2 + 3x + 9$

21. $f(x) = 3x^2 - 2x + 4$

22. $f(x) = -6x^2 + 3x - 2$

23. $f(x) = -4x^2 + 3x + \frac{1}{2}$

24. $f(x) = -20x^2 + 5x - \frac{1}{4}$

25. $f(x) = \frac{1}{5}x^2 + 4x + 10$

26. The World Time Company finds that its monthly profit P (in dollars) from the manufacture and sale of x units of its economy model digital alarm clock is given by the equation $P(x) = -x^2/100 + 12x - 1200$. Sketch the graph of the function P and, by finding the vertex of the parabola, find the number of clocks the company should manufacture and sell per month in order to maximize its profit. What is the maximum possible monthly profit?

27. The Easy-Open Company manufactures and sells electric can openers. The company finds that its monthly cost C (in dollars) to produce x can openers is given by the equation $C(x) = x^2/1000 + 2x + 6000$. If the company sells the can openers for $14 each and can sell as many units as it can produce per month, find an equation which gives the monthly profit P as a function of x (the profit P is total sales revenue minus production cost). Sketch the graph of the profit function P, and by finding the vertex of the parabola, find the number of can openers the company should manufacture per month to maximize its monthly profit. What is the maximum profit?

28. A farmer finds that the number N of quarts of strawberries he can produce per 50 square ft of planting area is a function of the number x of plants set out per 50 square ft as defined by the equation $N(x) = -x^2/10 + 12x - 250$, $x \geqslant 50$. Sketch the graph of the function N and, by finding the vertex of the parabola, find the number of plants the farmer should set out per 50 square ft in

order to produce the maximum number of strawberries. What is the maximum number of quarts of strawberries that can be produced per 50 square ft?

29. The increase I in pulse rate due to the injection of 0.5 cc of a certain drug is found to be a function of the time t (in minutes) after injection as given by the equation $I(t) = -t^2/30 + 2t + 5$, for $4 \leqslant t \leqslant 60$. Sketch the graph of the function I and, by finding the vertex of the parabola, find the time after injection when the increase in pulse rate is the greatest. What is the maximum increase in pulse rate?

30. The pollution index p in Pleasantville on an average day is approximated by the function $p(t) = -t^2/2 + 9t + \frac{39}{2}$, where t is time in hours with $t = 0$ corresponding to 8:00 AM and $0 \leqslant t \leqslant 16$. Sketch the graph of the function p and, by finding the vertex of the parabola, find the time of day when the pollution index is at a maximum. What is the pollution index at this time?

2.5.   Composition of Functions

We saw in Section 2.2 that we can define the sum, difference, product, and quotient of two functions. In this section we discuss another important method of combining functions, called *functional composition*.

Recall that in Example 1.4-3, the Paris Paint Company economists projected that the number of gallons of their premium quality paint that could be sold during the coming year was a function f of the price per gallon x as shown in Table 2.5-1. Now suppose that the number of employees to be hired by the company is a function g of the number y of gallons of paint the company expects to sell, as shown in Table 2.5-2. Then, if the paint is sold at $2.67 per gallon, the anticipated number of gallons of paint the company will sell is 250,000 (from Table 2.5-1), and consequently 179 employees should be hired (from Table 2.5-2). So once the price per gallon is determined, the anticipated number of gallons of paint the company will sell can be found from the function f (given in Table 2.5-1) and then the number of employees to be hired can be found from the function g (given in Table 2.5-2). Thus we have a new function h, whose values are determined by successively evaluating f and g, i.e.,

$$h(x) = g(f(x))$$

for x in the domain of f; this new function gives the number of employees to be hired as a function of the price per gallon of the paint. This function h is called the *composition* of g with f and is denoted by

$$h = g \circ f$$

The function $g \circ f$ is shown in Table 2.5-3.

The composition of a function g with a function f is denoted $g \circ f$.

TABLE 2.5-1

Price per gallon (dollars)	x	2.67	2.68	2.69	2.70	2.71
Number of gallons sold (thousands)	$f(x)$	250	247	240	212	200

TABLE 2.5-2

Number of gallons sold (thousands)	y	200	212	240	247	250
Number of employees hired	$g(y)$	167	169	175	178	179

TABLE 2.5-3

Price per gallon (dollars)	x	2.67	2.68	2.69	2.70	2.71
Number of employees	$(g \circ f)(x)$	179	178	175	169	167

Table 2.5-3 is constructed from Tables 2.5-1 and 2.5-2 by successively evaluating f and g at each value of x in the domain of f, i.e., $(g \circ f)(x) = g(f(x))$.

As another illustration of the concept of functional composition, let us suppose a contestant on a television quiz show has two opportunities to win money. First, he throws a dart at a circular target (1 ft in radius), and the distance x (in feet) of the dart from the center of the target determines his winnings according to the formula

$$\text{Dollars won} = \frac{1000}{10x + 1}$$

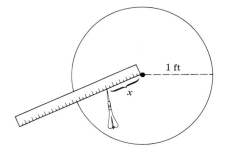

Thus if $x = \frac{9}{10}$ (of a foot), the contestant wins \$100, and if the contestant hits the bull's-eye ($x = 0$), he wins \$1000. The contestant is then given the opportunity to win three times this amount of money plus \$50 by correctly answering a bonus question. If we let f denote the function which gives the dollar winnings for throwing the dart x ft from the bull's-eye we have $f(x) = 1000/(10x + 1)$, and if g denotes the function which gives the dollar winnings for correctly answering the bonus question after winning \$y$ by throwing the dart, we have $g(y) = 3y + 50$. Tables 2.5-4 and 2.5-5 show these functions in tabular form.

TABLE 2.5-4

Any real number x such that $0 \leqslant x \leqslant 1$	x	0	$\frac{1}{10}$	$\frac{2}{5}$	$\frac{9}{10}$
$f(x)$	$1000/(10x + 1)$	1000	500	200	100

In Table 2.5-5, the domain of g actually consists only of those real numbers y between 0 and 1000 inclusive which are meaningful in terms of standard coinage, but we can ignore this restriction for our purposes here.

TABLE 2.5-5

Any real number y such that $0 \leqslant y \leqslant 1000$	y	100	200	300	700
$g(y)$	$3y + 50$	350	650	950	2150

Now, when a contestant's dart lands $\frac{2}{5}$ ft off center and he answers the bonus question correctly, we can compute his winnings by successively evaluating the functions f and g. First we evaluate $f(x)$ for $x = \frac{2}{5}$ and find that $f(\frac{2}{5}) = 200$; then we evaluate $g(200)$ and get $g(200) = 650$, so the total amount of money won is \$650. This same procedure can be followed for any distance x of the dart from the bull's-eye; we compute $f(x)$ to get the initial winnings and then compute

$$g(f(x)) = 3(f(x)) + 50$$

to find the total amount of money won for correctly answering the bonus question. Thus if $h(x)$ denotes the total winnings that result from correctly answering the bonus question after throwing the dart x ft from the bull's-eye, then h is the composition $g \circ f$ of g with f, i.e., for each x in the domain of f,

$$h(x) = (g \circ f)(x) = g(f(x)) = 3(f(x)) + 50$$

$$= 3[1000/(10x + 1)] + 50$$

The function $h = g \circ f$ is shown in tabular form in Table 2.5-6.

TABLE 2.5-6

Any real number x such that $0 \leqslant x \leqslant 1$	x	0	$\frac{1}{10}$	$\frac{2}{5}$	$\frac{9}{10}$
$(g \circ f)(x)$	$3[1000/(10x + 1)] + 50$	3050	1550	650	350

The definition of the composite function $g \circ f : (g \circ f)(x) = g(f(x))$

The domain of $g \circ f$ **consists of those numbers x in the domain of f for which $f(x)$ is in the domain of g. Note:** $f(x)$ **must be in the domain of g in order that it be possible to evaluate** $g(f(x)) = (g \circ f)(x).$

The illustrations above should indicate the utility of the notion of the composition $g \circ f$ of a function g with a function f defined by

$$(g \circ f)(x) = g(f(x))$$

The domain of $g \circ f$ consists of those real numbers x which are in the domain of f and which satisfy the requirement that $f(x)$ be in the domain of g.

┌─── **EXAMPLE 2.5-1** ───┐

Given $f(x) = x^2$ and $g(x) = 2x + 1$, find an equation that defines the composition function $g \circ f$ and compute the domain of $g \circ f$.

SOLUTION

The domain of f and g both consist of all real numbers, so for any x in the domain of f, the value of f at x (i.e., $f(x)$) is in the domain of g. Thus the domain of $g \circ f$ is also the set of all real numbers. For each real number x,

$$(g \circ f)(x) = g(f(x)) = g(x^2) = 2(x^2) + 1 = 2x^2 + 1$$

and this is an equation which defines $g \circ f$. Table 2.5-7 shows $g \circ f$ in tabular form, and Fig. 2.5-1 shows the graph of $g \circ f$.

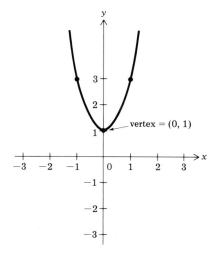

FIGURE 2.5-1 The graph of $(g \circ f)(x) = 2x^2 + 1$.

TABLE 2.5-7

All real numbers x	x	0	-2	2	4	-3
$(g \circ f)(x)$	$2x^2 + 1$	1	9	9	33	19

┌─── **EXAMPLE 2.5-2** ───┐

Given the functions f and g of Example 2.5-1, find an equation which defines the composition function $f \circ g$ and compute its domain.

SOLUTION

An equation defining $f \circ g$ is obtained by evaluating $f \circ g$ at x, so we have

$$(f \circ g)(x) = f(g(x)) = f(2x + 1)$$

$$= (2x + 1)^2 = 4x^2 + 4x + 1$$

Given two functions f and g, the composition function $g \circ f$ is usually not the same as the function $f \circ g$.

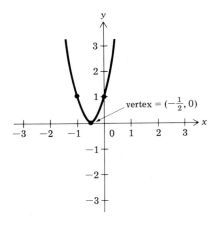

FIGURE 2.5-2   The graph of $(f \circ g)(x) = 4x^2 + 4x + 1.$

Since $g(x)$ is in the domain of f for each x in the domain of g, the domain of $f \circ g$ consists of all real numbers. Table 2.5-8 shows $f \circ g$ in tabular form, and Fig. 2.5-2 shows the graph of $f \circ g$. Note that the function $f \circ g$ is different from $g \circ f$ (see Example 2.5-1). This is usually the case; given functions f and g, the composition $g \circ f$ is usually different from $f \circ g$.

TABLE 2.5-8

All real numbers x	x	0	-2	2	4	-3
$(f \circ g)(x)$	$4x^2 + 4x + 1$	1	9	25	81	25

EXAMPLE 2.5-3

Given $f(x) = x + 2$ and $g(x) = \sqrt{x - 1}$, find equations which define $g \circ f$ and $f \circ g$ and compute the domains of each of these composition functions.

SOLUTION

An equation defining $f \circ g$ is

$$(f \circ g)(x) = f(g(x)) = f(\sqrt{x - 1}) = \sqrt{x - 1} + 2$$

The domain of f is all real numbers, and the domain of g consists of all real numbers greater than or equal to 1. Thus the domain of $f \circ g$ consists of those x in the domain of g for which $g(x)$ is in the domain of f, i.e., real numbers $x \geqslant 1$.

An equation defining $g \circ f$ is

$$(g \circ f)(x) = g(f(x)) = g(x + 2) = \sqrt{(x + 2) - 1} = \sqrt{x + 1}$$

The domain of $g \circ f$ consists of all real numbers x for which $f(x) \geqslant 1$, i.e., $x + 2 \geqslant 1$ or $x \geqslant -1$. The functions $f \circ g$ and $g \circ f$ are shown in tabular form in Tables 2.5-9 and 2.5-10, and their graphs are shown in Figs. 2.5-3 and 2.5-4.

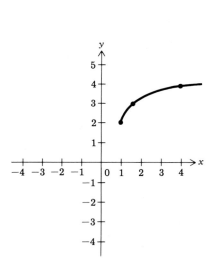

FIGURE 2.5-3   The graph of $(f \circ g)(x) = \sqrt{x - 1} + 2.$

TABLE 2.5-9

All real numbers $x \geqslant 1$	x	1	2	5	10
$(f \circ g)(x)$	$\sqrt{x-1}+2$	2	3	4	5

TABLE 2.5-10

All real numbers $x \geqslant -1$	x	-1	0	3	8
$(g \circ f)(x)$	$\sqrt{x+1}$	0	1	2	3

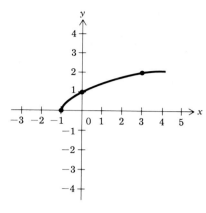

FIGURE 2.5-4   The graph of $(g \circ f)(x) = \sqrt{x+1}$.

EXAMPLE 2.5-4

The Carve-Ease Company manufactures and sells electric carving knives. The production cost C per unit (in dollars) for its deluxe model knife is a function of the number x of knives produced as given by the equation $C(x) = 5x/1000 + 6 + 400/x$. The selling price of each knife is a function of the production cost per knife C according to the equation $S(C) = C + 0.2C$. Find an equation which defines the selling price S as a function of the number x of knives produced, i.e., find an equation which defines the composition function $S \circ C$. What is the selling price per unit if 2000 knives are produced?

SOLUTION

We are given $C(x) = 5x/1000 + 6 + 400/x$ and $S(C) = C + 0.2C$. Thus if x knives are produced, the value of the composition function $S \circ C$ at x is given by the equation

$$(S \circ C)(x) = S(C(x))$$

$$= S\left(\frac{5x}{1000} + 6 + \frac{400}{x}\right)$$

$$= \left(\frac{5x}{1000} + 6 + \frac{400}{x}\right) + 0.2\left(\frac{5x}{1000} + 6 + \frac{400}{x}\right)$$

Simplifying this expression, we have

$$(S \circ C)(x) = \frac{6x}{1000} + 7.2 + \frac{480}{x}$$

The selling price per unit if 2000 knives are manufactured is the value of the function $S \circ C$ at $x = 2000$. Thus

$$(S \circ C)(2000) = \frac{6(2000)}{1000} + 7.2 + \frac{480}{2000}$$

$$= 12 + 7.2 + 0.24 = \$19.44$$

is the selling price per unit when 2000 knives are manufactured.

We shall frequently find it helpful to express a given function as the composition of two simpler functions. The following examples illustrate how this can be done.

EXAMPLE 2.5-5

Given $h(x) = \sqrt{x + 1}$, find functions f and g such that $h = g \circ f$.

SOLUTION

The equation defining the function h involves two successive operations: first, 1 is added to x to obtain $x + 1$, then the square root of $x + 1$ yields $h(x) = \sqrt{x + 1}$. Thus, letting

$$f(x) = x + 1 \quad \text{and} \quad g(x) = \sqrt{x}$$

we have

$$(g \circ f)(x) = g(f(x)) = g(x + 1) = \sqrt{x + 1} = h(x)$$

so $h = g \circ f$.

EXAMPLE 2.5-6

Given $h(x) = 1/(x + 2)^2$, find functions f and g such that $h = g \circ f$.

SOLUTION

The equation defining h involves the following two operations. First, 2 is added to x to obtain $x + 2$, then $x + 2$ is squared and divided into 1. Thus, letting

$$f(x) = x + 2 \quad \text{and} \quad g(x) = 1/x^2$$

we have

$$(g \circ f)(x) = g(f(x)) = g(x + 2) = \frac{1}{(x + 2)^2} = h(x)$$

so $h = g \circ f$. There are, however, other possibilities for f and g. If we let

$$f(x) = (x + 2)^2 \quad \text{and} \quad g(x) = 1/x$$

then

$$(g \circ f)(x) = g(f(x)) = \frac{1}{f(x)} = \frac{1}{(x + 2)^2} = h(x)$$

and again, $h = g \circ f$.

Exercises 2.5

In each of Exercises 1 through 10, two tables are given which define two functions f and g. Complete the table which describes the function $g \circ f$.

In Exercises 1 through 10, recall that when a function is defined entirely by a table (rather than by an equation), the domain of the function consists of the entries in the top row of the table.

1.

x	1	3	5	7	9
$f(x)$	6	2	8	4	10

y	2	4	6	8	10
$g(y)$	200	− 4	60	− 23	0

x	1	3	5	7	9
$(g \circ f)(x)$	60		− 23		

2.

x	− 2	− 1	0	1	2
$f(x)$	2.9	5.6	− 3.2	− 1	0

y	− 3.2	− 1	0	2.9	5.6
$g(y)$	0	π	100	− 2	0

x	− 2	− 1	0	1	2
$(g \circ f)(x)$			0		100

3.

x	100	200	300	400	500
$f(x)$	2.3	1	0	− 4	5.2

y	− 4	0	1	2.3	5.2
$g(y)$	3	− 4	2	1	− 1

x	100				
$(g \circ f)(x)$	1				

4.

x	1	2	3	4	5
$f(x)$	5	4	3	2	1

y	1	2	3	4	5
$g(y)$	-5	-4	-3	-2	-1

x	1				
$(g \circ f)(x)$	-1				

5.

x	-3	-1	0	1	3
$f(x)$	10	-10	5	-5	0

y	-10	-5	0	5	10
$g(y)$	2	2	2	2	2

x					
$(g \circ f)(x)$					

6.

x	1	3	5	7	9
$f(x)$	9	7	5	3	1

y	1	3	5	7	9
$g(y)$	9	7	5	3	1

x					
$(g \circ f)(x)$					

7.

x	2	4	6	8	10
$f(x)$	1	1	1	1	1

y	1	3	5	7	9
$g(y)$	0	2	4	6	8

x					
$(g \circ f)(x)$					

8.

x	-7	-3	0	3	5
$f(x)$	5	3	4	2	1

y	1	2	3	4	5
$g(y)$	1	2	3	4	5

x					
$(g \circ f)(x)$					

9.

x	10	11	12	13	14
$f(x)$	3	-3	π	-2	-1

y	0	2	π	-2	3
$g(y)$	100	200	300	400	500

x			
$(g \circ f)(x)$			

10.

x	4	9	16	25	36
$f(x)$	9	12	-4	-6	15

y	3	6	9	12	15
$g(y)$	π	$-\pi$	2π	-2π	3π

x			
$(g \circ f)(x)$			

In Exercises 11 through 23, for the given functions f and g, find equations which define the two composition functions $g \circ f$ and $f \circ g$, compute their domains, and sketch their graphs.

11. $f(x) = 3x - 1$, $g(x) = 2x + 1$
12. $f(x) = \frac{1}{2}x - 2$, $g(x) = -5x - 3$
13. $f(x) = x^3$, $g(x) = x$
14. $f(x) = x^3$, $g(x) = x^2$
15. $f(x) = x^2 - 2x + 1$, $g(x) = x - 1$
16. $f(x) = x - 2$, $g(x) = x^2 + 1$
17. $f(x) = -3$, $g(x) = x^2 + 2x + 1$
18. $f(x) = x^3 + 4$, $g(x) = 2$
19. $f(x) = \sqrt{x - 2}$, $g(x) = x + 1$
20. $f(x) = x - 3$, $g(x) = \sqrt{x + 4}$
21. $f(x) = |x|$, $g(x) = 3x - 1$
22. $f(x) = 2x^2 - 3$, $g(x) = |x|$
23. $f(x) = |x - 3|$, $g(x) = x^2 + 1$

In Exercises 24 through 32, express the given function as a composition of two other functions.

24. $h(x) = (x + 1)^5$

25. $h(x) = \sqrt{x^2 + 4}$

26. $h(x) = \dfrac{5}{(x - 7)^3}$

27. $h(x) = (x^4 - 2x)^4 + 10$

28. $s(t) = \sqrt{t^3} - 40$

29. $r(s) = \dfrac{1}{\sqrt{2s^2 - 1}}$

30. $t(z) = \left(\dfrac{1}{z^2 - 1}\right)^4$

31. $w(t) = (t^2 + 1)^3 - \dfrac{4}{t^2 + 1}$

32. $u(v) = \dfrac{\left(2v^3 + 2\right)^5 + 7}{2v^3 + 2}$

33. Let c be a fixed real number and let f be any function whose domain contains the number c. Let g be the constant function $g(x) = c$. Show that the composition functions $f \circ g$ and $g \circ f$ are each constant functions.

34. Let f be the function defined by the equation $f(x) = x$ (f is called the *identity function*). Let g be any function. Show that $g \circ f = g$ and $f \circ g = g$.

35. An ecologist finds that in a forest area near an industrial city, the percent D of diseased trees per acre is a function of the average daily pollution index p as defined by the equation $D(p) = 3p^2 / 400 + 4$. He also estimates that the average daily pollution index p is a function of the distance d (in miles) from the city as defined by the equation $p(d) = 65 - 2d$. Find an equation which defines D as a function of d, i.e., find an equation which defines the composition function $D \circ p$. What percent of trees per acre are diseased 5 mi from the city?

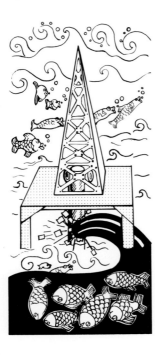

36. Oil seeping from an offshore oil well is spreading on the surface of the water in a circle. The radius r (in miles) of the circle is a function of time t (in hours) as defined by the equation $r(t) = 0.5 + 1.5t$, where $t = 0$ corresponds to the time at which the radius of the spill is $\frac{1}{2}$ mi. The area A of the oil spill is given by $A = \pi r^2$, where r is the radius of the spill. Find an equation which defines the area A as a function of t, i.e., find an equation which defines $A \circ r$. What is the area of the spill at time $t = 2$ hr?

37. The Lite-Rite Company manufactures and sells bedside reading lamps. For its economy model lamp, the production cost c per unit (in dollars) is a function of the number x of lamps produced as

given by the equation $c(x) = 3x/1000 + 4 + 150/x$. The selling price s of each lamp is a function of the production cost per lamp c according to the equation $s(c) = c + 0.15c$. Find an equation which defines the selling price s as a function of the number x of lamps produced, i.e., find an equation which defines $s \circ c$. What is the selling price of each lamp if 1500 lamps are produced?

38. The city of Pleasantville estimates that its population P during the next ten years will be approximated by the function $P(t) = 200t^2 + 40{,}000$ where t is time in years with $t = 0$ corresponding to the present year. The city also estimates that its average daily pollution index I is a function of the population of the city as given by the equation $I(P) = 30 + P/2000$. Find an equation which defines the pollution index I as a function of t, i.e., find an equation which defines the composition function $I \circ P$. What is the expected average daily pollution index for the city six years from now?

39. Researchers find that the percent concentration P of a certain drug in the bloodstream t min after oral administration of the drug is given by the function $P(t) = -t^2/600 + 13t/60 - 1$ for $5 \leqslant t \leqslant 110$. They also find that the increase in pulse rate R is a function of the percent concentration P of the drug in the bloodstream as given by the equation $R(P) = P^2/2 + 2P$ for $0 \leqslant P \leqslant 6$. Find an equation which defines the increase in pulse rate R as a function of t, i.e., find an equation which defines the composition function $R \circ P$. What is the increase in pulse rate 30 min after administration of the drug?

By *percent concentration* of a drug in the bloodstream we mean the percent of the bloodstream that is the drug.

40. A petroleum company is considering a site for the construction of a service station. Projected residential and commercial development of the area around the site leads the company to estimate that for the next five years, the average daily traffic count T on the highway fronting the property will be a function of time t (in months) as defined by $T(t) = -t^2 + 160t + 4100$, where $t = 0$ corresponds to the present time. The company also estimates that the number N of gallons of gasoline that will be sold at the projected station per day is a function of the daily traffic count T and is given by $N(T) = 1200 + T/3$. Find an equation which gives the number of gallons N that will be sold per day at the projected station as a function of t, i.e., find an equation which defines $N \circ T$. If the station is constructed, how many gallons of gasoline will be sold (assuming that the company's estimates are accurate) per day at the station 20 months from now?

Self Test   Part I

1. Find a rational number whose decimal representation is $1.0\overline{14}$.
2. Find two distinct real numbers x and y such that $|x| = |y| = 2$.
3. Find the distance between the point $P_1 = (-1, 0)$ and $P_2 = (1, 1)$.
4. Fill in the missing entries for the function f defined in tabular form by

	x	0	1	3	8
$f(x)$	$2/\sqrt{x+1}$		$\sqrt{2}$		

5. Find the equation of the line passing through the points $P_1 = (-1, 2)$ and $P_2 = (3, -1)$.
6. Let $f(x) = \frac{1}{2}x$ and $g(x) = \frac{1}{2}x - 1$. Find equations that define each of the functions $f + g$, $f - g$, fg, f/g, $f \circ g$, and $g \circ f$. Determine the domains of each of these functions.
7. Give the domain and sketch the graphs of the following functions.

$$f(x) = -2\sqrt{x} + 1$$

$$g(x) = \frac{1}{2}x^2 + x - 2$$

and

$$h(x) = \frac{2}{x} - 1$$

8. Find the slope of the line that is the graph of the linear function $f(x) = -2x + 3$ and find two points on this line.
9. The population P of the city of Pleasantville for each year since 1960 is approximated by the function $P(t) = -250t^2 + 5000t + 30{,}000$, where t is the time in years with $t = 0$ corresponding to 1960. Sketch the graph of the function P and, by finding the vertex of the parabola, find that year in which the population of Pleasantville was the greatest.
10. Purity Feeds, Inc. sells its premium grade poultry feed at \$.13 a pound. There is a delivery charge of \$.01 a pound and a handling charge of \$2 per order. Find the equation that expresses the cost C (in dollars) of an order in terms of the number of pounds x of the order. Then write the function C in tabular form; include in the table the values of $C(x)$, when x is 50, 100, 200, and 500. What is the cost of a 625-lb order?
11. The Timely Watch Company manufactures and sells wrist watches. For its premium quality watch the company finds that it can sell 1250 watches per year if they are priced at \$130 each and 1100 watches per year if they are priced at \$150 each. Assuming that the demand D (number of watches per year that can be sold) is a linear function of the selling price x, $110 \le x \le 165$, find an equation that defines $D(x)$. (Check your answer.) How many watches per year can the company sell if they are priced at \$145 each?

Limits and Continuity

PART II

Limits of Sequences

3.1. Introduction

The concept of a *limit* is fundamental to the development of the two central topics of the calculus, the derivative and the integral. For this reason, a thorough understanding of the notion of a limit is necessary before going any further. The word "limit" occurs frequently in ordinary language to express aspects of daily experience; examples such as "speed limit," "load limit," and "the limit of my patience" are commonplace. As we shall see, it is a refinement of the meanings associated with such common usages of the word "limit" that is profitably employed in the calculus.

3.2. Sequences

Before we talk about limits, we introduce a special class of functions called *sequences*. A *sequence* is a function whose domain consists of all positive integers.

Definition of a sequence
Recall that the positive integers are the numbers
$1, 2, 3, 4, 5, \ldots .$

EXAMPLE 3.2-1

The function s whose value at each positive integer n is given by the equation

The domain of a sequence must be the set of positive integers. Thus if a function s is specified to be a sequence, its domain is limited to the positive integers even though the rule defining s may be valid for other real numbers as well. For example, the domain of the *function* f defined by $f(x) = 1/x^2$ consists of all real numbers different from 0, while the domain of the *sequence* $s(n) = 1/n^2$ is the positive integers, even though the rule defining s is the same as the rule defining f. The graphs of f and s are shown in Figs. 3.2-1 and 3.2-2.

$$s(n) = \frac{1}{n^2}$$

is a sequence. (Note that we are specifying the domain of s as the positive integers. Actually, s could be defined at nonintegral values of n as well; for example, it would be meaningful to talk about $s(\frac{1}{2}) = 1/(\frac{1}{2})^2 = 4$, but we must exclude $\frac{1}{2}$ from the domain of s if s is to be a sequence.) The values of the sequence s at the integers 1, 2, 3, 7, and 12 are

$$s(1) = \frac{1}{1^2} = 1, \qquad s(2) = \frac{1}{2^2} = \frac{1}{4},$$

$$s(3) = \frac{1}{3^2} = \frac{1}{9}, \qquad s(7) = \frac{1}{7^2} = \frac{1}{49}, \quad \text{and}$$

$$s(12) = \frac{1}{12^2} = \frac{1}{144}$$

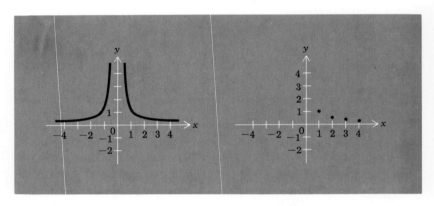

FIGURE 3.2-1 The graph of the function $f(x) = 1/x^2$.

FIGURE 3.2-2 The graph of the sequence $s(n) = 1/n^2$.

EXAMPLE 3.2-2

The function g whose value at each positive integer n is defined by

$$g(n) = (-1)^n$$

is a sequence (again, the domain is specified as the positive integers), and we see that $g(1) = (-1)^1 = -1$, $g(2) = (-1)^2 = 1$, $g(3) = (-1)^3 = -1$, $g(4) = (-1)^4 = 1$, etc.

Sequences are the mathematical counterpart of the commonplace practice of listing things in order. That is, for a sequence s, we can think of the value $s(1)$ as the first object in the list, $s(2)$ as the second object, $s(3)$ as the third object, $s(50)$ as the fiftieth object, and so on. To provide for the theoretical possibility of an infinitely long list, the domain of s is then taken to be *all* positive integers. For example, the manager of a recently opened store might study the sequence s, where $s(n)$ is the number of customers coming into the store on the nth day after the opening of the store.

As another example of a sequence, imagine a line of people in front of a theater; if we are willing to imagine further that the IQ of each individual in line is painted on his or her back and that the line is infinitely long, we have a sequence at hand. Simply take the value of the sequence s at the positive integer n to be the number painted on the back of the nth person in line. Thus the first person in line wears the number $s(1)$, the second person wears $s(2)$, and so on.

The value of the sequence s at 1, $s(1)$, is called the first *term* of the sequence. Similarly, $s(2)$ is the second term, $s(3)$ the third term, and in general $s(n)$ is the nth term.

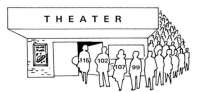

EXAMPLE 3.2-3

A business executive is offered the position of vice-president of a company for an annual salary of $80,000 per year plus a raise each year of $\frac{1}{20}$ of the preceding year's salary. Let s be the sequence whose value at each positive integer n is the salary the executive will receive after n yı. Find $s(1)$, $s(2)$, $s(3)$, and $s(4)$. Verify that the equation $s(n) = (1.05)^n (80,000)$ gives the value of the sequence s at n for each positive integer n.

$s(n)$, the value of the sequence s at n, is called the nth *term* of the sequence.

SOLUTION

After 1 yr, the executive will receive a raise of

$$\frac{1}{20}(80,000) = \$4000$$

and thus have a total salary of $s(1) = \$84,000$. After the second year, the raise will be

$$\frac{1}{20}(84,000) = \$4200$$

so the total salary will be $s(2) = \$88,200$. Similarly,

$$s(3) = 88,200 + \frac{1}{20}(88,200) = \$92,610$$

and

$$s(4) = 92,610 + \frac{1}{20}(92,610) = \$97,240.50$$

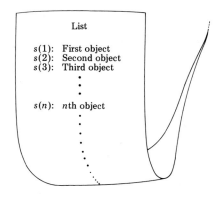

List

$s(1)$: First object
$s(2)$: Second object
$s(3)$: Third object
  ⋮
$s(n)$: nth object

To find the equation that defines $s(n)$ for any positive integer n, we note that for any integer $n > 1$, our sequence s has the property that

$$s(n) = s(n - 1) + \tfrac{1}{20} s(n - 1)$$
$$= s(n - 1)(1 + \tfrac{1}{20})$$
$$= (1.05)s(n - 1)$$

Thus, to get from any particular term $s(n - 1)$ of this sequence to the next term $s(n)$, we multiply $s(n - 1)$ by 1.05. But then, to get from the first term of this sequence to $s(n)$, we must multiply the first term by 1.05 $(n - 1)$ times, i.e., we must multiply $s(1) = 84{,}000 = (1.05)(80{,}000)$ by $(1.05)^{n-1}$ to get $s(n)$. This is the same as saying that $s(n) = (1.05)^n 80{,}000$.

In many texts the functional notation $s(n)$ for the nth term of a sequence s is abandoned in favor of subscript notation. Thus $s(1)$ is denoted by s_1, $s(2)$ by s_2, and, in general, $s(n)$ by s_n.

Although sequences are simply functions whose domains consist of all positive integers, several idiosyncrasies of presentation have developed. The value $s(n)$ of the sequence s at n is called the nth *term* of the sequence, and it has become customary to define a sequence by simply displaying its nth term in braces. Thus the sequence s defined by $s(n) = 1/n^2$ is often presented as

$$\left\{ \frac{1}{n^2} \right\}$$

Similarly,

$$\{(-1)^n\} \quad \text{and} \quad \{2^n\}$$

denote the sequences p and q defined by the equations

$$p(n) = (-1)^n \quad \text{and} \quad q(n) = 2^n$$

Definition of a line graph for a sequence

Furthermore, the graph of a sequence s is seldom presented in the traditional Cartesian fashion employed for other functions, locating the pairs $(n, s(n))$ in a Cartesian plane. Instead, the terms $s(n)$ of the sequence are simply located on a number line. This kind of graph is called a *line graph*; it provides a useful pictorial representation of a sequence. Care must be taken to label the terms of the sequence clearly on the line graph so that the order in which the terms appear is clear.

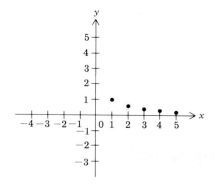

EXAMPLE 3.2-4

Sketch the Cartesian graph and the line graph of the sequence $s(n) = 1/n$.

SOLUTION

The Cartesian graph of $s(n) = 1/n$ consists of those points in a Cartesian plane whose coordinates are of the form $(n, s(n))$ for some positive integer n. Thus we locate the points $(1, 1)$, $(2, \frac{1}{2})$, $(3, \frac{1}{3})$, etc., as shown in Fig. 3.2-3. The line graph consists of the points on a number line which correspond to numbers of the form $s(n) = 1/n$ for some positive integer n. Thus we locate the points $s(1) = 1$, $s(2) = \frac{1}{2}$, $s(3) = \frac{1}{3}$, etc., on a number line as shown in Fig. 3.2-4 to obtain the line graph.

FIGURE 3.2-3   The Cartesian graph of the sequence $s(n) = 1/n$.

FIGURE 3.2-4   The line graph of the sequence $s(n) = 1/n$.

EXAMPLE 3.2-5

Construct the line graph of the sequence $x(n) = (\frac{1}{2})^n$.

SOLUTION

We first compute the first few terms of the sequence $x(n)$. We see that $x(1) = (\frac{1}{2})^1 = \frac{1}{2}$, $x(2) = (\frac{1}{2})^2 = \frac{1}{4}$, $x(3) = (\frac{1}{2})^3 = \frac{1}{8}$, $x(4) = (\frac{1}{2})^4 = \frac{1}{16}$, and notice that

$$\tfrac{1}{2}x(n) = \tfrac{1}{2}\left(\tfrac{1}{2}\right)^n = \left(\tfrac{1}{2}\right)^{n+1} = x(n + 1)$$

Thus once we have located $x(n)$, the next term $x(n + 1)$ is half as far from 0. The line graph is shown in Fig. 3.2-5.

In Example 3.2-5 (and elsewhere) we use x (and later, y, z, etc.) as the name of a function, rather than as a variable.

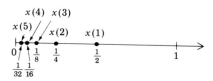

FIGURE 3.2-5   The line graph of $x(n) = (\frac{1}{2})^n$.

EXAMPLE 3.2-6

Construct the line graph of the sequence $\{1 - (\frac{1}{2})^n\}$.

SOLUTION

The sequence $\{1 - (\frac{1}{2})^n\}$ is a function y whose domain is the positive integers and whose value at the positive integer n is given by $y(n) = 1 - (\frac{1}{2})^n$. Thus the first few terms of the sequence are $y(1) = 1 - (\frac{1}{2})^1 = \frac{1}{2}$, $y(2) = 1 - (\frac{1}{2})^2 = \frac{3}{4}$, $y(3) = 1 - (\frac{1}{2})^3 = \frac{7}{8}$, and $y(4) = 1 - (\frac{1}{2})^4 = \frac{15}{16}$. Notice that since $(\frac{1}{2})^n$ is always positive, the nth term $y(n) = 1 - (\frac{1}{2})^n$ is always less than 1. Also, as the value of n increases, the value of $(\frac{1}{2})^n$ moves closer to 0 (as we saw in Example 3.2-5), and thus the value of $y(n) = 1 - (\frac{1}{2})^n$ moves closer to 1. Using this information, we construct the line graph as shown in Fig. 3.2-6.

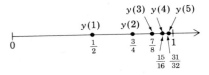

FIGURE 3.2-6 The line graph of the sequence $\{1 - (\frac{1}{2})^n\}$.

EXAMPLE 3.2-7

Construct the line graph of the sequence $\{(-1)^n/n\}$.

SOLUTION

The sequence $\{(-1)^n/n\}$ is a function z whose domain is the positive integers and whose value at the positive integer n is given by $z(n) = (-1)^n/n$. Thus the first few terms of the sequence are $z(1) = (-1)^1/1 = -1$, $z(2) = (-1)^2/2 = \frac{1}{2}$, $z(3) = (-1)^3/3 = -\frac{1}{3}$, $z(4) = (-1)^4/4 = \frac{1}{4}$, and $z(5) = (-1)^5/5 = -\frac{1}{5}$. Note that the terms alternate in sign; the terms corresponding to even values of n will be to the right of 0, while those corresponding to odd values of n will be to the left of 0. Also, note that for any positive integer n,

$$|z(n)| = \left| \frac{(-1)^n}{n} \right| = \frac{1}{n}$$

Recall that $(-1)^n$ equals 1 if n is an even positive integer and equals -1 if n is odd.

so the distance of $z(n)$ from 0 is $1/n$. Thus, as the value of n increases, the corresponding term $z(n)$ moves closer to 0 (see Example 3.2-4). The line graph is shown in Fig. 3.2-7.

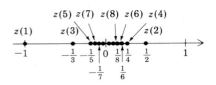

FIGURE 3.2-7 The line graph of the sequence $\{(-1)^n/n\}$.

Exercises 3.2

In each of Exercises 1 through 10, a sequence s is given. Write the equation which defines the function s, compute the first six terms of the sequence, and construct its line graph.

1. $\{(-1)^n\}$

2. $\left\{1 + \dfrac{1}{n}\right\}$

3. $\left\{1 - \dfrac{1}{n}\right\}$

4. $\left\{1 + \dfrac{(-1)^n}{n}\right\}$

5. $\left\{3 + \dfrac{(-1)^n}{n}\right\}$

6. $\{1 + (-1)^n\}$

7. $\left\{[1 + (-1)^n] + \dfrac{1}{n}\right\}$

8. $\left\{2[1 + (-1)^n] - \dfrac{1}{n}\right\}$

9. $\left\{(-1)^{n+1} \dfrac{4}{2^n}\right\}$

10. $\left\{(-1)^{n+1} \dfrac{2^n}{4}\right\}$

In Exercises 11 through 20, the first six terms of a sequence s are given in order, i.e., the first number given is $s(1)$, the second number is $s(2)$, etc. Find an equation which defines the nth term $s(n)$ of such a sequence.

11. $2, 2, 2, 2, 2, 2, \ldots$
12. $1, 2, 3, 4, 5, 6, \ldots$
13. $1, 4, 9, 16, 25, 36, \ldots$
14. $1, \frac{1}{2}, \frac{1}{4}, \frac{1}{8}, \frac{1}{16}, \frac{1}{32}, \ldots$
15. $1, \frac{1}{4}, \frac{1}{9}, \frac{1}{16}, \frac{1}{25}, \frac{1}{36}, \ldots$
16. $4 + \frac{1}{2}, 4 + \frac{1}{3}, 4 + \frac{1}{4}, 4 + \frac{1}{5}, 4 + \frac{1}{6}, \ldots$
17. $\frac{1}{2}, \frac{1}{4}, \frac{1}{6}, \frac{1}{8}, \frac{1}{10}, \frac{1}{12}, \ldots$
18. $-\frac{1}{2}, \frac{1}{4}, -\frac{1}{6}, \frac{1}{8}, -\frac{1}{10}, \frac{1}{12}, \ldots$
19. $\frac{1}{4}, -\frac{1}{6}, \frac{1}{8}, -\frac{1}{10}, \frac{1}{12}, -\frac{1}{14}, \ldots$
20. $1, -\frac{1}{3}, \frac{1}{5}, -\frac{1}{7}, \frac{1}{9}, -\frac{1}{11}, \ldots$

21. A ball is dropped from a height of 15 ft, and each time it bounces it rises $\frac{3}{4}$ of the distance it had previously fallen. Let s be the sequence whose value at each positive integer n is the height the ball rises after the nth bounce. Find $s(1)$, $s(2)$, $s(3)$, and $s(4)$. Using an argument like the one in Example 3.2-3, justify to yourself that the equation $s(n) = 15(\frac{3}{4})^n$ gives the value of the nth term of the sequence s for each positive integer n.

22. The Central Realty Corporation is considering the purchase of a tract of land. The land is currently worth $200,000, and the company estimates that its value will appreciate each year by an amount equal to $\frac{1}{10}$ of its value at the beginning of the year. Let s be the sequence whose value at each positive integer n is the worth of the land in n yrs. Find $s(1)$, $s(2)$, $s(3)$, and $s(4)$. Using an

argument like the one in Example 3.2-3, justify to yourself that the equation $s(n) = (1.1)^n(200,000)$ gives the value of the nth term of the sequence s for each positive integer n.

23. The Environmental Control Board of the city of Pleasantville decides that the city's population increase each year should be limited to an amount equal to $\frac{1}{20}$ of the population at the beginning of the year. Assume that the Board's recommendation is followed, and that the present population of the city is 160,000. Let s be the sequence whose value at each positive integer n is the population of the city n yr from now; find $s(1)$, $s(2)$, $s(3)$, and $s(4)$. Using an argument like the one in Example 3.2-3, justify to yourself that the equation $s(n) = (1.05)^n(160,000)$ gives the value of the nth term of the sequence s for each positive integer n.

24. Industrial Engineers, Inc. designed a pollution-control filter for the smokestacks of a local factory. It is feasible to install several filters in each smokestack, and each time the emissions from the factory pass through a filter, $\frac{1}{5}$ of the pollutants that are then in the emissions are removed. The pollution index of the unfiltered emissions is 220,000 ppm. Let s be the sequence whose value at each positive integer n is the pollution index of emissions from a smokestack in which n filters have been installed. Find $s(1)$, $s(2)$, $s(3)$, and $s(4)$. Using an argument like the one in Example 3.2-3, justify to yourself that the equation $s(n) = (\frac{4}{5})^n(220,000)$ gives the value of the nth term of the sequence s for each positive integer n.

25. A group of economic researchers is conducting a study on inflation and its future effect on the national economy. For purposes of the study, they agreed to operate under the assumption that the value of the dollar will decrease each year by an amount equal to $\frac{1}{20}$ of its value at the beginning of the year. Let s be the sequence whose value at each positive integer n is the value of $800,000, n$ years from the present time. Find $s(1)$, $s(2)$, $s(3)$, and $s(4)$. Using an argument like the one in Example 3.2-3, justify to yourself that the equation $s(n) = (0.95)^n(800,000)$ gives the value of the nth term of the sequence s for each positive intgeger n.

Recall that to convert a percent to a number, we divide by 100; thus $5\% = \frac{5}{100} = 0.05$.

26. Suppose $8000 is deposited in a bank that pays 5% interest per year compounded annually. The interest paid at the end of the first year is $(0.05)(8000) = \$400$, so the total in the account at the beginning of the second year is $8400; then the interest paid at the end of the second year is $(0.05)(8400) = \$420$, so the total in the account at the beginning of the third year is $8820, etc. Let s be the sequence whose value at each positive integer n is the amount in the account after n years. Then we have found that $s(1) = \$8400$ and $s(2) = \$8820$. Find $s(3)$ and $s(4)$. Using an argument like the one in Example 3.2-3, justify to yourself that the equation $s(n) = (1.05)^n(8000)$ gives the amount in the account after n years, for each positive integer n.

*3.3. The Limit Concept (*Optional*)

Now that we have defined sequences and seen some examples of them, the next step toward understanding the term "limit" as it is used in the calculus is to define what we mean by the phrase "the limit of a sequence." The following illustrations lay the groundwork for the next section, where the definition of "the limit of a sequence" is fully developed.

Suppose a man takes a 1-gal container and, after 1 min, pours $\frac{1}{2}$ gal of water into it. One minute later he adds $\frac{1}{4}$ gal of water, 1 min later he adds $\frac{1}{8}$ gal, and so on. Thus n min after the start of the process, he pours $(\frac{1}{2})^n$ gal of water into the container. Will he ever fill the container with water? As we shall soon see, the answer is no; but if he were to pour the same amounts of water into any smaller container, the container would eventually overflow.

Now suppose that a woman wishes to set up a trust fund so that her favorite charity may withdraw $1 million at the end of the first year, $.5 million at the end of the second year, $.25 million at the end of the third year, and so on. Thus at the end of the nth year, the charity is to withdraw $$(\frac{1}{2})^{n-1}$ million. How much money must the woman initially put into the trust fund (assuming that no interest is accrued) to guarantee that the fund will never be exhausted by the yearly withdrawals of the charity? The answer is $2 million.

For $n = 1$, $(\frac{1}{2})^{n-1} = (\frac{1}{2})^{1-1} = (\frac{1}{2})^0 = 1$; for $n = 2$, $(\frac{1}{2})^{n-1} = (\frac{1}{2})^{2-1} = (\frac{1}{2})^1 = \frac{1}{2}$; and so on.

The answers to the questions posed in the above illustrations can be verified by investigating the behavior of appropriate sequences. Before proceeding further, however, we need to recall from elementary algebra that if r is any real number, then

$$(1 - r)(1 + r) = 1 - r^2$$

$$(1 - r)(1 + r + r^2) = 1 - r^3$$

and

$$(1 - r)(1 + r + r^2 + r^3) = 1 - r^4$$

In fact, if n is any positive integer, then

$$(1 - r)(1 + r + r^2 + \cdots + r^{n-1}) = 1 - r^n$$

To see, for example, that $(1 - r)(1 + r + r^2) = 1 - r^3$, simply perform the multiplication:

$$
\begin{array}{r}
1 + r + r^2 \\
1 - r \\
\hline
-r - r^2 - r^3 \\
1 + r + r^2 \\
\hline
1 + 0 + 0 - r^3 = 1 - r^3
\end{array}
$$

Note that since $(1 - r)(1 + r + r^2 + \cdots + r^{n-1}) = 1 - r^n$, we have $1 + r + r^2 + \cdots + r^{n-1} = (1 - r^n)/(1 - r)$.

Now let us return to the man pouring water into a container. We define a sequence y where for each positive integer n, $y(n)$ is the amount of water in the container after n min. Thus $y(1) = \frac{1}{2}$ gal, $y(2) = \frac{1}{2} + \frac{1}{4} = \frac{1}{2} + (\frac{1}{2})^2$, $y(3) = \frac{1}{2} + \frac{1}{4} + \frac{1}{8} = \frac{1}{2} + (\frac{1}{2})^2 + (\frac{1}{2})^3$, and, in general,

In this section y and z are used as the names of functions (sequences) rather than as variables.

$$y(n) = \tfrac{1}{2} + \left(\tfrac{1}{2}\right)^2 + \left(\tfrac{1}{2}\right)^3 + \cdots + \left(\tfrac{1}{2}\right)^n$$

$$= \tfrac{1}{2}\left[1 + \tfrac{1}{2} + \left(\tfrac{1}{2}\right)^2 + \cdots + \left(\tfrac{1}{2}\right)^{n-1}\right]$$

Thus, using the formula $1 + r + r^2 + \cdots + r^{n-1} = (1 - r^n)/(1 - r)$ (with $r = \tfrac{1}{2}$), we have

$$y(n) = \tfrac{1}{2}\left[\frac{1 - \left(\tfrac{1}{2}\right)^n}{1 - \tfrac{1}{2}}\right] = \tfrac{1}{2}\left[\frac{1 - \left(\tfrac{1}{2}\right)^n}{\tfrac{1}{2}}\right] = 1 - \left(\tfrac{1}{2}\right)^n$$

Since $y(n)$ equals the total amount of water (in gallons) in the container after n min, and since $y(n) = 1 - \left(\tfrac{1}{2}\right)^n$, we see that no matter how large the integer n may be, $y(n)$ is always less than 1 gal, so the container will never be full. The line graph for the sequence $y = \{1 - \left(\tfrac{1}{2}\right)^n\}$ (see Fig. 3.3-1 and Example 3.2-6) reflects the fact that more and more water is added to the water in the container but the total amount of water in the container at any time is less than 1 gal.

On the other hand, if the same amounts of water were poured into a container of capacity less than 1 gal, the container would eventually overflow. For example, if a container of capacity 0.991 gal were used, it would overflow after 7 min, since $y(7) = 1 - \left(\tfrac{1}{2}\right)^7 = 127/128 > 0.991$. Similarly, a container of capacity 0.99971 gal would overflow after 16 min, since $y(16) = 1 - \left(\tfrac{1}{2}\right)^{16} = \tfrac{65535}{65536} > 0.99971$. Thus there is a *limit* to how small the container can be and still never overflow; evidently, that limit is 1 gal. There is also a *limit* to the amount of water collected in the container: again, 1 gal. We extend this language to the sequence y (whose nth term $y(n)$ is the amount of water in the container after n min) and say that "the limit of the sequence $\{y(n)\}$ is 1." It is cumbersome, however, to write out this entire phrase, so we will use the notation

$$\lim_{n\to\infty} y(n) = 1$$

instead. If you were watching the line graph of $\{y(n)\}$ (see Fig. 3.3-1) being sketched as the process of pouring the water took place, you would first see the point $y(1) = \tfrac{1}{2}$ marked on the line; next $y(2) = \tfrac{3}{4}$, and then $y(3) = \tfrac{7}{8}$, would be marked. As time went on (as n got larger), you would notice that the points being marked on the line approached the number 1 on the line. This geometric effect, which reflects the fact that the limit of the sequence $\{y(n)\}$ is 1, is often described by saying that "$y(n)$ approaches 1 (or gets close to 1) as n approaches infinity." Indeed, the notation $\lim_{n\to\infty} y(n) = 1$ is often translated this way.

We have yet to consider our second illustration, the trust fund. Again, we use an appropriate sequence to verify that $2 million must initially be placed in the fund to insure that the fund will never be exhausted. Recall that the charity is to withdraw $\$\left(\tfrac{1}{2}\right)^{n-1}$ million at the end of the nth year. Let $z(n)$ be the total amount withdrawn after n yr. Then $z(1) = 1$ million, $z(2) = 1 + \tfrac{1}{2}$, $z(3) = 1 + \tfrac{1}{2} + \tfrac{1}{4}$, and in general,

After n min, we are pouring in half the amount of water then needed to fill the container. This should relieve whatever element of surprise there may be in discovering that the container is never filled.

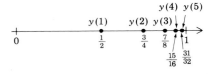

FIGURE 3.3-1

The notation $\lim_{n\to\infty} y(n) = 1$ means "the limit of the sequence $\{y(n)\}$ is 1."

The notation $\lim_{n\to\infty} y(n) = 1$, which means "the limit of the sequence $\{y(n)\}$ is 1," is often translated by saying that "$y(n)$ approaches 1 as n approaches infinity." This reflects the geometric behavior of the terms $y(n)$ of the sequence on the line graph, since they all "jam up," or cluster, by the number 1 on the line.

$$z(n) = 1 + \tfrac{1}{2} + \tfrac{1}{4} + \cdots + \left(\tfrac{1}{2}\right)^{n-1}$$

But

$$1 + \tfrac{1}{2} + \left(\tfrac{1}{2}\right)^2 + \cdots + \left(\tfrac{1}{2}\right)^{n-1} = \left(1 - \left(\tfrac{1}{2}\right)^n\right) / \left(1 - \tfrac{1}{2}\right) = 2\left[1 - \left(\tfrac{1}{2}\right)^n\right]$$

so

$$z(n) = 2\left[1 - \left(\tfrac{1}{2}\right)^n\right]$$

To see that $1 + \tfrac{1}{2} + \left(\tfrac{1}{2}\right)^2 + \cdots + \left(\tfrac{1}{2}\right)^{n-1} = (1 - \left(\tfrac{1}{2}\right)^n)/(1 - \tfrac{1}{2})$, we use the formula $(1 + r + r^2 + \cdots + r^{n-1}) = (1 - r^n)/(1 - r)$ with $r = \tfrac{1}{2}$.

Since $1 - \left(\tfrac{1}{2}\right)^n$ is always less than 1, $z(n) = 2[1 - \left(\tfrac{1}{2}\right)^n]$ is always less than 2, so the total amount of money drawn out of trust after n yr is always less than \$2 million, regardless of the value of n. Now, computing the first few terms of the sequence $\{z(n)\}$, we have $z(1) = 1$, $z(2) = \tfrac{3}{2}$, $z(3) = \tfrac{7}{4}$, $z(4) = \tfrac{15}{8}$, and $z(5) = \tfrac{31}{16}$, so the line graph of the sequence z is as shown in Fig. 3.3-2. Again, if we were watching someone sketch the line graph as the years went on, we would notice that as the sequence values were marked on the line they would be located successively farther and farther to the right and appear to "jam up near" or "get arbitrarily close to" 2. If the woman had initially put only a \$1.5 million into the trust, then after 3 years the trust would be overdrawn, since $z(3) = \tfrac{7}{4} > \tfrac{6}{4} = 1.5$. If \$1$\tfrac{7}{8}$ million were initially put into the fund, it would be overdrawn in 5 years, since $z(5) = \tfrac{31}{16} > \tfrac{30}{16} = 1\tfrac{7}{8}$. In fact, if any amount less than \$2 million were initially put into the trust, the account would eventually be overdrawn. Thus there is a *limit* to how small the initial deposit may be, namely, \$2 million, which coincides with the *limit* of the total amount of money to be withdrawn by the charity. As before, we extend this language to the sequence $\{z(n)\}$ and say that "the limit of the sequence $\{z(n)\}$ is 2," and write $\lim_{n \to \infty} z(n) = 2$.

Since $1 - \left(\tfrac{1}{2}\right)^n < 1$, $2(1 - \left(\tfrac{1}{2}\right)^n) < 2 \cdot 1 = 2$.

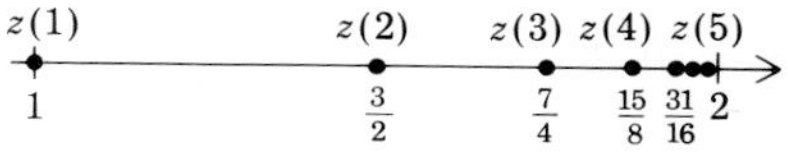

FIGURE 3.3-2

Recall that $z(n)$ is the total amount of money withdrawn after n yr.

Recall that the notation $\lim_{n \to \infty} z(n) = 2$ means "the limit of the sequence $\{z(n)\}$ is 2."

Exercises 3.3

In each of Exercises 1 through 12, an equation is given that defines the nth term $s(n)$ of a sequence s. Simplify the equation for $s(n)$ by using the formula $(1 + r + r^2 + \cdots + r^{n-1}) = (1 - r^n)/(1 - r)$, compute the first four terms of the sequence, and sketch the line graph. Estimate the number on the line at which the terms of the sequence appear to "jam up" or "cluster."

1. $s(n) = 2[1 + \tfrac{1}{2} + \left(\tfrac{1}{2}\right)^2 + \cdots + \left(\tfrac{1}{2}\right)^{n-1}]$
2. $s(n) = 3[1 + \tfrac{1}{2} + \left(\tfrac{1}{2}\right)^2 + \cdots + \left(\tfrac{1}{2}\right)^{n-1}]$
3. $s(n) = 2[1 + \tfrac{1}{3} + \left(\tfrac{1}{3}\right)^2 + \cdots + \left(\tfrac{1}{3}\right)^{n-1}]$
4. $s(n) = 4[1 + \tfrac{1}{3} + \left(\tfrac{1}{3}\right)^2 + \cdots + \left(\tfrac{1}{3}\right)^{n-1}]$
5. $s(n) = 3[1 + \tfrac{1}{4} + \left(\tfrac{1}{4}\right)^2 + \cdots + \left(\tfrac{1}{4}\right)^{n-1}]$
6. $s(n) = 4[1 + \tfrac{1}{5} + \left(\tfrac{1}{5}\right)^2 + \cdots + \left(\tfrac{1}{5}\right)^{n-1}]$
7. $s(n) = 8[1 + \tfrac{1}{5} + \left(\tfrac{1}{5}\right)^2 + \cdots + \left(\tfrac{1}{5}\right)^{n-1}]$

8. $s(n) = \frac{9}{5}[1 + \frac{1}{10} + (\frac{1}{10})^2 + \cdots + (\frac{1}{10})^{n-1}]$

9. $s(n) = 2[\frac{1}{3} + (\frac{1}{3})^2 + \cdots + (\frac{1}{3})^n]$

10. $s(n) = 3[\frac{1}{4} + (\frac{1}{4})^2 + \cdots + (\frac{1}{4})^n]$

11. $s(n) = 6[(\frac{1}{2})^2 + (\frac{1}{2})^3 + \cdots + (\frac{1}{2})^{n+1}]$

12. $s(n) = 90[(\frac{1}{10})^2 + (\frac{1}{10})^3 + \cdots + (\frac{1}{10})^{n+1}]$

13. A woman takes a 1-gal container and, after 1 min, pours $\frac{1}{3}$ gal of water into the container. After another minute, she pours $\frac{1}{9}$ gal of water in, after another minute $\frac{1}{27}$ gal, etc., so that n min after the process starts, she pours in $(\frac{1}{3})^n$ gal of water. Let s be the sequence whose value $s(n)$ at each positive integer n is the total number of gallons of water in the container after n min. Show that $s(n) = \frac{1}{2}[1 - (\frac{1}{3})^n]$. Construct the line graph of the sequence $\{s(n)\}$. What is the limit to the amount of water collected in the container?

14. As in Exercise 13, a woman pours water into a 1-gal container, but now suppose that she first pours in $\frac{1}{5}$ gal, then 1 min later $\frac{1}{25}$ gal, 1 min later $\frac{1}{125}$ gal, etc., so that n min after the process starts, she pours in $(\frac{1}{5})^n$ gal of water. Let t be the sequence whose value $t(n)$ at each positive integer n is the total number of gallons of water in the container after n min. Show that $t(n) = \frac{1}{4}[1 - (\frac{1}{5})^n]$. Construct the line graph of the sequence $\{t(n)\}$. What is the limit to the amount of water collected in the container?

15. A man puts $1 million in trust for his favorite charity, which is to draw out $\$\frac{1}{4}$ million at the end of the first year, $\$\frac{1}{16}$ million at the end of the second year, $\$\frac{1}{64}$ million at the end of the third year, etc., so that at the end of the nth year, the charity will withdraw $\$(\frac{1}{4})^n$ million. Let p be the sequence whose value $p(n)$ at each positive integer n is the total amount of money withdrawn by the charity after n years. Show that $p(n) = \frac{1}{3}[1 - (\frac{1}{4})^n]$. Construct the line graph of the sequence $\{p(n)\}$. What is the limit to the total amount of money the charity will receive from the trust?

16. A man wishes to put enough money into a trust fund so that his charity may withdraw $\$\frac{2}{3}$ million at the end of 1 year, $\$\frac{4}{9}$ million at the end of 2 years, $\$\frac{8}{27}$ million at the end of the third year, etc., so that at the end of the nth year, the charity is to withdraw $\$(\frac{2}{3})^n$ million. Let q be the sequence whose value $q(n)$ at each positive integer n is the total amount of money to be withdrawn by the charity after n years. Show that $q(n) = 2[1 - (\frac{2}{3})^n]$. Construct the line graph of the sequence $\{q(n)\}$. How much money must the man initially put in trust in order that the account will never be overdrawn?

17. A ball is thrown 8 ft straight up into the air and is allowed to bounce indefinitely. Each time it bounces it rises $\frac{1}{2}$ the distance it had previously fallen. Let t be the sequence whose value $t(n)$ at

each positive integer n is the height the ball rises after the nth bounce. Verify that for each positive integer n, $t(n) = 8(\frac{1}{2})^n$. Let s be the sequence whose value $s(n)$ at each positive integer n is the total distance traveled by the ball when it hits the ground the nth time. Then $s(1) = 16$, $s(2) = 16 + 2t(1)$, $s(3) = 16 + 2t(1) + 2t(2)$, and in general, $s(n) = 16 + 2t(1) + 2t(2) + \cdots + 2t(n - 1)$. Show that for each positive integer n, $s(n) = 32[1 - (\frac{1}{2})^n]$. Construct the line graph of the sequence $\{s(n)\}$. What is the limit to the total distance traveled by the ball?

3.4.   The Limit of a Sequence

The primary purpose of this section is to develop a practical definition of the statement "the limit of the sequence $\{s(n)\}$ is the number L." First let us examine some particular sequences and their line graphs; for, as we shall soon see, the value of the limit L of a sequence (if the sequence has a limit) can often be determined by the line graph.

EXAMPLE 3.4-1

Let s be the sequence defined by

$$s(n) = \frac{1}{n}$$

Then $s(1) = 1$, $s(2) = \frac{1}{2}$, $s(3) = \frac{1}{3}$, etc., and the line graph of s is shown in Fig. 3.4-1. Notice that the terms $s(n)$ of the sequence get closer and closer to (or "jam up at") 0 as n becomes larger and larger. Thus, when n is large, $s(n)$ is very close to 0. In fact, as mathematicians like to put it, the terms $s(n)$ get *arbitrarily close* to 0; this means that there is no positive number that is smaller than *all* the terms $s(n)$ of this sequence. For this reason we say that the limit of the sequence $\{1/n\}$ is 0, and we write

$$\lim_{n\to\infty} \frac{1}{n} = 0$$

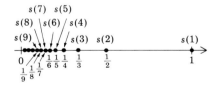

FIGURE 3.4-1   The line graph of the sequence s, where $s(n) = 1/n$.

EXAMPLE 3.4-2

Let p be the sequence defined by

$$p(n) = \frac{(-1)^n}{n}$$

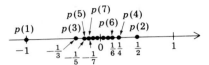

FIGURE 3.4-2 The line graph of the sequence p, where $p(n) = (-1)^n/n$.

Then $p(1) = -1$, $p(2) = \frac{1}{2}$, $p(3) = \frac{-1}{3}$, etc., and the line graph of p is shown in Fig. 3.4-2. Again, the terms $p(n)$ of the sequence move closer and closer to (or "jam up at") 0 as n gets larger and larger (as n approaches infinity). Even though some terms of this sequence are positive and some are negative, both positive and negative terms eventually get very close to 0. Thus the limit of the sequence $\{(-1)^n/n\}$ is 0, and we write

$$\lim_{n\to\infty} \frac{(-1)^n}{n} = 0$$

In each of the two examples above, the terms of the sequences s and p drew steadily closer to 0 as n took on larger and larger values; that is, for each positive integer n, $s(n + 1)$ was closer to 0 than $s(n)$, and $p(n + 1)$ was closer to 0 than $p(n)$. However, it is not necessary that the terms of a sequence behave in this way for the limit of the sequence to be 0, as the next example illustrates.

EXAMPLE 3.4-3

Recall that a positive integer n is even if it is of the form $n = 2k$ for some integer k and odd if it is of the form $n = 2k + 1$ for some integer k.

Let q be the sequence defined by the equations

$$q(n) = \frac{1}{2(n + 1)} \qquad \text{if } n \text{ is odd}$$

and

$$q(n) = \frac{1}{2(n - 1)} \qquad \text{if } n \text{ is even}$$

Then the first few terms of the sequence are $q(1) = \frac{1}{4}$, $q(2) = \frac{1}{2}$, $q(3) = \frac{1}{8}$, $q(4) = \frac{1}{6}$, $q(5) = \frac{1}{12}$, $q(6) = \frac{1}{10}$, $q(7) = \frac{1}{16}$, $q(8) = \frac{1}{14}$, . . . , and the line graph of q is shown in Fig. 3.4-3. Notice that the terms $q(n)$ of the sequence do not steadily get closer to 0 as n takes on larger and larger values; for example, $q(4)$ is not closer to 0 than $q(3)$. Nonetheless, the terms $q(n)$ of the sequence do "jam up" or "cluster" at 0 as more and more terms are marked on the line (as n gets larger). Thus for large values of n, $q(n)$ is close to 0, and the limit of this sequence is again 0, i.e., $\lim_{n\to\infty} q(n) = 0$. This fact is sometimes expressed by stating that the sequence $\{q(n)\}$ *converges* to 0.

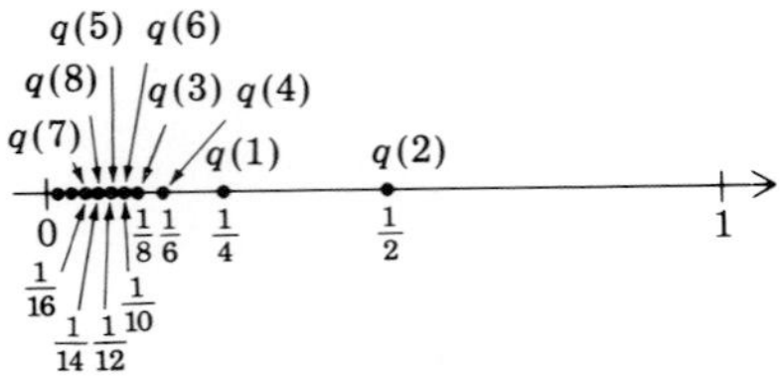

FIGURE 3.4-3 The line graph of the sequence q, where $q(n) = 1/2(n + 1)$ if n is odd and $q(n) = 1/2(n - 1)$ if n is even.

This geometric property of the terms of a sequence, "jamming up" or "clustering" about exactly one point on the line, will serve as our criterion for a sequence to have a limit. We say that *the limit of the sequence $\{s(n)\}$ is the number L* (and write $\lim_{n \to \infty} s(n) = L$) provided that there is exactly one point L on the line at which the terms $s(n)$ of the sequence cluster or jam up. In this case, we also say that the sequence $\{s(n)\}$ *converges to L*. (As later examples will show, it is possible for the terms of a sequence to cluster near two distinct points on the line; when this happens, the sequence does not have a limit.)

Definition of the limit of a sequence

Alternative phrases for "the limit of the sequence $\{s(n)\}$ is L" are "the sequence $\{s(n)\}$ approaches L as n approaches infinity" and "the sequence $\{s(n)\}$ approaches L as n increases indefinitely." These phrases all have the same meaning and are all represented symbolically by the notation $\lim_{n \to \infty} s(n) = L$.

EXAMPLE 3.4-4

The sequence r defined by the equation

$$r(n) = (-1)^n$$

does not have a limit. The terms of the sequence are $r(1) = -1$, $r(2) = 1$, $r(3) = -1$, $r(4) = 1, \ldots$, and thus the terms of the sequence alternate between -1 and 1. So the tendency is not for the terms $r(n)$ to jam up or cluster at one particular point on the line; rather, they pop back and forth from -1 to 1 indefinitely (see Fig. 3.4-4).

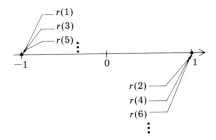

FIGURE 3.4-4 The line graph of the sequence r, where $r(n) = (-1)^n$.

EXAMPLE 3.4-5

Let a be a fixed real number, and let c be the sequence defined by $c(n) = a$ for each positive integer n. Such a sequence is called a *constant sequence*. The limit of the constant sequence c is the number a, since the terms $c(n)$ of the sequence certainly jam up at the point a (in fact, they pile up on a, since $c(n) = a$ for all n). Thus $\lim_{n \to \infty} c(n) = a$ (see Fig. 3.4-5).

Any sequence c defined by an equation of the form $c(n) = a$ for some fixed number a is called a *constant sequence*. Every constant sequence has a limit, and the limit is the number a, i.e., $\lim_{n \to \infty} c(n) = a$.

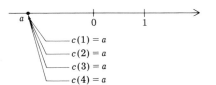

FIGURE 3.4-5 The line graph of the constant sequence c, where $c(n) = a$ for each positive integer n.

EXAMPLE 3.4-6

Let x be the sequence $\{1 + (-1)^n(n - 1)/n\}$, i.e., x is defined by the equation

$$x(n) = 1 + \frac{(-1)^n(n - 1)}{n}$$

for each positive integer n. Then $x(1) = 1$, $x(2) = 1 + \frac{1}{2} = \frac{3}{2}$, $x(3) = 1 - \frac{2}{3} = \frac{1}{3}$, $x(4) = 1 + \frac{3}{4} = \frac{7}{4}$, $x(5) = 1 - \frac{4}{5} = \frac{1}{5}$, $x(6) = 1 + \frac{5}{6} = \frac{11}{6}, \ldots$, and the line graph of the sequence x is as shown in Fig. 3.4-6. Notice that the terms $x(n)$ of the sequence x jam up near two different points on the line, namely, 0 and 2; consequently, the sequence $\{x(n)\}$ does not have a limit, since the terms do not cluster around exactly one point, as the definition of limit requires.

Note in Fig. 3.4-6 that when n is odd, the terms $x(n)$ jam up at 0, while when n is even, the terms $x(n)$ jam up at 2. Nonetheless, there is no single point L on the line such that the terms $x(n)$ *all* cluster around L.

FIGURE 3.4-6

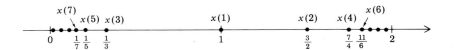

While our definition of the limit of a sequence will be adequate for our purposes, the technically impeccable definition offered in more advanced courses is as follows. The sequence $s(n)$ is said to have limit L (L a real number) providing that given any specified real number $\epsilon > 0$ there exists a positive integer n_0 (depending on ϵ) such that $|s(n) - L| < \epsilon$ whenever $n \geq n_0$.

Frequently it is possible to find the limit of a sequence s without relying on its line graph. This is done by analyzing the equation that gives the nth term $s(n)$ of the sequence to see what happens to the values of $s(n)$ as n becomes large. If the values get arbitrarily close to (jam up at) some number L, then L is the limit. The following examples illustrate this technique.

EXAMPLE 3.4-7

Find the limit of the sequence $\{(2n + 3)/(3n + 1)\}$.

SOLUTION

The sequence $\{(2n + 3)/(3n + 1)\}$ is a function s, whose value at each positive integer n is given by the equation

$$s(n) = \frac{2n + 3}{3n + 1}$$

It is not clear from this equation what happens to the values of $s(n)$ as n becomes large; however, if we divide the numerator and denominator by n, we have

$$s(n) = \frac{(2n + 3)/n}{(3n + 1)/n} = \frac{2 + (3/n)}{3 + (1/n)}$$

Now, as n becomes large, $3/n$ becomes very small (close to 0) and $2 + (3/n)$ becomes very close to 2. Similarly, $3 + (1/n)$ approaches 3; thus it is reasonable to believe that

$$\lim_{n \to \infty} \frac{2n + 3}{3n + 1} = \lim_{n \to \infty} \frac{2 + (3/n)}{3 + (1/n)} = \frac{2}{3}$$

The line graph of $\{(2n + 3)/(3n + 1)\}$, shown in Fig. 3.4-7, also indicates that $\frac{2}{3}$ is the correct limit.

Recall that dividing the numerator and denominator of a fraction by the positive integer n does not change the value of the fraction. Such algebraic "tricks" are frequently helpful when analyzing the behavior of the terms $s(n)$ of a sequence as n gets large.

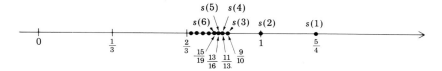

FIGURE 3.4-7

EXAMPLE 3.4-8

Find the limit of the sequence $\{(6n^2 - 7n + 2)/(4n^2 - 1)\}$.

SOLUTION

The sequence $\{(6n^2 - 7n + 2)/(4n^2 - 1)\}$ is a function s, whose value at each positive integer n is given by the equation

$$s(n) = \frac{6n^2 - 7n + 2}{4n^2 - 1}$$

If we divide numerator and denominator by n^2 (the highest power of n occuring in the denominator), we have

Sequences of the kind investigated in Examples 3.4-7 and 3.4-8 can usually be handled by dividing numerator and denominator by the largest power of n occuring in the denominator. Then the result that $\{1/n^p\}$ has limit 0 (p a positive integer) is usually brought into play.

$$s(n) = \frac{(6n^2 - 7n + 2)/n^2}{(4n^2 - 1)/n^2} = \frac{6 - (7/n) + (2/n^2)}{4 - (1/n^2)}$$

Now as n becomes large, $7/n$ and $2/n^2$ become quite small (close to 0), so

$$6 - (7/n) + (2/n^2)$$

approaches 6. Similarly, $4 - (1/n^2)$ approaches 4, so it is reasonable to believe that

$$\lim_{n \to \infty} \frac{6n^2 - 7n + 2}{4n^2 - 1} = \lim_{n \to \infty} \frac{6 - (7/n) + (2/n^2)}{4 - (1/n^2)} = \frac{6}{4} = \frac{3}{2}$$

An alternative approach to determining the behavior of this sequence is afforded by the fact that $6n^2 - 7n + 2$ and $4n^2 - 1$ factor into $(2n - 1)(3n - 2)$ and $(2n - 1)(2n + 1)$, respectively. Thus

$$s(n) = \frac{6n^2 - 7n + 2}{4n^2 - 1} = \frac{(2n - 1)(3n - 2)}{(2n - 1)(2n + 1)} = \frac{3n - 2}{2n + 1}$$

At this point we can divide numerator and denominator by n and proceed as in Example 3.4-7.

EXAMPLE 3.4-9 (*Optional.*)

A group of psychologists conducted a learning experiment with rats. In the experiment, 20 of 100 rats successfully found their way through a maze on the first trial. On each succeeding trial, a reinforcement was given to the rats which increased their likelihood of getting through the maze. The data indicated that on the nth trial ($n > 1$), the number of rats finding their way through the maze increased by approximately $\frac{1}{4}$ the number that failed on the previous trial. Let s be the sequence whose nth term $s(n)$ is the number of rats finding their way through the maze on the nth trial. Find the equation which defines $s(n)$ and find $\lim_{n \to \infty} s(n)$.

SOLUTION

We are given that for each positive integer n,

$$s(n + 1) = s(n) + \tfrac{1}{4}\big[100 - s(n)\big]$$

We may rewrite this equation as

$$s(n + 1) = 100 - 100 + s(n) + \tfrac{1}{4}(100 - s(n))$$

$$= 100 - (100 - s(n)) + \tfrac{1}{4}(100 - s(n))$$

$$= 100 - \Big[(100 - s(n)) - \tfrac{1}{4}(100 - s(n))\Big]$$

$$= 100 - (100 - s(n))(1 - \tfrac{1}{4})$$

Thus $s(2) = 100 - (100 - s(1))(1 - \tfrac{1}{4})$, and since $s(3) = 100 - (100 - s(2))(1 - \tfrac{1}{4})$, we have

$$s(3) = 100 - \Big[100 - \big(100 - (100 - s(1))(1 - \tfrac{1}{4})\big)\Big](1 - \tfrac{1}{4})$$

$$= 100 - \Big[100 - 100 + (100 - s(1))(1 - \tfrac{1}{4})\Big](1 - \tfrac{1}{4})$$

$$= 100 - \Big[(100 - s(1))(1 - \tfrac{1}{4})\Big](1 - \tfrac{1}{4})$$

$$= 100 - (100 - s(1))(1 - \tfrac{1}{4})^2$$

Similarly,

$$s(4) = 100 - (100 - s(3))(1 - \tfrac{1}{4})$$

$$= 100 - \Big[100 - \big(100 - (100 - s(1))(1 - \tfrac{1}{4})^2\big)\Big](1 - \tfrac{1}{4})$$

$$= 100 - \Big[100 - 100 + (100 - s(1))(1 - \tfrac{1}{4})^2\Big](1 - \tfrac{1}{4})$$

$$= 100 - \Big[(100 - s(1))(1 - \tfrac{1}{4})^2\Big](1 - \tfrac{1}{4})$$

$$= 100 - (100 - s(1))(1 - \tfrac{1}{4})^3$$

Thus we see that a pattern is developing; it appears that

$$s(n) = 100 - (100 - s(1))(1 - \tfrac{1}{4})^{n-1}$$

Since we are given that $s(1) = 20$, we have

$$s(n) = 100 - (100 - 20)(\tfrac{3}{4})^{n-1} = 100 - 80(\tfrac{3}{4})^{n-1}$$

and this is the equation that defines the nth term of the

The number of rats failing to find their way through the maze on the nth trial is $100 - s(n)$, that is, 100 minus the number successfully completing the maze on the nth trial.

Adding and subtracting 100.

Inserting parentheses.

Factoring out $(100 - s(n))$.

Substituting $s(2) = 100 - (100 - s(1))(1 - \tfrac{1}{4})$.

Substituting $s(3) = 100 - (100 - s(1))(1 - \tfrac{1}{4})^2$.

Note that the equation we have found here for $s(n)$ generalizes to other experiments in learning theory. In general, if M rats (or people, etc.) participate in an experiment with M_1 correct responses on the first trial, and if on each succeeding trial the number of correct responses increases by a fraction p/q of the number that failed on the previous trial, then the number of correct responses $s(n)$ on the nth trial is given by the equation $s(n) = M - (M - M_1)(1 - p/q)^{n-1}$.

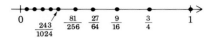

$$\overset{\bullet\bullet\bullet\bullet\ \ \bullet\ \ \ \bullet\ \ \ \ \ \bullet\ \ \ \ \ \ \ \ \ \ \bullet}{\underset{0 \quad \underset{1024}{243} \quad \underset{256}{81} \ \ \underset{64}{27} \ \ \underset{16}{9} \quad \ \ \underset{4}{3} \qquad \quad 1}{\rule{0pt}{0pt}}}$$

FIGURE 3.4-8   The line graph of the sequence $\{(\frac{3}{4})^{n-1}\}$. The first few terms of this sequence are

$$\left(\tfrac{3}{4}\right)^{1-1} = \left(\tfrac{3}{4}\right)^{0} = 1,$$

$$\left(\tfrac{3}{4}\right)^{2-1} = \left(\tfrac{3}{4}\right)^{1} = \tfrac{3}{4},$$

$$\left(\tfrac{3}{4}\right)^{3-1} = \left(\tfrac{3}{4}\right)^{2} = \tfrac{9}{16},$$

$$\left(\tfrac{3}{4}\right)^{4-1} = \left(\tfrac{3}{4}\right)^{3} = \tfrac{27}{64},$$

$$\left(\tfrac{3}{4}\right)^{5-1} = \left(\tfrac{3}{4}\right)^{4} = \tfrac{81}{256},$$

$$\left(\tfrac{3}{4}\right)^{6-1} = \left(\tfrac{3}{4}\right)^{5} = \tfrac{243}{1024},$$

sequence s. Since $\lim_{n \to \infty}(\frac{3}{4})^{n-1} = 0$ (as shown by the line graph of the sequence $\{(\frac{3}{4})^{n-1}\}$ in Fig. 3.4-8), $80(\frac{3}{4})^{n-1}$ becomes very small (close to 0) as n gets large, which means that $100 - 80(\frac{3}{4})^{n-1}$ approaches 100:

$$\lim_{n \to \infty} s(n) = \lim_{n \to \infty} \left[100 - 80\left(\tfrac{3}{4}\right)^{n-1}\right] = 100$$

This means that as more and more trials are conducted, the number of rats successfully finding their way through the maze approaches 100.

Exercise 3.4.

In each of Exercises 1 through 12, a sequence s is given. Write the equation that defines the function s, compute the first six terms of the sequence, and sketch its line graph. State whether or not the sequence has a limit L as n approaches infinity, and if the sequence does have a limit, estimate the value of L.

1. $\left\{\dfrac{1}{4n}\right\}$ 2. $\left\{\dfrac{-1}{6n}\right\}$

3. $\left\{\dfrac{(-1)^n}{3n}\right\}$ 4. $\left\{\dfrac{(-1)^n}{n^2}\right\}$

5. $\left\{1 + \dfrac{1}{n^2}\right\}$ 6. $\left\{1 + \dfrac{(-1)^n}{n^2}\right\}$

7. $\{7.25 - (1)^n\}$ 8. $\{-5 - (-1)^n\}$

9. $\{3 + (-1)^n\}$ 10. $\left\{-2 + \dfrac{(-1)^n}{n}\right\}$

11. $\left\{-1 - \dfrac{(-1)^n}{n}\right\}$ 12. $\left\{4 + \dfrac{(-1)^n}{n^2}\right\}$

As in Exercises 1 through 12, find (if it exists):

13. $\displaystyle\lim_{n \to \infty} ([1 + (-1)^n] + 2)$ 14. $\displaystyle\lim_{n \to \infty} (n[1 + (-1)^n])$

15. $\displaystyle\lim_{n \to \infty} \left(\dfrac{1 + (-1)^n}{n}\right)$ 16. $\displaystyle\lim_{n \to \infty} \left(2 + \dfrac{(-1)^n(n + 1)}{n}\right)$

17. $\displaystyle\lim_{n \to \infty} \left(\dfrac{2}{3} + \dfrac{(-1)^n n}{2^n}\right)$ 18. $\displaystyle\lim_{n \to \infty} \left((-1)^n + \dfrac{n}{2^n}\right)$

In each of Exercises 19 through 32, write the equation that defines the nth term $s(n)$ of the given sequence s, and analyze this equation to see what happens to the values of $s(n)$ as n becomes large (see Examples 3.4-7 and 3.4-8). Based on your analysis of the behavior of the values of $s(n)$ as n gets large, state what the limit of the sequence is.

Recall that in analyzing the behavior of the values of $s(n)$, it may be helpful to modify the form of $s(n)$, i.e., to divide (or multiply) the numerator and denominator of $s(n)$ by some quantity (see Example 3.4-7).

19. $\left\{ \dfrac{4}{2n+1} \right\}$

20. $\left\{ \dfrac{4 - 1/n}{2 + 1/n} \right\}$

21. $\left\{ \dfrac{2 - 1/n}{2 + n} \right\}$

22. $\left\{ \dfrac{4n + 1}{2n - 4} \right\}$

23. $\left\{ \dfrac{2 - n}{1 + 2n} \right\}$

24. $\left\{ \dfrac{6n - 5}{1 - 3n} \right\}$

25. $\left\{ \dfrac{3n + 2}{5n - 1} \right\}$

26. $\left\{ \dfrac{7n - 6}{11n + 4} \right\}$

27. $\left\{ \dfrac{n^2 - 1}{2n^2 - 2} \right\}$

28. $\left\{ \dfrac{2n^2 + 4n}{n^2 + 1} \right\}$

29. $\left\{ \dfrac{7n - 5n^2}{n^2 + 3n} \right\}$

30. $\left\{ \dfrac{4n - 7}{2n^2 - 3n + 1} \right\}$

31. $\left\{ \dfrac{3n^2 + 4n - 1}{7n^2 - n + 4} \right\}$

32. $\left\{ \dfrac{-4n^3 + 5n - 2}{-2n^3 - 6n^2 + 1} \right\}$

In each of Exercises 33 through 37, a sequence s is defined by the two given equations; one defines $s(n)$ when n is odd, and the other defines $s(n)$ when n is even. Compute the first six terms of the sequence and find $\lim_{n \to \infty} s(n)$ if the limit exists.

33.

$$s(n) = \begin{cases} \dfrac{1}{n^2} & \text{if } n \text{ is odd} \\[2mm] \dfrac{1}{(n + 2)^2} & \text{if } n \text{ is even} \end{cases}$$

34.

$$s(n) = \begin{cases} 2 + \dfrac{(-1)^n}{n} & \text{if } n \text{ is odd} \\[2mm] 2 - \dfrac{1}{n^2 + 2} & \text{if } n \text{ is even} \end{cases}$$

35.

$$s(n) = \begin{cases} 3 - \dfrac{1}{n^2} & \text{if } n \text{ is odd} \\[2mm] 3\left(1 + \dfrac{1}{n + 1}\right) & \text{if } n \text{ is even} \end{cases}$$

36.

$$s(n) = \begin{cases} 5 & \text{if } n \text{ is odd} \\ 5 - 2/n & \text{if } n \text{ is even} \end{cases}$$

37.

$$s(n) = \begin{cases} \dfrac{2n^2 - 4n^3 - 2}{n^3 - 1} & \text{if } n \text{ is odd} \\ 3 - \dfrac{n+1}{n} & \text{if } n \text{ is even} \end{cases}$$

38. In Exercise 3.2-24, a sequence s was found whose nth term $s(n) = (\frac{4}{5})^n (220{,}000)$ is the pollution index of smoke emitted from a smokestack in which n pollution-control filters have been installed. Find $\lim_{n \to \infty} s(n)$. What does the value of this limit mean?

39. In Exercise 3.2-25, a sequence s was found whose nth term $s(n) = (0.95)^n (800{,}000)$ is the value of $\$800{,}000$ n yr from the present time (due to inflation). Find $\lim_{n \to \infty} s(n)$. What does the value of this limit mean?

In Exercise 40, if $\$P_n$ is put on deposit, after n years the amount A in the account is $A = P_n(1.05)^n$ (see Exercise 3.2-26). Setting $A = \$8000$ and solving for P_n, we find that $P_n = 8000/(1.05)^n$.

40. If a bank pays 5% interest compounded annually, and if you wish to have $\$8000$ in your account n years from now, then the amount P_n of money you must now deposit (assuming that no other deposits are made) in your account is given by the equation $P_n = 8000/(1.05)^n$. Find $\lim_{n \to \infty} P_n$. What does the value of this limit mean?

41. A company purchases a piece of heavy machinery for $\$25{,}000$. Suppose the estimated resale value of the machinery n yr from now is given by the equation $s(n) = 4000 + (25{,}000)(\frac{3}{5})^n$. Find $\lim_{n \to \infty} s(n)$; this limit can be considered the "scrap value" of the machinery.

42. Suppose a portable electronic calculator operates properly for 10 hr on a new set of batteries. Suppose further that after the batteries are recharged, the calculator operates for 9 hr, and after each successive recharge the calculator operates $\frac{9}{10}$ as long as it did on the previous recharge. Let s be the sequence whose nth term $s(n)$ is the total length of time the calculator will operate before the batteries must be recharged the nth time. Find the equation that defines the nth term $s(n)$ of the sequence s, and find $\lim_{n \to \infty} s(n)$. How long will the calculator operate on one set of batteries, assuming that an infinite number of recharges are possible?

Hint: In Exercise 42, use the formula $(1 + r + r^2 + \cdots + r^{n-1}) = (1 - r^n)/(1 - r)$ to find the equation which defines $s(n)$.

43. (See Example 3.4-9.) In a certain experiment, 10 of 60 rats find their way through a maze on the first trial. On each succeeding trial, a reinforcement is given which increases their likelihood of getting through the maze so that on the nth trial $(n > 1)$, the number of rats getting through is increased by $\frac{1}{6}$ of the number

that failed on the previous trial. Let s be the sequence whose nth term $s(n)$ is the number of rats getting through on the nth trial. Find the equation that defines $s(n)$ and find $\lim_{n \to \infty} s(n)$.

*3.5. Sequences with Infinite Limits (*Optional*)

All the sequences considered in the previous section had the property that their terms jammed up at, or got arbitrarily close to, one or more points on the line. Frequently, however, sequences are encountered whose terms do not jam up at *any* point on the line. Such sequences, of course, do not have any real number L as a limit; nonetheless, they play an important role in the development of the calculus, and we need to examine some of their properties. Let us begin with an illustration which shows how sequences like this might arise.

Suppose we have an unlimited supply of batteries with which to operate a portable radio, and suppose the radio will operate for 10 hr on 1 battery (we assume that the radio never wears out). Then the radio will operate for 20 hr on 2 batteries, for 30 hr on 3 batteries, and in general, for $10n$ hr on n batteries. Thus the sequence s that assigns to each positive integer n the number of hours the radio will operate on n batteries is defined by $s(n) = 10n$. The line graph of s is shown in Fig. 3.5-1. Note that the terms $s(n)$ of the sequence continue indefinitely to the right on the line (so they become larger and larger) and do not appear to jam up at any point on the line, so the sequence does not have any number L as a limit. However, if we pick any point M on the line, then eventually one of the terms $s(1) = 10$, $s(2) = 20$, $s(3) = 30$, $s(4) = 40$, . . . of the sequence will be larger than M; and once we find a

If the terms of a sequence jam up at exactly one point, the sequence has a limit; if the terms jam up at more than one point or if they do not jam up at all, then the sequence does not have a number L as a limit.

The first few terms of the sequence s are

$$s(1) = 10, s(2) = 20,$$

$$s(3) = 30, s(4) = 40,$$

$$s(5) = 50, \ldots .$$

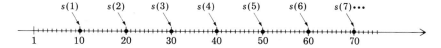

FIGURE 3.5-1

term of the sequence that is to the right of M on the line, *all* the remaining terms of s are to the right of M. Thus, only a finite number of the terms of s are to the left of M, no matter how large the number M may be. When a sequence s has this property (namely, that for any number M on the line that we might choose, only a finite number of the terms $s(n)$ of the sequence lie to the left of M) we say that the sequence $\{s(n)\}$ has limit *plus infinity* and write

$$\lim_{n \to \infty} s(n) = +\infty$$

Since, in our illustration here, each term $s(n)$ is the number of hours the radio will operate on n batteries, the assertion that $\lim_{n \to \infty} s(n) = +\infty$ means that we can operate the radio as long as we wish, provided that we have an unlimited supply of batteries (in other words, the terms $s(n)$ become arbitrarily large as n increases indefinitely).

Definition of $\lim_{n \to \infty} s(n) = +\infty$

The symbol $+\infty$ does not represent any number. Thus when we say "the sequence $\{s(n)\}$ has limit plus infinity," we are only asserting that the terms $s(n)$ become arbitrarily large as n increases; we are not asserting that the terms $s(n)$ jam up near any particular number on the line. In fact, if $\lim_{n \to \infty} s(n) = +\infty$, then the sequence $\{s(n)\}$ cannot have any number L as a limit.

An alternative phrase for "the sequence $\{s(n)\}$ has limit plus infinity" is "the terms $s(n)$ of the sequence s become arbitrarily large as n increases indefinitely."

EXAMPLE 3.5-1

Find the limit of the sequence y defined by the equation $y(n) = n^2$.

SOLUTION

The first few terms of the sequence are $y(1) = 1$, $y(2) = 4$, $y(3) = 9$, $y(4) = 16$, $y(5) = 25, \ldots$, and the line graph is shown in Fig. 3.5-2. Evidently the terms $y(n)$ do not jam up at any point on the line; rather, they continue indefinitely to the right. Moreover, if we pick any number M on the line, then there are only a finite number of terms to the left of M. For example, if $M = 27$, there are five terms to the left of M, and if $M = 200$, there are 14 terms to the left of M, namely $y(1)$, $y(2), \ldots, y(14)$. Thus the limit of the sequence y (that is, the sequence $\{n^2\}$) is plus infinity, and we write

$$\lim_{n \to \infty} n^2 = +\infty$$

FIGURE 3.5-2

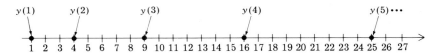

EXAMPLE 3.5-2

Verify that the sequence $\{(-1)^n n^2\}$ has neither $+\infty$ nor a number L as a limit.

SOLUTION

The sequence $\{(-1)^n n^2\}$ is a function z whose value at each positive integer n is given by the equation $z(n) = (-1)^n n^2$. Thus, the first few terms of the sequence are $z(1) = -1$, $z(2) = 4$, $z(3) = -9$, $z(4) = 16$, $z(5) = -25$, $z(6) = 36$, $z(7) = -49, \ldots$, and the line graph is shown in Fig. 3.5-3. Note that the terms of the sequence do not jam up at any point on the line, so the sequence does not have any number L as a limit. Moreover, if we pick any number M, there are infinitely many terms to the left of M, since the terms continue indefinitely to the left as well as to the right. Thus the sequence does not have $+\infty$ as a limit either so the sequence has neither a number nor $+\infty$ as a limit.

For any number M, there are infinitely many terms of the sequence $\{z(n)\} = \{(-1)^n n^2\}$ to the left of M, since the odd terms $z(n)$ (the terms $z(n)$ when n is odd) are negative numbers and continue indefinitely far to the left on the line as n (odd) becomes larger and larger.

FIGURE 3.5-3

Now let us consider for a moment the sequence r defined by the equation $r(n) = -n^2$. The first few terms of this sequence are $r(1) = -1$, $r(2) = -4$, $r(3) = -9$, $r(4) = -16$, etc., and the line graph is shown in Fig. 3.5-4. Notice that the terms $r(n)$ of this sequence continue farther and farther to the left on the line as n takes on larger and larger

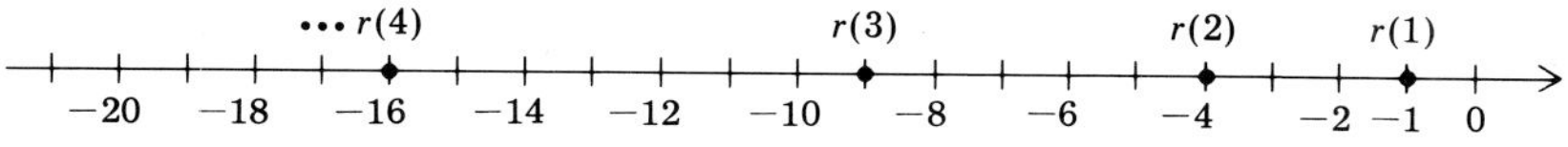

FIGURE 3.5-4

values. The behavior of the terms $r(n)$ is similar to the behavior of the terms of a sequence whose limit is $+\infty$, but here the terms are moving indefinitely far to the *left* as n increases instead of to the right. Thus we mirror the definition given above for a sequence having limit $+\infty$ and say that a sequence s has limit *minus infinity* if for any number M on the line that we choose, only a finite number of terms $s(n)$ lie to the *right* of M, and we write this symbolically as

Definition of $\lim_{n \to \infty} s(n) = -\infty$

$$\lim_{n \to \infty} s(n) = -\infty$$

According to this definition and Fig. 3.5-4, we see that for the sequence r, where $r(n) = -n^2$, we have $\lim_{n \to \infty} r(n) = -\infty$.

EXAMPLE 3.5-3

Find the limit of the sequence y defined by $y(n) = 1 - n$.

In Fig. 3.5-5, since the terms $y(n)$ do not jam up at any point on the line, the sequence y has no *number L* as a limit; however, since $\lim_{n \to \infty} y(n) = -\infty$, we do say that the sequence y has a *limit*. Thus a sequence is said to have a *limit* provided either that it has a number L as a limit or that it has limit $+\infty$ or $-\infty$.

SOLUTION

The first few terms of the sequence y are $y(1) = 0$, $y(2) = -1$, $y(3) = -2$, $y(4) = -3$, $y(5) = -4$, etc., and the line graph is shown in Fig. 3.5-5. Evidently the terms $y(n)$ do not jam up at any point on the line, but rather continue indefinitely to the left. Moreover, if M is any number on the line, then only a finite number of the terms $y(n)$ lie to the right of M; for example, if $M = 1$, then no terms are to the right of M, or if $M = -9$, then nine terms are to the right of M. Thus the limit of the sequence y is minus infinity:

$$\lim_{n \to \infty} (1 - n) = -\infty$$

FIGURE 3.5-5

EXAMPLE 3.5-4

Verify that the sequence $\{(-1)^n n^2\}$ has no limit.

SOLUTION

We have already seen, in Example 3.5-2, that this sequence has neither a number L nor $+\infty$ as a limit. Moreover, given any number M, we see from Fig. 3.5-3 that there are infinitely many terms of the sequence to the right of M, since the even terms (values of $z(n) = (-1)^n n^2$ for even integers n) continue indefinitely to the right (become larger and larger) as n gets larger. Thus the sequence does not have $-\infty$ as a limit either, so it has no limit.

As was the case for sequences with finite limits, it is frequently possible to determine whether a sequence s has $+\infty$ or $-\infty$ as a limit without relying on its line graph by analyzing the equation which gives the nth term $s(n)$ of the sequence. If the values of $s(n)$ become arbitrarily large (and stay large), then $+\infty$ is the limit; if the values become arbitrarily large and negative, then $-\infty$ is the limit. The following examples illustrate this technique.

To say that a number is "large and negative" means that it is a negative number whose absolute value (distance from 0) is large.

EXAMPLE 3.5-5

Find the limit of the sequence $\{(2n^2 + 3)/(3n + 1)\}$.

SOLUTION

The sequence $\{(2n^2 + 3)/(3n + 1)\}$ is a function q whose value at n is given by

$$q(n) = \frac{2n^2 + 3}{3n + 1}$$

Again, to see how the values of $q(n)$ behave as n becomes large, we divide the numerator and denominator by n; then

$$q(n) = \frac{(2n^2 + 3)/n}{(3n + 1)/n} = \frac{2n + (3/n)}{3 + (1/n)}$$

Now as n becomes large, $3 + (1/n)$ approaches 3, and $2n + (3/n)$ grows larger and larger. Thus the numerator becomes very large and the denominator approaches 3; thus the quotient must become larger and larger, i.e., we see that

$$\lim_{n \to \infty} \frac{2n^2 + 3}{3n + 1} = \lim_{n \to \infty} \frac{2n + (3/n)}{3 + (1/n)} = +\infty$$

Since we are dividing larger and larger numbers by numbers that are very close to 3, the quotient must become larger and larger.

Again, the line graph (Fig. 3.5-6) also indicates that $+\infty$ is the correct limit.

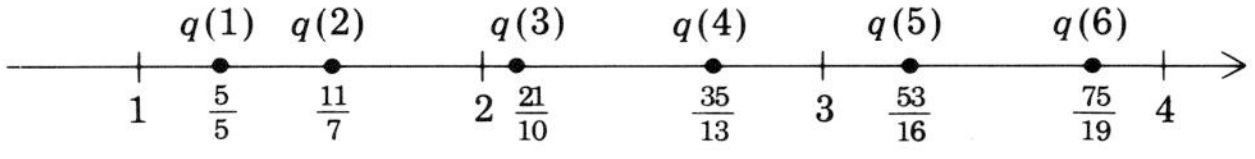

FIGURE 3.5-6

EXAMPLE 3.5-6

Find the limit of the sequence $\{(n^2 + 1)/(3 - 2n)\}$.

SOLUTION

Let w be the sequence $\{(n^2 + 1)/(3 - 2n)\}$; then for each positive integer n,

$$w(n) = \frac{n^2 + 1}{3 - 2n}$$

Dividing numerator and denominator by n, we see that

$$w(n) = \frac{(n^2 + 1)/n}{(3 - 2n)/n} = \frac{n + (1/n)}{(3/n) - 2}$$

Thus as n becomes large, the numerator $n + 1/n$ also becomes very large, while the denominator $(3/n) - 2$ approaches -2 (since $3/n$ becomes very small). The quotient, then, becomes very large and negative; thus we see that

$$\lim_{n \to \infty} w(n) = \lim_{n \to \infty} \frac{n^2 + 1}{3 - 2n} = -\infty$$

and the line graph (Fig. 3.5-7) again indicates this.

FIGURE 3.5-7

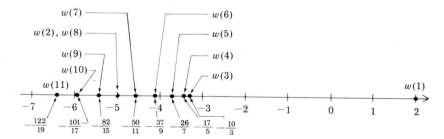

Exercises 3.5

In each of Exercises 1 through 14, a sequence s is given. Write the equation that defines the function s, compute the first six terms of the sequence, sketch the line graph, and state whether or not the sequence has a limit as n approaches infinity. If the sequence has a limit, state whether the limit is a number L, $+\infty$, or $-\infty$; if the limit is a number L, estimate the value of L.

Recall that a sequence $\{s(n)\}$ is said to have a *limit* provided either that it has a number L as a limit or that it has limit $+\infty$ or $-\infty$.

1. $\{2 + n\}$

2. $\left\{4 - \dfrac{n}{2}\right\}$

3. $\left\{2n + \dfrac{1}{n}\right\}$

4. $\left\{-2n + \dfrac{(-1)^n}{n}\right\}$

5. $\left\{3 - \dfrac{(-1)^n}{n}\right\}$

6. $\{-4 + (-1)^n\}$

7. $\{(-1)^{n+1} 2n\}$

8. $\left\{-n^2 + \dfrac{1}{n}\right\}$

9. $\left\{n - \dfrac{1}{n^2}\right\}$

10. $\left\{2n + \dfrac{(-1)^n}{n}\right\}$

11. $\left\{-2^n + \dfrac{(-1)^n}{n}\right\}$

12. $\{n^2[1 + (-1)^n]\}$

13. $\{[1 + (-1)^n] - 2n\}$

14. $\{2(n - n^2) + 1\}$:

As in Exercises 1 through 14, find (if it exists):

15. $\lim\limits_{n \to \infty} \left((-1)^{n+1} 2^n - n\right)$

16. $\lim\limits_{n \to \infty} \left(\dfrac{1 + (-1)^n}{2^n - n^2}\right)$

17. $\lim\limits_{n \to \infty} \left(2^n + \dfrac{(-1)^n(n - 1)}{n}\right)$

18. $\lim\limits_{n \to \infty} \left((-1)^n n - \dfrac{n + 1}{n}\right)$

In each of Exercises 19 through 24, analyze the equation that defines the nth term $s(n)$ of the given sequence s to see what happens to the values of $s(n)$ as n becomes large (see Examples 3.5-5 and 3.5-6). From your analysis of the behavior of the values of $s(n)$ as n gets large, state the limit of the sequence.

In analyzing the behavior of $s(n)$, it may be helpful to modify the form of $s(n)$ by dividing (or multiplying) the numerator and denominator of $s(n)$ by some quantity (see Examples 3.5-5 and 3.5-6).

19. $\left\{\dfrac{2n + 1}{4}\right\}$

20. $\left\{\dfrac{1 - 2n}{2}\right\}$

21. $\left\{\dfrac{2n + 1}{1 + (1/n)}\right\}$

22. $\left\{\dfrac{2 + n^2}{2n + 1}\right\}$

23. $\left\{\dfrac{3n^2 - 2}{1 - n}\right\}$

24. $\left\{\dfrac{5n^2 - 4n + 1}{-2n + 1}\right\}$

As in Exercises 19 through 24, find (if it exists):

25. $\lim\limits_{n \to \infty} \dfrac{-2n^2 + 6n - 1}{4n + 3}$

26. $\lim\limits_{n \to \infty} \dfrac{-2n^3 - 6n}{-4n^2 + 1}$

27. $\lim\limits_{n \to \infty} \dfrac{5 - 1/n}{1/n^2}$

28. $\lim\limits_{n \to \infty} \dfrac{-4n^2 + 1/n}{n}$

In Exercises 29 through 33, a sequence s is defined by the two given equations; one defines $s(n)$ when n is odd, and the other defines $s(n)$ when n is even. Compute the first six terms of the sequence and find $\lim_{n \to \infty} s(n)$ if the limit exists.

29.

$$s(n) = \begin{cases} 6n + 1 & \text{if } n \text{ is odd} \\ 2n & \text{if } n \text{ is even} \end{cases}$$

30.

$$s(n) = \begin{cases} -n^2 + 1 & \text{if } n \text{ is odd} \\ 10 - n & \text{if } n \text{ is even} \end{cases}$$

31.

$$s(n) = \begin{cases} 3 - 2n & \text{if } n \text{ is odd} \\ 2n - 3 & \text{if } n \text{ is even} \end{cases}$$

32.

$$s(n) = \begin{cases} \dfrac{n^2 + n}{n} & \text{if } n \text{ is odd} \\[2ex] \dfrac{2n^2 - n}{n} & \text{if } n \text{ is even} \end{cases}$$

33.

$$s(n) = \begin{cases} \dfrac{2n^2 - 1}{4n} + 2 & \text{if } n \text{ is odd} \\[2ex] 4 + \dfrac{(-1)^n}{n^2} & \text{if } n \text{ is even} \end{cases}$$

34. In Exercise 3.2-26, a sequence s was found whose nth term $s(n) = (1.05)^n (8000)$ is the amount in a bank account after n years, where \$8000 is the initial deposit and the bank pays 5% interest per year compounded annually. Find $\lim_{n \to \infty} s(n)$. What is the meaning of this limit?

35. In Exercise 3.2-22, a sequence s was found whose nth term $s(n) = (1.1)^n (200{,}000)$ is the value of a tract of land n years from now. Find $\lim_{n \to \infty} s(n)$. What is the meaning of this limit?

Limits of Functions; Continuity; Exponential and Logarithmic Functions

4.1. Limits of Functions

Now that we have defined and discussed limits of sequences, we are ready to consider the more general topic of limits of functions. Our initial objective is to explore, for a given function f and a given number a, the meaning of the statement "the limit of $f(x)$ as x approaches a is the number L." To do this, we shift our attention from the value of the function f at $x = a$ to the values of f at points x *near a*. Even if f is not defined at $x = a$, it is often possible to investigate the values of f at points x *near a*.

EXAMPLE 4.1-1

Suppose

$$f(x) = \frac{1000}{x + 1}$$

gives the reading on a thermometer when the thermometer is x in. away from a flame. Suppose further that the thermometer will immediately burst if it touches the flame. Then $f(0)$

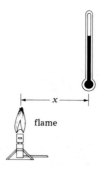

x

flame

= 1000 does not give the reading on the thermometer when the thermometer is 0 in. ($x = 0$) away from the flame; for the thermometer then touches the flame, bursts, and shows no reading at all. However, it is still reasonable to ask whether the temperature (the value of $f(x)$) gets close to 1000 when the thermometer is close to the flame (when x is close to 0). To answer this question we compute $f(x)$ for several values of x near 0 and interpret our results. The data (in Table 4.1-1) support the conclusion that the temperature approaches 1000 as the thermometer gets close to the flame; in fact, $f(x)$ gets arbitrarily close to 1000 when x is close to 0. We will say here that the *limit* of $f(x)$, as x approaches 0, is 1000, and we will write $\lim_{x \to 0} f(x) = 1000$.

TABLE 4.1-1

Distance from flame (inches)	x	1	$\frac{1}{10}$	$\frac{1}{100}$	$\frac{1}{1000}$	$\frac{1}{10,000}$
Temperature	$f(x) = \dfrac{1000}{(x + 1)}$	500	909.09	990.09	999.00	999.90

1 is not in the domain of $f(x)$ $= (x^2 - 1)/(x - 1)$, since division by 0 is not defined.

EXAMPLE 4.1-2

Let $f(x) = (x^2 - 1)/(x - 1)$; then 1 is not in the domain of f, and it is nonsense to ask for the value of f at $x = 1$. It is possible, however, to compute $f(x)$ for any number x other than 1, so we can ask whether there is some number which $f(x)$ is always close to if x is close (but not equal) to 1. We compute $f(x)$ for several numbers near 1 and record the results in Table 4.1-2 below. The work is simplified by noting that for $x \neq 1$,

$$f(x) = \frac{x^2 - 1}{x - 1} = \frac{(x + 1)(x - 1)}{x - 1} = x + 1$$

It is apparent from Table 4.1-2 that when x is close to 1, $f(x)$ is close to 2. We express this by saying that the limit of $f(x)$, as x approaches 1, is 2. Here we write

$$\lim_{x \to 1} f(x) = 2 \quad \text{or} \quad \lim_{x \to 1} \frac{x^2 - 1}{x - 1} = 2$$

TABLE 4.1-2

x	0.9	1.1	0.99	1.01	0.999	1.001
$f(x) = \dfrac{(x^2 - 1)}{(x - 1)}$	1.9	2.1	1.99	2.01	1.999	2.001

The previous examples highlight the essence of the limit question. For a given function f and number a, we say that the limit of $f(x)$ as x approaches a is the number L providing that L is the only number with the property that $f(x)$ gets arbitrarily close to L when x is close to a. We do not care what value f has at a itself, or even whether a is in the domain of f; our only concern is the behavior of f at points *near a*. The notation $\lim_{x \to a} f(x) = L$ means "the limit of $f(x)$ as x approaches a is the number L."

The definition of the notation $\lim_{x \to a} f(x) = L$

EXAMPLE 4.1-3

Let $f(x) = 2x^2 + 1$; determine whether there is a number L such that $\lim_{x \to 2} (2x^2 + 1) = L$.

SOLUTION

As before, we compute $f(x)$ for several numbers x near 2 and record the results in Table 4.1-3.

TABLE 4.1-3

x	1.9	2.1	1.99	2.01	1.999	2.001
$f(x) = 2x^2 + 1$	8.22	9.82	8.92	9.08	8.992	9.008

From this table we see that when x is close to 2, $f(x) = 2x^2 + 1$ is close to 9. Thus $L = 9$, i.e.,

$$\lim_{x \to 2} (2x^2 + 1) = 9$$

$\lim_{x \to 2}(2x^2 + 1) = 9$ has the same meaning as $\lim_{t \to 2}(2t^2 + 1) = 9$ and $\lim_{z \to 2}(2z^2 + 1) = 9$.

The graph of $f(x) = 2x^2 + 1$ is a parabola opening upward with vertex at $(0, 1)$ as shown in Fig. 4.1-1. Note that the graph of f passes smoothly through the point $(2, 9)$ and that there are no "holes" or "jumps" in it. Consequently, when x is close to 2, the point $(x, f(x))$ on the graph is close to the point $(2, 9)$, so $f(x)$ will be close to 9. This again indicates that $\lim_{x \to 2} f(x) = 9$.

FIGURE 4.1-1

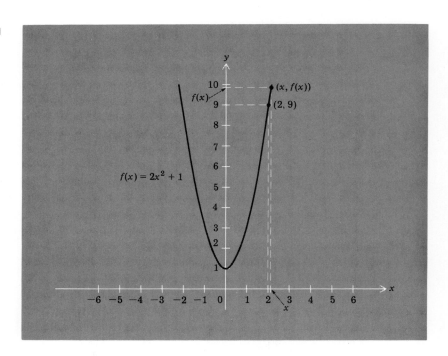

The graph of a function is often helpful in determining whether a function has a limit L at a point a. The following examples illustrate this.

EXAMPLE 4.1-4

Suppose that the cost of a telephone call from Miami to New Orleans is given by the function f, where

$$f(x) = \begin{cases} \$0.60 & \text{if } 0 < x \leqslant 3 \\ \$0.80 & \text{if } 3 < x \leqslant 4 \end{cases}$$

and x is the length of the call in minutes. Thus the cost of a 2-min call is \$0.60, as is the cost of a $2\frac{1}{2}$-min call or a 3-min call. A $3\frac{1}{2}$-min call, however, costs \$0.80, as does a call $3\frac{1}{4}$ min long or one 3 min and 1 sec long. The graph of f is shown in Fig. 4.1-2. Note that the graph of f has a "jump" at 3; if x is close to 3 but to the left of 3, the value of $f(x)$ is 60, whereas if x is close to 3 but to the right of 3, the value of $f(x)$ is 80. Thus there is no *one* number that $f(x)$ is always close to when x is close to 3: sometimes $f(x)$ is close to 60 and sometimes it is close to 80. This means that there is no number L such that $\lim_{x \to 3} f(x) = L$. In this case, we say that the limit of $f(x)$ as x approaches 3 does not exist.

The notation

$$f(x) = \begin{cases} \$0.60 & \text{if } 0 < x \leqslant 3 \\ \$0.80 & \text{if } 3 < x \leqslant 4 \end{cases}$$

defines the value $f(x)$ of f at any number x for which $0 < x \leqslant 4$. This is because if x is a number for which $0 < x \leqslant 4$, then either $0 < x \leqslant 3$ or $3 < x \leqslant 4$; if $0 < x \leqslant 3$, then $f(x) = 60$; if $3 < x \leqslant 4$, then $f(x) = 80$. Thus $f(2) = 60$, $f(2.9) = 60$, $f(3) = 60$, $f(3.5) = 80$, $f(3.001) = 80$, etc.

FIGURE 4.1-2

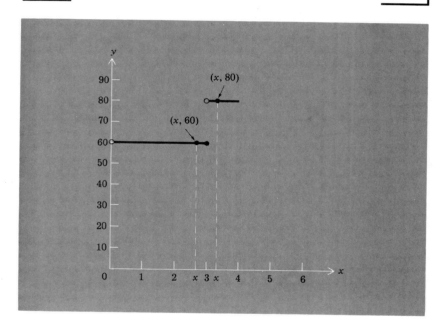

EXAMPLE 4.1-5

Let $g(x) = \sqrt{x}$ and determine whether there is a number L such that $\lim_{x \to 1} g(x) = L$.

SOLUTION

We will determine this from the graph of $g(x) = \sqrt{x}$, shown in Fig. 4.1-3. If x is a number close to 1 and to the left of 1, then we can see from Fig. 4.1-3(a) that $g(x) = \sqrt{x}$ is close to

$\lim_{x \to 1} \sqrt{x} = 1$ can also be expressed by $\lim_{w \to 1} \sqrt{w} = 1$ or $\lim_{r \to 1} \sqrt{r} = 1$.

(and less than) $1 = \sqrt{1}$. Similarly, if x is a number close to 1 and to the right of 1, we can see from Fig. 4.1-3(b) that $g(x) = \sqrt{x}$ is close to (and larger than) $1 = \sqrt{1}$. In any event, if x is close to 1 it is apparent that $\sqrt{x}$ is close to $\sqrt{1} = 1$, so $L = 1$, i.e., $\lim_{x \to 1} \sqrt{x} = 1$.

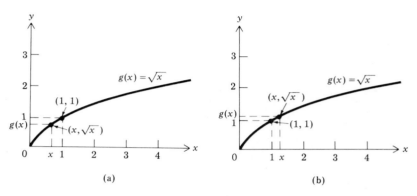

(a)   (b)

FIGURE 4.1-3

EXAMPLE 4.1-6

A foreign country agrees to supply the U.S. with

$$f(x) = \frac{x - 1}{\sqrt{x} - 1} - 1$$

million tons of sugar at a price of \$$x$ thousand per ton. Thus if the U.S. is willing to pay \$4000 per ton, then the country will supply $f(4) = (4 - 1)/(\sqrt{4} - 1) - 1 = 3 - 1 = 2$ million tons of sugar. It seems reasonable to ask what quantity L of sugar will be supplied to the U.S. at \$1000 per ton. However, since $x = 1$ is not in the domain of f, we cannot use the given formula. But if we find out how much sugar will be supplied to the U.S. at a price of \$$x$ thousand per ton, where x is *close* to 1, then this amount should be *close* to the amount L of sugar supplied at \$1000 per ton, i.e.,

$$L = \lim_{x \to 1} f(x) = \lim_{x \to 1} \left(\frac{x - 1}{\sqrt{x} - 1} - 1 \right)$$

To find L, i.e., to evaluate

$$\lim_{x \to 1} \left(\frac{x - 1}{\sqrt{x} - 1} - 1 \right)$$

we first note that

$$x - 1 = (\sqrt{x} - 1)(\sqrt{x} + 1)$$

so that

$$\frac{x - 1}{\sqrt{x} - 1} = \frac{(\sqrt{x} - 1)(\sqrt{x} + 1)}{\sqrt{x} - 1} = \sqrt{x} + 1 \qquad \text{if } x \neq 1$$

Thus $f(x) = (x - 1)/(\sqrt{x} - 1) - 1 = (\sqrt{x} + 1) - 1 = \sqrt{x}$ if $x \neq 1$.

That is, the graph of f and the graph of g, where $g(x) = \sqrt{x}$, are identical except that the point $(1, 1)$ is missing from the graph of f (see Fig. 4.1-4). Since in evaluating the limit at 1 we are not interested in what happens *at* 1, and since the above functions have the same values *near* 1, the limit $\lim_{x \to 1} f(x)$ is the same as $\lim_{x \to 1} g(x)$ (which is 1, as we saw in Example 4.1-5). Thus the U.S. should expect to be able to buy $L = 1$ million tons of sugar at \$1000 per ton.

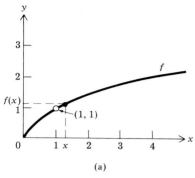

(a)

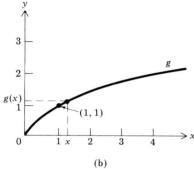

(b)

FIGURE 4.1-4

The above example points out a valuable aid in evaluating limits. If any two functions f and g have the property that

$$f(x) = g(x) \qquad \text{for all } x \neq a$$

and if

$$\lim_{x \to a} g(x) = L$$

then necessarily

$$\lim_{x \to a} f(x) = L$$

As was the case with sequences, a more precise definition of the limit of a function is used in advanced courses. The definition above will suffice for our purposes; however, we state the following definition for the sake of completeness. A function f is said to have limit L (L a real number) at the number a, provided that given any $\epsilon > 0$ there exists a number $\delta > 0$ such that whenever $0 < |x - a| < \delta$ we have $|f(x) - L| < \epsilon$.

Exercises 4.1

In Exercises 1 through 38, find the indicated limits.

1. $\lim\limits_{x \to 2} (x - 4)$

2. $\lim\limits_{x \to 3} (x - 1)$

3. $\lim\limits_{x \to -1} 2x$

4. $\lim\limits_{x \to 1/2} 4x$

5. $\lim\limits_{x \to 2} (2x - 1)$

6. $\lim\limits_{x \to 1} (\frac{1}{3} x + 2)$

7. $\lim\limits_{x \to -1} (-2x + 1)$

8. $\lim\limits_{x \to 1/2} (x - \frac{1}{2})$

9. $\lim\limits_{x \to -1/4} (2x + 1)$

10. $\lim\limits_{x \to 0} (x^2 - 2)$

11. $\lim\limits_{x \to -2} (x^2 - 1)$

12. $\lim\limits_{x \to 0} (x^2 + 2)$

13. $\lim\limits_{x \to 0} (2x^2 - 1)$

14. $\lim\limits_{x \to -1} (2x^2 + 2)$

15. $\lim\limits_{x \to 2} (2x^2 - x + 1)$

16. $\lim\limits_{x \to 1} (x^2 + x - 1)$

17. $\lim\limits_{x \to 0} (3x^2 + x - 1)$

18. $\lim\limits_{x \to 0} (4x^2 - 2x + 2)$

19. $\lim\limits_{x \to 0} |x|$

20. $\lim\limits_{x \to -3} 2|x + 3|$

21. $\lim\limits_{x \to 2} |x - 1|$

22. $\lim\limits_{x \to -3} |x + 2|$

23. $\lim\limits_{x \to -2} (2|x| - 1)$

24. $\lim\limits_{x \to -1} (|x| - 2)$

25. $\lim\limits_{x \to 0} |x^2 - 2|$

26. $\lim\limits_{x \to 1} \dfrac{1}{x + 2}$

27. $\lim\limits_{x \to 2} \left(\dfrac{1}{x} + 3 \right)$

28. $\lim\limits_{x \to 2} \dfrac{1}{x - 1}$

29. $\lim\limits_{x \to 1} \dfrac{1}{2x + 3}$

30. $\lim\limits_{x \to -1} \dfrac{3}{x - 1}$

31. $\lim\limits_{x \to -2} \dfrac{1}{|x + 1|}$

32. $\lim\limits_{x \to -1} \dfrac{3}{|x|}$

33. $\lim\limits_{x \to 2} \sqrt{x + 2}$

34. $\lim\limits_{x \to 4} \sqrt{x - 3}$

35. $\lim\limits_{x \to 2} \dfrac{1}{\sqrt{x^2 + 3}}$

36. $\lim\limits_{x \to 1} (2\sqrt{x + 3} - 1)$

37. $\lim\limits_{x \to 1} \dfrac{2}{\sqrt{x^2 + 3}}$

38. $\lim\limits_{x \to 4} \left(\dfrac{1}{\sqrt{x}} + 2 \right)$

39. Let f be the function defined by

$$f(x) = \begin{cases} 1 & \text{if } 0 \leqslant x \leqslant 2 \\ 0 & \text{if } 2 < x \leqslant 3 \end{cases}$$

Show that $\lim_{x \to 2} f(x)$ does not exist.

40. Let g be the function defined by

$$g(x) = \begin{cases} -1 & \text{if } x < 0 \\ 1 & \text{if } x \geqslant 0 \end{cases}$$

Show that $\lim_{x \to 0} g(x)$ does not exist.

41. Let h be the function defined by

$$h(x) = \begin{cases} \frac{1}{2} & \text{if } x < 1 \\ 0 & \text{if } x \geqslant 1 \end{cases}$$

Show that $\lim_{x \to 1} h(x)$ does not exist.

42. Let q be the function defined by

$$q(x) = \begin{cases} -1 & \text{if } 0 \leqslant x < 3 \\ \frac{1}{2} & \text{if } 3 < x \leqslant 4 \end{cases}$$

Show that $\lim_{x \to 3} q(x)$ does not exist.

43. A bacteriologist finds that the population P of a certain colony of bacteria at time t (in hours) is given by the equation $P(t) = 100(4t^2 + t)$. Find $\lim_{t \to 2} P(t)$.

44. A chemical plant discharges liquid waste into a river. On an average work day, the number of gallons v of waste deposited in the river at the end of t hr of plant operation is given by $v(t) = 600t^2$, where $t = 0$ corresponds to the opening of the plant at 8:00 AM and $0 \leqslant t \leqslant 10$. Find $\lim_{t \to 3} v(t)$.

45. The C & C Candy Company finds that the number d of boxes of its premium quality candy that it can sell per day is given by $d(x) = 100/x + 250$, where x is the selling price of the candy per box (in dollars) and $1 \leqslant x \leqslant 3$. Find $\lim_{x \to 2} d(x)$.

46. Medical researchers, studying the contraction characteristics of a particular muscle, extended the muscle by subjecting it to a stretching force. They found that the extension of the muscle beyond its normal length was a function f of time t (for $0 < t < 3$) given by $f(t) = (3 - t)/(1 + t^2)$, where f is measured in centimeters and t is the time elapsed (in seconds) after initial application of the force. Find $\lim_{t \to 2} f(t)$.

47. The monthly service charge S (in dollars) for a checking account at a certain bank is given by the function

$$S(x) = \begin{cases} 0.5 & \text{if } 1 \leqslant x \leqslant 5 \\ 0.75 & \text{if } 5 < x \leqslant 10 \end{cases}$$

where x is the number of checks written on the account during the month, $1 \leqslant x \leqslant 10$. Using this definition to define $S(x)$ for all x between 1 and 10, show that $\lim_{x \to 5} S(x)$ does not exist.

48. The postage P (in cents) required for a first-class letter of weight x oz, $0 < x \leqslant 2$, is given by the equation

$$P(x) = \begin{cases} 10 & \text{if } 0 < x \leqslant 1 \\ 20 & \text{if } 1 < x \leqslant 2 \end{cases}$$

Show that $\lim_{x \to 1} P(x)$ does not exist.

Recall that the function d in Exercise 45 is called the *demand* function.

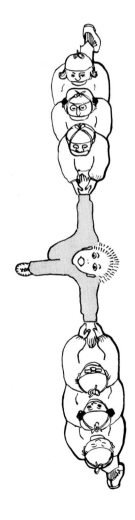

49. If $100 is placed in a bank that pays 5% interest compounded annually, the amount A (in dollars) of money in the account x yr after the account is opened is given by the function

$$A(x) = \begin{cases} 100 & \text{if } 0 < x < 1 \\ 105 & \text{if } 1 \leqslant x < 2 \\ 111.25 & \text{if } 2 \leqslant x < 3 \end{cases}$$

for $0 < x < 3$. Find $\lim_{x \to 1.5} A(x)$. Show that $\lim_{x \to 2} A(x)$ does not exist.

4.2.   Functions with Infinite Limits

Thus far our definition and examples of the notation $\lim_{x \to a} f(x) = L$ deal only with the case where L and a are both numbers. Frequently, however, limits are encountered where L or a (or both) are replaced by the symbols $+\infty$ or $-\infty$. In this section we describe how such limits are defined. Let us begin with an illustration which will lead us to limits of the type $\lim_{x \to a} f(x) = +\infty$ and $\lim_{x \to +\infty} f(x) = L$.

────── **EXAMPLE 4.2-1** ──────

Recall that a demand function d for a commodity is a function whose value $d(x)$ is the amount of the commodity that can be sold when x is the selling price per unit.

The B and B Brewing Company estimates that the demand function d for its premium quality beer is given by the equation

$$d(x) = \frac{10}{x}$$

where x is the selling price per gallon and $d(x)$ is the number of gallons that can be sold (in thousands of gallons). Thus if the beer is priced at $x = \$1$ per gallon, the company will sell $d(1) = \frac{10}{1}$ or 10,000 gal; if $x = \$2$, then $d(2) = \frac{10}{2}$ or 5000 gal will be sold; and so on. The graph of $d(x) = 10/x$ is shown in Fig. 4.2-1, and from this graph we see that if the price x is small, then the demand $d(x)$ is large, but as the price x becomes larger (as x moves to the right on the x axis) the demand $d(x)$ rapidly decreases ($d(x)$ moves toward 0 on the y axis). Computing $d(x)$ for some values of x close to 0 (see Table 4.2-1), we note that $d(x)$ becomes very large as x moves closer and closer to 0. This corresponds to the fact that the graph rises indefinitely near the y axis (see Fig. 4.2-1). The limit of $d(x)$ as x approaches 0 does not exist as a number L (the values $d(x)$ do not get close to any number L as x gets

Ordinarily, when examining $\lim_{x \to 0} d(x)$ we need to consider values of x near 0, both to the left of 0 and to the right of 0. In this case, however, only numbers to the right of 0 are in the domain of d.

close to 0), but we would like to describe the behavior of d near 0 by saying that $d(x)$ increases indefinitely, or that the limit of $d(x)$ as x approaches 0 is $+\infty$. Thus we write $\lim_{x \to 0} d(x) = +\infty$.

That $\lim_{x \to 0} d(x) = +\infty$ means that as the price of beer approaches 0, the demand $d(x)$ for beer grows arbitrarily large.

TABLE 4.2-1

x	$1/10$	$1/100$	$1/1000$	$1/10,000$	$1/10^n$
$d(x) = 10/x$	100	1000	10,000	100,000	10^{n+1}

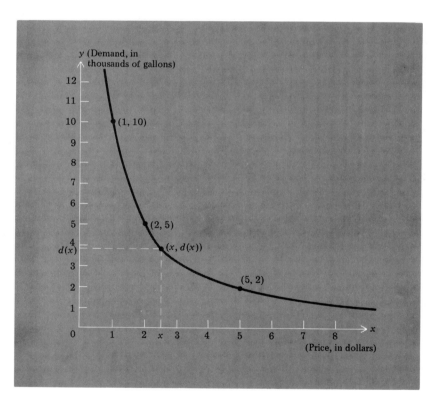

FIGURE 4.2-1 The graph of the demand function $d(x) = 10/x$. Note that since x represents the price of the commodity, only positive values of x are meaningful; thus the domain of d consists of positive values of x.

If f is any function and a any number, we agree to write

$$\lim_{x \to a} f(x) = +\infty$$

Definition of $\lim_{x \to a} f(x) = +\infty$

if the values $f(x)$ get arbitrarily large (larger than *any* number we might choose) as x moves nearer to a. Similarly, if the values $f(x)$ are all negative near a and $|f(x)|$ gets arbitrarily large (larger than *any* number we might choose) as x approaches a, then we say f has limit minus infinity at a and write

Definition of $\lim_{x \to a} f(x) = -\infty$

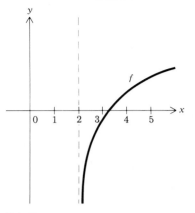

FIGURE 4.2-2   The graph of a function f such that $\lim_{x \to 2} f(x) = -\infty$.

That $\lim_{x \to +\infty} d(x) = 0$ means that as the price x per gallon of beer grows larger and larger, the demand $d(x)$ for beer approaches 0.

Definition of $\lim_{x \to +\infty} f(x) = L$

Definition of $\lim_{x \to -\infty} f(x) = L$

FIGURE 4.2-3   (a) the graph of a function f for which $\lim_{x \to +\infty} f(x) = 2$. (b) the graph of a function f for which $\lim_{x \to -\infty} f(x) = -1$.

$$\lim_{x \to a} f(x) = -\infty$$

(see Fig. 4.2-2).

Now let us return to our demand function $d(x) = 10/x$ and examine the behavior of the demand d as the selling price x per gallon grows larger and larger. To do this, we construct a table of values for $d(x)$, where x takes on large values (see Table 4.2-2).

TABLE 4.2-2

x	10	100	1000	10,000	100,000	10^n
$d(x) = 10/x$	1	$1/10$	$1/100$	$1/1000$	$1/10,000$	$1/10^{n-1}$

We see here that as x gets larger and larger $d(x)$ gets ever closer to 0. To express this fact, that $d(x)$ gets very close to 0 as x becomes very large, we say that d has limit 0 as x approaches $+\infty$ and write

$$\lim_{x \to +\infty} d(x) = 0$$

If f is any function and L any number, we agree to write

$$\lim_{x \to +\infty} f(x) = L$$

if the values $f(x)$ get close to L as x increases indefinitely. Similarly, we write

$$\lim_{x \to -\infty} f(x) = L$$

if the values $f(x)$ get close to L when x is negative and $|x|$ increases indefinitely (see Fig. 4.2-3).

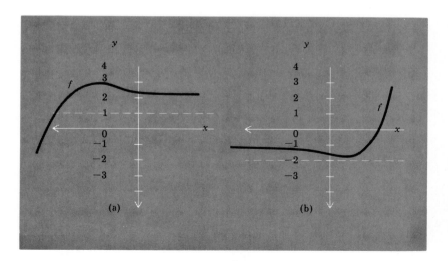

(a)   (b)

We conclude this section with an illustration which leads us to a limit $\lim_{x \to a} g(x) = L$, where a and L are both replaced by the symbol $+\infty$, i.e., a limit of the form

$$\lim_{x \to +\infty} g(x) = +\infty$$

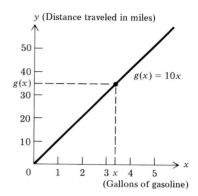

y (Distance traveled in miles)

$g(x) = 10x$

(Gallons of gasoline)

FIGURE 4.2-4

EXAMPLE 4.2-2

Let us suppose that a particular automobile will travel 10 mi on 1 gal of gasoline. It will then travel 5 mi on $\frac{1}{2}$ gal, 20 mi on 2 gal, and in general, $10x$ mi on x gal of gasoline. Thus if g is the function whose value $g(x)$ at each x ($x \geqslant 0$) is the number of miles the car will travel on x gal of gasoline, then $g(x) = 10x$, and the graph of g is shown in Fig. 4.2-4. Evidently, from the graph of g, as the number x of gallons of gasoline used becomes very large, the distance $g(x)$ that the car will travel also becomes very large. Thus we say that the distance g the car will travel (assuming that the car never wears out) has limit $+\infty$ as x (number of gallons of gasoline used) approaches $+\infty$, and we write

$$\lim_{x \to +\infty} g(x) = +\infty$$

In Fig. 4.2-4, the function $g(x) = 10x$ is linear (has form $g(x) = ax + b$ with $a = 10$, $b = 0$), so its graph is a straight line.

$\lim_{x \to +\infty} g(x) = +\infty$ means that the car will travel indefinitely far on an unlimited supply of gasoline.

More generally, for any function f, we say "the limit of $f(x)$ as x approaches $+\infty$ is $+\infty$" and write

$$\lim_{x \to +\infty} f(x) = +\infty$$

Definition of $\lim_{x \to +\infty} f(x) = +\infty$.

provided that $f(x)$ increases indefinitely (becomes arbitrarily large) as x increases indefinitely (becomes arbitrarily large). In a similar manner, we may define the limits

$$\lim_{x \to +\infty} f(x) = -\infty, \quad \lim_{x \to -\infty} f(x) = +\infty, \quad \text{and} \quad \lim_{x \to -\infty} f(x) = -\infty$$

as follows.

(1) $\lim_{x \to +\infty} f(x) = -\infty$ provided that $f(x) < 0$ for all x to the right of some number M, and $|f(x)| \cdot$becomes arbitrarily large as x becomes arbitrarily large.

Definition of $\lim_{x \to +\infty} f(x) = -\infty$

(2) $\lim_{x \to -\infty} f(x) = +\infty$ provided that $f(x)$ becomes arbitrarily large as $|x|$ becomes arbitrarily large (for $x < 0$).

Definition of $\lim_{x \to -\infty} f(x) = +\infty$

Definition of $\lim\limits_{x \to -\infty} f(x) = -\infty$

(3) $\lim\limits_{x \to -\infty} f(x) = -\infty$ provided that $f(x) < 0$ for all x to the left of some number M, and $|f(x)|$ becomes arbitrarily large as $|x|$ becomes arbitrarily large (for $x < 0$).

Recall that graphs of quadratic functions are called parabolas (see Section 2.4).

For example, if the graph of f is a parabola that opens upward, then $\lim\limits_{x \to +\infty} f(x) = +\infty$ and $\lim\limits_{x \to -\infty} f(x) = +\infty$. If the graph of f is a parabola that opens downward, then $\lim\limits_{x \to +\infty} f(x) = -\infty$ and $\lim\limits_{x \to -\infty} f(x) = -\infty$ (see Fig. 4.2-5).

FIGURE 4.2-5   (a) $\lim\limits_{x \to +\infty} f(x)$ $= +\infty$ and $\lim\limits_{x \to -\infty} f(x) = +\infty$. (b) $\lim\limits_{x \to +\infty} f(x) = -\infty$ and $\lim\limits_{x \to -\infty} f(x) = -\infty$.

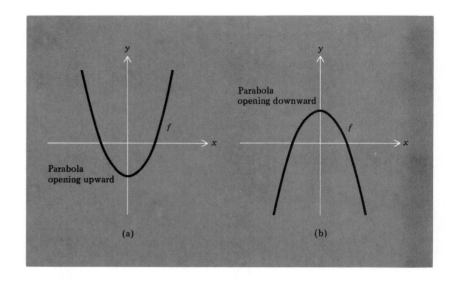

Parabola opening upward

Parabola opening downward

(a)

(b)

EXAMPLE 4.2-3

Let g be the function defined by

$$g(x) = \frac{2}{|x - 2|}$$

Sketch the graph of g and find $\lim\limits_{x \to 2} g(x)$ from the graph.

SOLUTION

Note that 2 is not in the domain of $g(x) = 2/|x - 2|$, since the denominator is 0 at $x = 2$. Nonetheless, $\lim\limits_{x \to 2} g(x)$ exists and equals $+\infty$.

To sketch the graph, we choose some sample values for x and find several points $(x, g(x))$ on the graph and connect these points with a smooth curve. When $x = 1$, $g(1) = 2/|1 - 2|$ $= 2$, so $(1, 2)$ is on the graph of g. Similarly, we find that the points $(0, 1)$, $(-1, \frac{2}{3})$, $(-2, \frac{1}{2})$, $(3, 2)$, $(4, 1)$, $(5, \frac{2}{3})$, $(\frac{3}{2}, 4)$, and $(\frac{5}{2}, 4)$ are on the graph. Plotting these points (and others, if necessary) and joining them with a smooth curve, we see that the graph of g is as shown in Fig. 4.2-6. Now, from the graph, we see that

To help in sketching the graph of $g(x) = 2/|x - 2|$, note that if x is close to 2, $|x - 2|$ is close to 0, so $g(x)$ is large. Also, if x is very large (positive or negative), $|x - 2|$ is large and positive, so $g(x)$ is positive and close to 0. This kind of analysis of the defining equation of a function will often allow one to determine the limit of a function without graphing the function.

$$\lim_{x \to 2} g(x) = \lim_{x \to 2} \frac{2}{|x - 2|} = +\infty$$

FIGURE 4.2-6

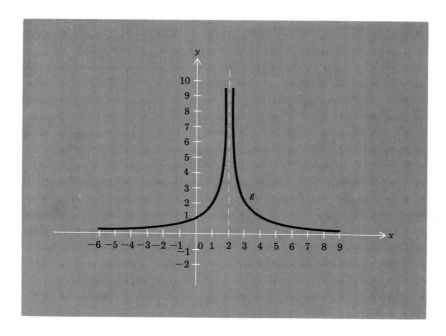

Exercises 4.2

In Exercises 1 through 41, carefully sketch the graph of the given function, and find the value of the indicated limit from the graph. Attempt to determine the limit from an examination of the defining equation before graphing the function.

1. $f(x) = 4 - x^2$, $\displaystyle\lim_{x \to +\infty} f(x)$

2. $f(x) = \frac{1}{2} - 2x^2$, $\displaystyle\lim_{x \to -\infty} f(x)$

3. $f(x) = x^3 - 1$, $\displaystyle\lim_{x \to -2} f(x)$

4. $f(x) = 2x^3 + 2$, $\lim_{x \to 1} f(x)$

5. $f(x) = (\frac{1}{2})x^3 - 1$, $\lim_{x \to -\infty} f(x)$

6. $f(x) = -x^3 + 2$, $\lim_{x \to +\infty} f(x)$

7. $f(x) = 1 - x^3$, $\lim_{x \to -1} f(x)$

8. $f(x) = x^2 + 2x + 1$, $\lim_{x \to 2} f(x)$

9. $f(x) = -4x^2 + 8x - 1$, $\lim_{x \to 1} f(x)$

10. $f(x) = 2x^2 - 2x - 2$, $\lim_{x \to \frac{1}{2}} f(x)$

11. $f(x) = -6x^2 - 6x + 2$, $\lim_{x \to -\frac{1}{2}} f(x)$

12. $g(x) = x^4 + 1$, $\lim_{x \to +\infty} g(x)$

13. $h(x) = x^5 - 2$, $\lim_{x \to -\infty} h(x)$

14. $p(x) = \frac{1}{4} x$, $\lim_{x \to +\infty} p(x)$

15. $q(x) = 4/x$, $\lim_{x \to -\infty} q(x)$

16. $t(x) = |x|$, $\lim_{x \to -\infty} t(x)$

17. $h(x) = |x| - 4$, $\lim_{x \to 0} h(x)$

18. $s(x) = \sqrt{x + 1}$, $\lim_{x \to +\infty} s(x)$

19. $p(x) = -\sqrt{-x}$, $\lim_{x \to -\infty} p(x)$

20. $h(x) = \dfrac{1}{x^2}$, $\lim_{x \to 0} h(x)$

21. $q(x) = \dfrac{1}{\sqrt{x - 1}}$, $\lim_{x \to 1} q(x)$

22. $s(x) = \dfrac{-1}{|x - 1|}$, $\lim_{x \to 1} s(x)$

23. $p(x) = |x| - 3$, $\lim_{x \to +\infty} p(x)$

24. $t(x) = 1 - |x|$, $\lim_{x \to -\infty} t(x)$

25. $h(x) = \dfrac{x - 1}{x - 1}$, $\lim_{x \to 1} h(x)$

26. $q(x) = \dfrac{x^2 - 4}{x + 2}$, $\lim_{x \to -2} q(x)$

27. $p(x) = \dfrac{|x + 1|}{x + 1}$, $\lim_{x \to +\infty} p(x)$

28. Let f be the constant function defined by $f(x) = c$, where c is some real number. Find $\lim_{x \to +\infty} f(x)$ and $\lim_{x \to -\infty} f(x)$.

29. A psychologist estimates that the percent p of certain material remembered by a patient, t mos after the patient learns it, is given by $p(t) = 100/(1 + t)$. Sketch the graph of the function p and find

$\lim_{t \to +\infty} p(t)$ from the graph. Interpret the value of this limit in the context of the problem.

30. The Quik-Lite Company estimates that its daily cost C (in dollars) for producing x units of its disposable cigarette lighters is given by $C(x) = (\frac{1}{2})x^2 + x + 2$. Sketch the graph of C and find $\lim_{x \to +\infty} C(x)$ from the graph. Interpret the value of this limit in the context of the problem.

31. The Dandy Detergent Company estimates that the daily demand d (in hundreds) for its 5-lb box of detergent is given by $d(x) = 4/x$, where x is the selling price (in dollars) of the detergent per box. Sketch the graph of d and find $\lim_{x \to +\infty} d(x)$ from the graph. Interpret the value of this limit in the context of the problem.

The demand d is the number of boxes of detergent the company can sell per day.

32. The Gro-Rite Fertilizer Company estimates that the daily demand d for its 25-lb bag of all-purpose lawn food is given by $d(x) = (2/x) + 4$ for $0 < x \leqslant 6$, where d is measured in hundreds of bags and x is the selling price (in dollars) per bag. Sketch the graph of d and find $\lim_{x \to 0} d(x)$ from the graph. Interpret the value of this limit in the context of the problem.

The domain of any demand function only contains values of x for which $x \geqslant 0$. In Exercise 32, the domain is further restricted to $0 < x \leqslant 6$.

33. A sales representative's contract specifies a weekly salary of $\$C$, where $C(x) = 120 + x/10$ and x is the amount of the representative's weekly sales. Sketch the graph of C and find $\lim_{x \to 0} C(x)$ and $\lim_{x \to +\infty} C(x)$ from the graph. Interpret the values of these limits in the context of the problem.

Note that the domain of C consists only of values of x such that $x \geqslant 0$.

4.3. The Algebra of Limits

Recall from Section 2.2 that if we are given two functions f and g, we may produce from them several new functions, namely, their sum $f + g$, their difference $f - g$, their product $f \cdot g$, and their quotient f/g. The limits of each of these functions as x approaches a can usually be easily determined once we know the limit of f at a and the limit of g at a. The rules governing the limit of the sum, difference, product, and quotient of two functions are given in Table 4.3-1.

These rules are valid only for functions f and g where $\lim_{x \to a} f(x) = L$, $\lim_{x \to a} g(x) = M$, and both L and M are *numbers*, i.e., neither L nor M is $\pm \infty$. The rules *are* valid, however, for $a = \pm \infty$.

TABLE 4.3-1

(1) $\lim\limits_{x \to a} (f \pm g)(x) = \lim\limits_{x \to a} f(x) \pm \lim\limits_{x \to a} g(x)$

(2) $\lim\limits_{x \to a} (f \cdot g)(x) = \lim\limits_{x \to a} f(x) \cdot \lim\limits_{x \to a} g(x)$

(3) $\lim\limits_{x \to a} (f/g)(x) = \lim\limits_{x \to a} f(x) / \lim\limits_{x \to a} g(x)$, provided that $\lim\limits_{x \to a} g(x) \neq 0$

(4) $\lim\limits_{x \to a} cf(x) = c \lim\limits_{x \to a} f(x)$ for any constant c

In rule (3), $\lim_{x \to a} g(x)$ must be $\neq 0$ so that the fraction $\lim_{x \to a} f(x) / \lim_{x \to a} g(x)$ does not have denominator 0.

EXAMPLE 4.3-1

Find $\lim_{x \to 1}(2x + \sqrt{x}\,)$.

SOLUTION

For any number a, the statement "x is close to a when x is close to a" is obviously true. Thus $\lim_{x \to a} x = a$.

Since the statement "x is close to 1 when x is close to 1" is obviously correct, we have

$$\lim_{x \to 1} x = 1$$

Thus by rule (4) of Table 4.3-1, we see that

$$\lim_{x \to 1} 2x = 2 \lim_{x \to 1} x = 2 \cdot 1 = 2$$

Also, from Example 4.1-5,

$$\lim_{x \to 1} \sqrt{x} = 1$$

Now, letting $f(x) = 2x$ and $g(x) = \sqrt{x}$, we have

$$(f + g)(x) = f(x) + g(x)$$

so

$$\lim_{x \to 1} (2x + \sqrt{x}\,) = \lim_{x \to 1} \left[\, f(x) + g(x)\,\right]$$

$$= \lim_{x \to 1} (f + g)(x) = \lim_{x \to 1} f(x) + \lim_{x \to 1} g(x)$$

(using rule (1) of Table 4.3-1)

$$= \lim_{x \to 1} 2x + \lim_{x \to 1} \sqrt{x} = 2 + 1 = 3$$

EXAMPLE 4.3-2

Find $\lim_{x \to 1}(2x^2/\sqrt{x}\,)$.

SOLUTION

This limit is of the form $\lim_{x \to a}(f/g)(x)$, where $f(x) = 2x^2$ and $g(x) = \sqrt{x}$, since

$$(f/g)(x) = \frac{f(x)}{g(x)} = \frac{2x^2}{\sqrt{x}}$$

Since $\lim_{x \to 1} x = 1$, we find using rules (2) and (4) of Table 4.3-1 that

$$\lim_{x \to 1} f(x) = \lim_{x \to 1} 2x^2 = 2 \lim_{x \to 1} x^2 = 2 \lim_{x \to 1} (x \cdot x)$$

$$= 2 \lim_{x \to 1} x \cdot \lim_{x \to 1} x = 2 \cdot 1 \cdot 1 = 2$$

From Example 4.1-5, we also have

$$\lim_{x \to 1} g(x) = \lim_{x \to 1} \sqrt{x} = 1$$

Thus by rule (3) of Table 4.3-1, we have

$$\lim_{x \to 1} \left(\frac{2x^2}{\sqrt{x}} \right) = \lim_{x \to 1} (f/g)(x) = \frac{\lim_{x \to 1} f(x)}{\lim_{x \to 1} g(x)} = \frac{2}{1} = 2$$

EXAMPLE 4.3-3

Find $\lim_{x \to +\infty} 6/x^2$.

SOLUTION

From rule (4) of Table 4.3-1, we have

$$\lim_{x \to +\infty} \frac{6}{x^2} = 6 \lim_{x \to +\infty} \frac{1}{x^2},$$

and since $1/x^2 = (1/x)(1/x)$, we now use rule (2) to write

$$6 \lim_{x \to +\infty} \frac{1}{x^2} = 6 \lim_{x \to +\infty} \left[\left(\frac{1}{x} \right) \left(\frac{1}{x} \right) \right] = 6 \left(\lim_{x \to +\infty} \frac{1}{x} \right) \left(\lim_{x \to +\infty} \frac{1}{x} \right)$$

In Example 4.3-3, note that we cannot use rule (3) to write $\lim_{x \to +\infty} 6/x^2 = \lim_{x \to +\infty} 6/\lim_{x \to +\infty} x^2$, since $\lim_{x \to +\infty} x^2 = +\infty$ and rules (1) through (4) are applicable only when the limits of all functions involved are finite.

The graph of the function $f(x) = 1/x$ is shown in Fig. 4.3-1, and from the graph we see that $\lim_{x \to +\infty} 1/x = 0$ (alternatively, we could make a table of the values $f(x)$ as x gets large to discover that $\lim_{x \to +\infty} f(x) = \lim_{x \to +\infty} (1/x) = 0$). Thus

$$\lim_{x \to +\infty} \frac{6}{x^2} = 6\left(\lim_{x \to +\infty} \frac{1}{x} \right)\left(\lim_{x \to +\infty} \frac{1}{x} \right) = 6 \cdot 0 \cdot 0 = 0$$

FIGURE 4.3-1　The graph of $f(x) = 1/x$.

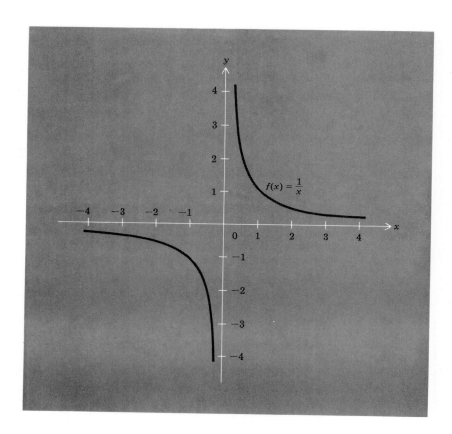

$$f(x) = \frac{1}{x}$$

EXAMPLE 4.3-4

Find $\lim_{x \to 2} (2x^3 - 3x)$.

SOLUTION

To evaluate this limit we use several of rules (1) through (4) in succession, indicating at each step which rule is used. Thus

$$\lim_{x \to 2} (2x^3 - 3x) = \lim_{x \to 2} 2x^3 - \lim_{x \to 2} 3x$$

(by rule (1))

$$= 2 \lim_{x \to 2} x^3 - 3 \lim_{x \to 2} x$$

(by rule (4))

$$= 2 \lim_{x \to 2} (x \cdot x^2) - 3 \lim_{x \to 2} x = 2 \lim_{x \to 2} x \lim_{x \to 2} x^2 - 3 \lim_{x \to 2} x$$

(by rule (2))

$$= 2 \lim_{x \to 2} x \lim_{x \to 2} (x \cdot x) - 3 \lim_{x \to 2} x$$

$$= 2 \lim_{x \to 2} x \lim_{x \to 2} x \lim_{x \to 2} x - 3 \lim_{x \to 2} x$$

(by rule (2)); so we see, since $\lim_{x \to 2} x = 2$, that

$$\lim_{x \to 2} (2x^3 - 3x) = 2(2)(2)(2) - 3(2) = 16 - 6 = 10$$

Recall that for any a, $\lim_{x \to a} x = a$.

It should be evident from the examples above that the rules (1) through (4) are useful tools for evaluating limits; they frequently allow us to express the limit of a complicated function as a combination of limits of simpler functions. The following illustration provides some insight into the kinds of functions that arise in practice and demonstrates the value of the techniques of this section in evaluating limits of such functions.

EXAMPLE 4.3-5 (*Optional.*)

A fertilizer retailer expects to sell 10,000 tons of fertilizer during the coming year. It costs $5 to keep 1 ton of fertilizer in storage for a year, and each ton of fertilizer delivered costs the retailer $15. In addition, the retailer must pay a fixed shipping cost of $10 for every shipment of fertilizer. From this information, the retailer wishes to determine the total cost G for receiving the 10,000 tons in equal shipments of x tons each, i.e., to describe G as a function of x, and then determine which value of x will make G as small as possible.

SOLUTION

The total cost G to the retailer will be the number of shipments times the cost per shipment plus the annual storage cost, i.e., $G(x) = f(x)g(x) + s(x)$.

To find the equation for $G(x)$, let $f(x)$ be the number of shipments received at x tons each, $g(x)$ the cost (in dollars) per shipment, and $s(x)$ the annual storage cost (in dollars). Then $G(x) = f(x)g(x) + s(x)$, so to find $G(x)$ we need only determine equations for $f(x)$, $g(x)$, and $s(x)$. Clearly,

$$f(x) = \frac{10{,}000}{x}$$

The number $f(x) \cdot g(x)$ is called the *annual reordering cost* of ordering shipments of x tons each.

since 10,000 tons must be delivered in shipments of x tons each; also, $g(x)$ is the fixed cost per shipment ($10) plus the cost per ton ($15) times the number of tons x per shipment, so

$$g(x) = 10 + 15x$$

The average inventory for a time period is half the sum of the initial and final inventory.

The computation of annual average inventory here is analogous to the computation of average speed over a distance. If the distance is broken into four parts and the average speed over each part is 50 mph, then the average speed over the entire distance is also 50 mph.

If x tons are received in each shipment, then there are $f(x)$ shipments in 1 yr; hence each shipment is exhausted after $1/f(x)$ of a year.

To determine $s(x)$, we first find the average tonnage of fertilizer on hand during the year (the *average inventory*) and then multiply this number by the annual storage cost per ton, which is $5. The average inventory is found by assuming a uniform demand for fertilizer throughout the year, so if 10,000 tons are received in one initial shipment, then the average inventory for the year (since the inventory at the end of the year is then 0 tons) is $(10{,}000 + 0)/2 = 5000$ tons. If 5000 tons are ordered initially and then another 5000 tons are ordered 6 months later, then the average inventory for each 6-month period is $(5000 + 0)/2$ (since the inventory at the end of each 6-month period is 0 tons); thus the average inventory for the year is $5000/2 = 2500$ tons. Similarly, if x tons are received initially, and if when (after some period of time, namely, $1/f(x)$ of a year) this supply is exhausted, another x tons are received, and then another x tons, etc., we see that the average inventory over each time period is $(x + 0)/2$, so the average annual inventory is $x/2$. Thus the annual storage cost $s(x)$ (average annual inventory times $5 a ton) is given by the equation

$$s(x) = 5\left(\frac{x}{2}\right) = \frac{5x}{2}$$

Now, since $G(x) = f(x) \cdot g(x) + s(x)$, we have

$$G(x) = \frac{10,000}{x}(10 + 15x) + \frac{5x}{2}$$

$$= \frac{100,000 + 150,000x}{x} + \frac{5x}{2}$$

$$= \frac{100,000}{x} + \frac{5x}{2} + 150,000$$

In the next chapter we will show that this cost G to the retailer is smallest when $x = 200$; so to minimize the cost, the retailer should order $f(200) = 10,000/200 = 50$ shipments of fertilizer of 200 tons each. For the present, let us find $\lim_{x \to 200} G(x)$, the limit of the total cost G as the number of tons x per shipment approaches 200. To do this, we successively apply rules (1) through (4) as follows, indicating at each step which rule is used. We have

$$\lim_{x \to 200} G(x) = \lim_{x \to 200} \left(\frac{100,000}{x} + \frac{5}{2}x + 150,000 \right)$$

$$= \lim_{x \to 200} \frac{100,000}{x} + \lim_{x \to 200} \frac{5}{2}x + \lim_{x \to 200} 150,000$$

(by rule (1))

$$= \frac{\lim_{x \to 200} 100,000}{\lim_{x \to 200} x} + \lim_{x \to 200} \frac{5}{2}x + \lim_{x \to 200} 150,000$$

(by rule (3))

For any constant function $h(x) = c$ and any number a, it is clear that $h(x)$ is close to c when x is close to a. Thus $\lim_{x \to a} h(x) = \lim_{x \to a} c = c$.

$$= \frac{100,000}{\lim_{x \to 200} x} + \lim_{x \to 200} \frac{5}{2}x + 150,000$$

$$= \frac{100,000}{\lim_{x \to 200} x} + \frac{5}{2} \lim_{x \to 200} x + 150,000$$

(by rule (4)). Thus we need only evaluate $\lim_{x \to 200} x$, and the value of this limit is 200. Therefore

Recall that for any a, $\lim_{x \to a} x = a$.

$$\lim_{x \to 200} G(x) = \frac{100,000}{200} + \frac{5}{2}(200) + 150,000$$

$$= 500 + 500 + 150,000$$

$$= \$151,000$$

This means that if the number of tons per shipment is close to 200, then the total cost G to the retailer for receiving the 10,000 tons of fertilizer is very close to $151,000.

Exercises 4.3

In Exercises 1 through 42, use rules (1) through (4) of this section to express the limit of the given function in terms of limits of the form $\lim_{x\to a} x$ and $\lim_{x\to a} c$, where c is a constant. Then use the fact that $\lim_{x\to a} x = a$ and $\lim_{x\to a} c = c$ to evaluate the limit of the function.

1. $\lim_{x\to 2} 2x$

2. $\lim_{x\to -3} -4x$

3. $\lim_{x\to 9} \frac{1}{3} x$

4. $\lim_{x\to -5} -\frac{1}{10} x$

5. $\lim_{x\to 2} (-3x + 1)$

6. $\lim_{x\to -1} (7x + 2)$

7. $\lim_{x\to 2} (-5x - 1)$

8. $\lim_{x\to 4} (3x - 4)$

9. $\lim_{x\to -4} \left(\frac{x}{4} - 2 \right)$

10. $\lim_{x\to 5} \left(\frac{-x}{5} + 2 \right)$

11. $\lim_{x\to -3} \left(\frac{x}{3} - \frac{3}{2} \right)$

12. $\lim_{x\to 7} \left(\frac{-x}{7} - \frac{3}{5} \right)$

13. $\lim_{x\to -3} (x^2 + 2)$

14. $\lim_{x\to -2} (-x^2 + 1)$

15. $\lim_{x\to 3} (2x^2 + 4)$

16. $\lim_{x\to -5} (-3x^2 - 1)$

17. $\lim_{x\to 2} \left(\frac{-x^2}{2} - \frac{1}{2} \right)$

18. $\lim_{x\to 2} \left(\frac{x^2}{7} - \frac{3}{2} \right)$

19. $\lim_{x\to 2} \left(\frac{2}{x} - 1 \right)$

20. $\lim_{x\to 10} \left(\frac{5}{x} + 1 \right)$

21. $\lim_{x\to 3} \left(\frac{-6}{x} + 2 \right)$

22. $\lim_{x\to -3} \left(\frac{-3}{x} - 5 \right)$

23. $\lim_{x\to 2} \left(\frac{-3}{2x} - \frac{1}{2} \right)$

24. $\lim_{x\to -1} \left(\frac{7}{5x} - \frac{3}{2} \right)$

25. $\lim_{x\to -2} (4x^3 + 2x - 1)$

26. $\lim_{x\to 2} (-3x^3 - 2x + 1)$

27. $\lim_{x\to -1} (-3x^2 - x + 2)$

28. $\lim_{x\to 2} (4x^4 - x + 2)$

29. $\lim_{x\to 3} \left(\frac{5}{x^2} + 2x - 1 \right)$

30. $\lim_{x\to 2} \left(\frac{-7}{x^3} - 2x + 1 \right)$

31. $\lim_{x\to -2} \left(\frac{-4}{x^3} + \frac{5}{x^2} + \frac{1}{x} \right)$

32. $\lim_{x\to -1} \left(\frac{-3}{2x^2} + \frac{6}{x} + 2 \right)$

33. $\lim_{x\to -1} \left(\frac{3}{2x^3} - \frac{4}{3x} - 7x^2 \right)$

34. $\lim_{x\to 1} \left(\frac{-5}{3x^4} - \frac{7}{x^3} + 2x^2 \right)$

35. $\lim_{x\to 2} \frac{6x^2 + 2x}{4 - 3x}$

36. $\lim_{x\to -1} \frac{-3x^3 + 2x}{3x^2 - 1}$

37. $\lim_{x\to 1} \frac{-x^3 + 2x - 1}{3 - x^2}$

38. $\lim_{x\to 1} \frac{-2x^3 + 3x}{2x^2 - 3}$

39. $\lim_{x \to 0} (2x^3 + x)(x^3 + 6x + 1)$

40. $\lim_{x \to -1} (2x^3 - x + 1)(x^4 + 3x - 1)$

41. $\lim_{x \to 0} (-4x^3 + 2x^2 - x)(x^3 + 2x)$

42. $\lim_{x \to -1} (-3x^2 - 4x + 1)(2x^2 - x - 1)$

In Exercises 43 through 65, use rules (1) through (4) of this section to express the limit of the given function in terms of limits of simpler functions and evaluate the limit.

43. $\lim_{x \to 1} (4\sqrt{x} + 2)$

44. $\lim_{x \to 1} (-2\sqrt{x} + 3)$

45. $\lim_{x \to 1} \dfrac{3x^2}{\sqrt{x}}$

46. $\lim_{x \to 1} \dfrac{2x + \sqrt{x}}{\sqrt{x}}$

47. $\lim_{x \to 1} \dfrac{\sqrt{x} + 1}{\sqrt{x}}$

48. $\lim_{x \to +\infty} \dfrac{-3}{x^2}$

49. $\lim_{x \to +\infty} \dfrac{6}{x^3}$

50. $\lim_{x \to -\infty} \left(\dfrac{10}{x^2} - 1 \right)$

51. $\lim_{x \to -\infty} \left(\dfrac{-4}{x^3} + 2 \right)$

52. $\lim_{x \to +\infty} \left(\dfrac{2}{x^3} + \dfrac{1}{2} \right)$

53. $\lim_{x \to +\infty} \left(\dfrac{-3}{x^3} + \dfrac{4}{x^4} \right)$

54. $\lim_{x \to -\infty} \left(\dfrac{4}{x^4} - \dfrac{3}{x^3} \right)$

55. $\lim_{x \to 4} \left(|-x| + \dfrac{1}{\sqrt{x}} \right)$

56. $\lim_{x \to 1} |x - 2|\sqrt{x}$

57. $\lim_{x \to -2} \dfrac{1 - |x|}{1 + |x|}$

58. $\lim_{x \to 2} \dfrac{\sqrt{x} - 1}{|-x + 2|}$

59. $\lim_{x \to 4} \dfrac{2\sqrt{x} + \dfrac{1}{\sqrt{x}}}{|x + 1|}$

60. $\lim_{x \to 4} \dfrac{\dfrac{3\sqrt{x}}{2} + \dfrac{1}{x}}{1 - |x|}$

61. $\lim_{x \to 4} (\sqrt{x} + \tfrac{1}{2})|x + 1|$

62. $\lim_{x \to 1} (3\sqrt{x} - 2/\sqrt{x})(4 + |x|)$

63. $\lim_{x \to 4} \dfrac{|-x + 2|\sqrt{x}}{1 - 2|x|}$

64. $\lim_{x \to 1} \left(|x^2 + 2|\sqrt{x} + \dfrac{1}{\sqrt{x}(x + 2)} \right)$

65. $\lim_{x \to 4} \left(\dfrac{x + 1}{|2x - 1|} + \dfrac{2\sqrt{x}}{|x|} \right)$

66. Recall from Section 2.4 that a polynomial function p is a function of the form $p(x) = a_n x^n + a_{n-1} x^{n-1} + \cdots + a_1 x + a_0$, where n is some nonnegative integer and $a_n, a_{n-1}, \ldots, a_1, a_0$ are fixed real numbers with $a_n \neq 0$. Use rules (1) through (4) of this section to verify that for any number a, $\lim_{x \to a} p(x)$ can always be expressed in terms of $\lim_{x \to a} x$.

In Exercises 67 through 70, use rules (1) through (4) of this section to evaluate the indicated limit.

67.   A bacteriologist estimates that the population P of a certain colony of bacteria at time t (in hours) is given by $P(t) = 300t^2 + 100t + 50$. Find $\lim_{t \to 3} P(t)$.

68.   The Super Spud Potato Chip Company finds that the number d of bags of potato chips that it can sell per day is given by $d(x) = 1000/x + 350$, where x is the selling price per bag (in cents) and $45 \leqslant x \leqslant 60$. Find $\lim_{x \to 50} d(x)$.

69.   The extension f (in centimeters) of a muscle beyond its normal length t sec after it is subjected to a stretching force is found to be given by $f(t) = (2 - t)/(2 + t^2)$ for $0 < t < 2$. Find $\lim_{t \to 1} f(t)$.

70.   The total amount V (in gallons) of waste discharged into a river on an average work day by a paper manufacturing plant during t hr of plant operation is given by $V(t) = 400t^2 + 200t + 300$ where $t = 0$ corresponds to the opening of the plant at 8:00 AM and $0 \leqslant t \leqslant 10$. Find $\lim_{t \to 4} V(t)$.

4.4.   Continuous Functions

As we have seen, rules (1) through (4) of the previous section are frequently helpful in evaluating the limit of a given function. However, even with these rules the task of finding the limit may still require considerable effort. Fortunately, for most of the functions f that we will encounter, there is a very simple method for evaluating $\lim_{x \to a} f(x)$: In most cases all that is required is to compute $f(a)$, the value of the function f at a. When f is given by an expression in x, this amounts to substituting a for x in the expression and completing the calculations. Any such function f for which $\lim_{x \to a} f(x) = f(a)$ is said to be *continuous* at the point a; and any function f which is continuous at each point a of its domain is called a *continuous function*.

Definition of a continuous function f at a point a

Definition of a continuous function

EXAMPLE 4.4-1

Let $f(x) = x$. Show that f is continuous at each point a in its domain.

SOLUTION

Let a be any point in the domain of $f(x) = x$ (i.e., any real number). We want to show that $\lim_{x \to a} f(x) = f(a)$. Now the statement "$x$ is close to a when x is close to a" is certainly correct. Since $f(x) = x$, this statement is exactly the same as "$f(x)$ is close to a when x is close to a," which means that

$$\lim_{x \to a} f(x) = a$$

But since $f(a) = a$, we have

$$\lim_{x \to a} f(x) = f(a)$$

Since this reasoning applies to any point a in the domain of f, f is continuous at each point a, so f is a continuous function.

If $f(x) = x$, then $\lim_{x \to a} f(x) = f(a)$ or $\lim_{x \to a} x = a$.

There is a useful connection between the graph of a function f near the point $(a, f(a))$ and the continuity of f at a. Consider the graph shown in Fig. 4.4-1 and assume that we know this to be the graph of some given function f. Now the graph of f passes from the left-hand side of the line $x = a$ to the right-hand side [through the point $(a, f(a))$] with no holes or gaps in the graph. This means that if x is close to a, then $(x, f(x))$ is close to $(a, f(a))$ and $f(x)$ is close to $f(a)$, i.e., $\lim_{x \to a} f(x) = f(a)$. So this is what a function f that is continuous at a (see Fig. 4.4-1) must look like. Another, perhaps simpler, way of visualizing continuity is this: if we can trace the graph of f from a point of the graph that lies to the left of the line $x = a$ to a point of the graph that lies to the right of the line $x = a$ without lifting our pencil from the paper, then f is continuous at a. Consequently, if we can trace the entire graph of f without lifting our pencil from the paper, then f is a continuous function, i.e., f is continuous at each point a of its domain.

For a function f to be continuous at a, a must be in the domain of f, since $f(a)$ must exist.

FIGURE 4.4-1

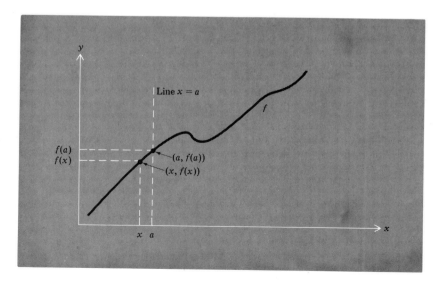

To say that f is continuous at a means that $\lim_{x \to a} f(x)$ can be evaluated by simply substituting $x = a$ in the expression defining $f(x)$, since $\lim_{x \to a} f(x) = f(a)$.

EXAMPLE 4.4-2

If $f(x)$ is a linear function, then its graph is a straight line, which (having no holes or gaps) can certainly be traced without lifting the pencil from the paper. Thus every linear function is continuous, so limits of linear functions can be evaluated by substitution. For instance, if $f(x) = 3x - 7$, then

$$\lim_{x \to 2} (3x - 7) = 3(2) - 7 = -1$$

and

$$\lim_{x \to 0} (3x - 7) = 3(0) - 7 = -7$$

The graph of $f(x) = x^2$ is a parabola and so can be traced without lifting the pencil from the paper; consequently $f(x) = x^2$ is continuous, so $\lim_{x \to a} x^2 = a^2$. By rule (2) of Table 4.3-1,

$$\lim_{x \to a} x^3 = \lim_{x \to a} (x^2 \cdot x) = \left(\lim_{x \to a} x^2 \right)\left(\lim_{x \to a} x \right) = a^2 \cdot a = a^3$$

which means that $g(x) = x^3$ is also a continuous function. Again using rule (2), we see that

Rule (2) of Section 4.3 states that for any functions f and g, $\lim_{x \to a} f \cdot g = \lim_{x \to a} f \cdot \lim_{x \to a} g$.

$$\lim_{x \to a} x^4 = \lim_{x \to a} (x^3 \cdot x) = \left(\lim_{x \to a} x^3 \right)\left(\lim_{x \to a} x \right) = a^3 \cdot a = a^4$$

so $h(x) = x^4$ is likewise a continuous function.

Evidently, by repeating this kind of argument often enough, we have that for any positive integer n, $\lim_{x \to a} x^n = a^n$; thus any function f of the form $f(x) = x^n$ is continuous at each number a. Consequently, if $t(x) = x^5 - 3x^2 + 1$, and if we wish to evaluate $\lim_{x \to 2} t(x)$, we may write (using the rules of Section 4.3)

$$\lim_{x \to 2} (x^5 - 3x^2 + 1) = \lim_{x \to 2} x^5 - \lim_{x \to 2} 3x^2 + \lim_{x \to 2} 1$$

$$= \lim_{x \to 2} x^5 - 3 \lim_{x \to 2} x^2 + \lim_{x \to 2} 1$$

$$= 2^5 - 3(2^2) + 1 = 21$$

Recall from Section 2.4 that a polynomial function p is any function of the form $p(x) = a_n x^n + a_{n-1} x^{n-1} + \cdots + a_1 x + a_0$, where each a_i is a real number and $a_n \neq 0$.

Since $t(2) = 2^5 - 3(2^2) + 1 = 21$, we have in fact shown that $t(x) = x^5 - 3x^2 + 1$ is continuous at $a = 2$.

These calculations suggest that any polynomial function

$$p(x) = a_n x^n + a_{n-1} x^{n-1} + \cdots + a_1 x + a_0$$

is continuous at each real number a. To verify this, we show that $\lim_{x \to a} p(x) = p(a)$ as follows.

$$\lim_{x \to a} p(x) = \lim_{x \to a} \left(a_n x^n + a_{n-1} x^{n-1} + \cdots + a_1 x + a_0 \right)$$

$$= \lim_{x \to a} a_n x^n + \lim_{x \to a} a_{n-1} x^{n-1} + \cdots + \lim_{x \to a} a_1 x + \lim_{x \to a} a_0$$

$$= a_n \lim_{x \to a} x^n + a_{n-1} \lim_{x \to a} x^{n-1} + \cdots + a_1 \lim_{x \to a} x + \lim_{x \to a} a_0$$

$$= a_n \cdot a^n + a_{n-1} \cdot a^{n-1} + \cdots + a_1 \cdot a + a_0 = p(a).$$

Thus every polynomial function $p(x)$ is continuous at each number a, so to evaluate $\lim_{x \to a} p(x)$ we need only compute $p(a)$.

If p is any polynomial function and a is any number, then $\lim_{x \to a} p(x) = p(a)$.

EXAMPLE 4.4-3

Find $\lim_{x \to 3}(2x^3 - 3x^2 + x - 1)$.

SOLUTION

This limit is of the form $\lim_{x \to a} p(x)$, where $a = 3$ and p is the polynomial function $p(x) = 2x^3 - 3x^2 + x - 1$. Thus $\lim_{x \to 3} p(x) = p(3)$, and since $p(3) = 2(3^3) - 3(3^2) + 3 - 1 = 29$, we have $\lim_{x \to 3}(2x^3 - 3x^2 + x - 1) = 29$.

EXAMPLE 4.4-4

Find $\lim_{x \to -2}(\sqrt{2}\, x^4 - \frac{1}{2} x^2 - 1)$.

SOLUTION

Again, this limit is of the form $\lim_{x \to a} p(x)$, where $a = -2$ and p is the polynomial function $p(x) = \sqrt{2}\, x^4 - \frac{1}{2} x^2 - 1$; thus $\lim_{x \to -2} p(x) = p(-2)$. Since $p(-2) = \sqrt{2}\,(-2)^4 - \frac{1}{2}(-2)^2 - 1 = 16\sqrt{2} - 3$, we have $\lim_{x \to -2}(\sqrt{2}\, x^4 - \frac{1}{2} x^2 - 1) = 16\sqrt{2} - 3$.

Definition of a rational function

A function r defined by an equation of the form $r(x) = p(x)/q(x)$, where p and q are each polynomials, is called a *rational* function. If a is any number for which $\lim_{x \to a} q(x) = q(a) \neq 0$, then by rule (3) of the previous section we have

Rule (3) of Section 4.3 states that for functions f and g, $\lim_{x \to a} f/g$ $= \lim_{x \to a} f / \lim_{x \to a} g$ provided $\lim_{x \to a} g \neq 0$.

$$\lim_{x \to a} r(x) = \lim_{x \to a} \frac{p(x)}{q(x)} = \frac{\lim_{x \to a} p(x)}{\lim_{x \to a} q(x)} = \frac{p(a)}{q(a)} = r(a)$$

so r is continuous at each point a for which $q(a) \neq 0$.

EXAMPLE 4.4-5

Find $\lim_{x \to -1}[(3x^7 + x^4 - 2x^2)/(14x^3 + 12x^2 + 3)]$.

SOLUTION

This limit is of the form

$$\lim_{x \to a} r(x)$$

where $a = -1$ and $r(x) = p(x)/q(x)$ with $p(x) = 3x^7 + x^4 - 2x^2$ and $q(x) = 14x^3 + 12x^2 + 3$. Thus r is a rational function, and since $q(-1) = 14(-1)^3 + 12(-1)^2 + 3 = 1 \neq 0$, we have

If r is any rational function (i.e., $r(x) = p(x)/q(x)$, where p and q are polynomials), then $\lim_{x \to a} r(x)$ $= r(a)$ provided $q(a) \neq 0$.

$$\lim_{x \to -1} r(x) = r(-1) = \frac{p(-1)}{q(-1)} = \frac{-4}{1} = -4$$

As we have shown, any polynomial function p is continuous at each point a, and any rational function $r = p/q$ is continuous at each point a for which $q(a) \neq 0$. Many other kinds of functions are also continuous; in fact, most of the functions in this book possess this convenient property. Remember, as a general rule, that any function whose graph can be drawn without taking the pencil off the paper is continuous. Thus $g(x) = \sqrt{x}$, $h(x) = |x|$, and $s(x) = \sqrt{|x|}$, though not polynomial or rational functions, are nevertheless continuous (see Figs. 4.4-2 through 4.4-4).

Since most functions f in this book are continuous, most of the limits we will encounter can now be easily evaluated, since if f is continuous at a, $\lim_{x \to a} f(x) = f(a)$.

Occasionally it may still be necessary to evaluate the limit of a function f at a point a where f is not continuous; if so, we must resort to our earlier methods (Sections 4.1, 4.2, and 4.3) or employ other tech-

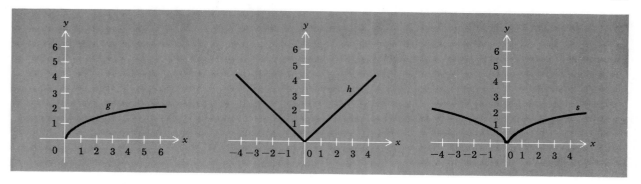

FIGURE 4.4-2 The graph of $g(x) = \sqrt{x}$ can be drawn without lifting the pencil from the paper.

FIGURE 4.4-3 The graph of $h(x) = |x|$ can be drawn without lifting the pencil from the paper.

FIGURE 4.4-4 The graph of $s(x) = \sqrt{|x|}$ can be drawn without lifting the pencil from the paper.

niques to evaluate the limit. Frequently, even though a function f may not be continuous at a point a, it is possible to find a function g which *is* continuous at a and whose limit at a is the same as that of f; if so, then $\lim_{x \to a} f(x) = \lim_{x \to a} g(x) = g(a)$.

For example, suppose we wish to evaluate $\lim_{x \to 0}(x^2 + x)/x$. The function $f(x) = (x^2 + x)/x$ is not continuous at $a = 0$, since 0 is not in the domain of the function, so we cannot write $\lim_{x \to 0} f(x) = f(0)$. Also, we cannot use rule (3) of Section 4.3 to write

$$\lim_{x \to 0}\left(\frac{x^2 + x}{x}\right) = \frac{\lim_{x \to 0}(x^2 + x)}{\lim_{x \to 0} x}$$

since $\lim_{x \to 0} x = 0$, so we must resort to other methods to evaluate $\lim_{x \to 0}(x^2 + x)/x$. Since $(x^2 + x)/x = x(x + 1)/x$, we can cancel x in the numerator and denominator (when $x \neq 0$) to obtain

$$\frac{x^2 + x}{x} = \frac{x + 1}{1} = x + 1 \quad (x \neq 0)$$

Thus the values of the functions

$$f(x) = \frac{x^2 + x}{x} \quad \text{and} \quad g(x) = x + 1$$

are identical for any $x \neq 0$. This means that the graphs of f and g are the same except that g is defined at 0 while f is not (see Fig. 4.4-5), i.e., the graph of f has a hole in it, while the graph of g does not. This means that $\lim_{x \to 0} f(x)$ must be the same as $\lim_{x \to 0} g(x)$(see Section 4.1), and since g is a polynomial,

$$\lim_{x \to 0} g(x) = g(0) = 0 + 1 = 1$$

Note that if a function f is not continuous at a point a, then $\lim_{x \to a} f(x)$ may or may not exist.

Note that $(x^2 + x)/x = p(x)/q(x)$ is a rational function, but it is not continuous at 0, since $q(0) = 0$. For a function f to be continuous at a point a we must have $\lim_{x \to a} f(x) = f(a)$, which means that f must be defined at a; thus a must be in the domain of the function.

FIGURE 4.4-5 (a) Graph of $f(x)$ $= (x^2 + x)/x.$ (b) Graph of $g(x)$ $= x + 1.$

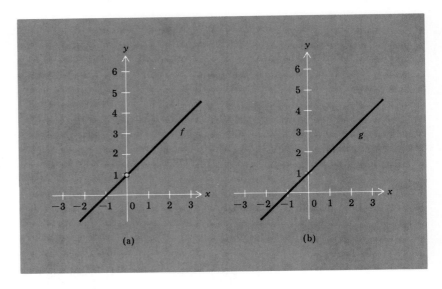

(a)

(b)

Thus

$$\lim_{x \to 0} f(x) = \lim_{x \to 0} \frac{x^2 + x}{x} = 1$$

As we saw in Section 4.1, it is true in general that if f and g are two functions whose values are identical for each $x \neq a$ (that is, $f(x) = g(x)$ for $x \neq a$) and if $\lim_{x \to a} g(x) = L$, then also $\lim_{x \to a} f(x) = L$. This fact is often useful when attempting to evaluate $\lim_{x \to a} f(x)$ for a function f which is not continuous at a; in particular, we have

$$\lim_{x \to a} \frac{(x - a)\,g(x)}{(x - a)} = \lim_{x \to a} g(x)$$

which is helpful in evaluating the limits of rational functions at points where the denominator is 0.

There are three ways in which a function f may fail to be continuous at a point a:

(1)   If a is not in the domain of f so that $f(a)$ is not defined,
(2)   If $\lim_{x \to a} f(x)$ exists but is different from $f(a)$.
(3)   If $\lim_{x \to a} f(x)$ does not exist.

> Recall that for f to be continuous at a, $\lim_{x \to a} f(x)$ must exist and be equal to $f(a)$.

The function $f(x) = (x^2 + x)/x$ (see Fig. 4.4-5(a)) fails to be continuous at $a = 0$ because 0 is not in the domain of f. The following illustrations show how functions might arise which fail to be continuous at a number a because of reasons (2) and (3) above.

Suppose a contestant on a television quiz show is asked to fill a pouch from 10 lb of available gold dust until the person thinks the pouch weighs 5 lb. The pouch is then weighed, and the contestant receives $1000 if the weight is exactly 5 lb; if the weight is not 5 lb, the contestant is given a $100 participation prize. Thus the amount of money a contestant receives is a function f of the weight x of the pouch, where $f(x)$ is given by

$$f(x) = \begin{cases} 100 & \text{if} \quad 0 \leqslant x \leqslant 10,\ x \neq 5 \\ 1000 & \text{if} \quad x = 5 \end{cases}$$

The restriction that x be between 0 and 10 reflects the fact that only 10 lb of gold dust is available.

The graph of f is shown in Fig. 4.4-6, and from the graph we see that f is not continuous at 5, since we must lift our pencil from the paper when tracing the graph from the left of 5 to the right of 5. The reason f is not continuous at 5 is that (as in (2) above) $\lim_{x \to 5} f(x) \neq f(5)$; we know that $f(5) = 1000$, but $\lim_{x \to 5} f(x) = 100$ since for any point $x \neq 5$ we have $f(x) = 100$. In particular, then, when x is close to but not equal to 5, we have $f(x) = 100$, which is certainly close to 100. Note that if we were to change the value of f at 5 from 1000 to 100, then f would be continuous at 5, since then we would have $\lim_{x \to 5} f(x) = f(5) = 100$. This would amount to dropping the point $(5, 1000)$ down to fill in the hole in the graph of f at $(5, 100)$ (see Fig. 4.4-6). Of course, if this change were made, then f would no longer describe the quiz-show winnings.

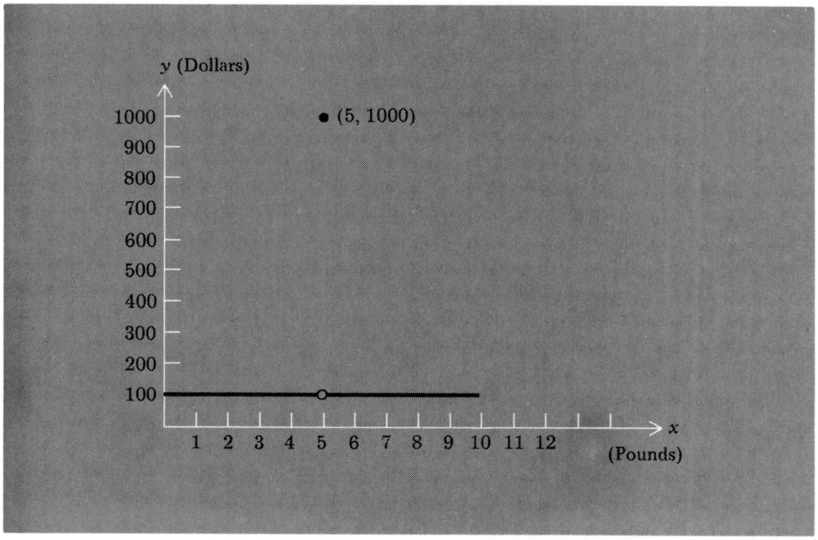

FIGURE 4.4-6

y (Dollars)

1000 ● (5, 1000)
900
800
700
600
500
400
300
200
100 ○

1 2 3 4 5 6 7 8 9 10 11 12   *x*
(Pounds)

Now suppose a 1-lb bucket is suspended from the ceiling by a length of wire that will support a load of exactly 5 lb before breaking. Suppose further that the top of the bucket is 6 ft from the ceiling and the bottom is 1 ft from the floor (see Fig. 4.4-7). Let g be the function whose value $g(x)$ is the distance from the top of the bucket to the ceiling after x lb of sand have been poured into the bucket. Then $g(x)$ is given by the rule

$$g(x) = \begin{cases} 6 & \text{if} \quad 0 \leqslant x \leqslant 4 \\ 7 & \text{if} \quad x > 4 \end{cases}$$

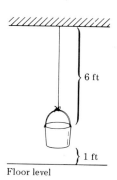

6 ft

1 ft

Floor level

FIGURE 4.4-7

since the wire will break if more than 4 lb of sand are poured into the bucket (remember that the bucket weighs 1 lb). The graph of g is shown in Fig. 4.4-8; from the graph we see that g is not continuous at 4, since we must lift our pencil when tracing the graph from the left of 4 to the right of 4. The reason g is not continuous is that (as in (3) above) $\lim_{x \to 4} g(x)$ does not exist; this limit does not exist because if x is close to 4 but to the left of 4, $g(x) = 6$, but if x is close to 4 but to the right of 4, $g(x) = 7$. Sometimes $g(x)$ is close to 6 when x is close to 4, and sometimes $g(x)$ is close to 7 when x is close to 4; there is not exactly one number L such that $g(x)$ is close to L if x is close to 4, so $\lim_{x \to 4} g(x)$ does not exist. Note the gap in the graph at $(4, 6)$. Although g is defined at 4 (since $g(4) = 6$), it cannot be true that $\lim_{x \to 4} g(x) = g(4)$, since $\lim_{x \to 4} g(x)$ does not exist; hence g is not continuous at 4.

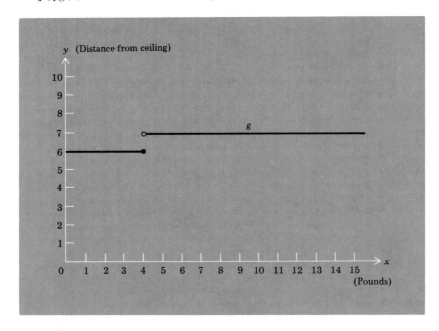

FIGURE 4.4-8

We have seen from the above examples that the graph of a function f can be quite helpful in providing information about $\lim_{x \to a} f(x)$. This information is summarized for certain (common) types of graphs in Fig. 4.4-9.

To close this section, we mention some general points about the notion of continuity that are frequently helpful. First, when functions arise from everyday situations, they are frequently presented in tabular form as discussed in Section 1.4; when their graphs are constructed, it is common practice to plot a finite number of points on the graph (from the information in the table) and to connect these points with a smooth curve. The resulting graph is the graph of a continuous function, so there is a tacit assumption that "most" functions arising from everyday experiences are continuous. This is indeed a strong assumption, but it is so

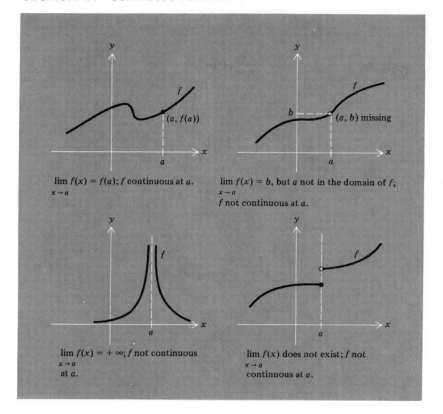

lim $f(x) = f(a)$; f continuous at a.
$x \to a$

lim $f(x) = b$, but a not in the domain of f;
$x \to a$
f not continuous at a.

lim $f(x) = +\infty$; f not continuous
$x \to a$
at a.

lim $f(x)$ does not exist; f not
$x \to a$
continuous at a.

FIGURE 4.4-9

often justified that it is seldom commented on. We shall follow this convention for the remainder of the text and operate under an implicit assumption of continuity; we shall only point out those few instances when functions that are not continuous arise.

Our final remark pertains to various combinations of continuous functions. Recall from Section 4.3 that if f and g are functions having finite limits at a, then

$$\lim_{x \to a} (f + g)(x) = \lim_{x \to a} f(x) + \lim_{x \to a} g(x)$$

Thus if f and g are each continuous at a, then

$$\lim_{x \to a} f(x) = f(a) \quad \text{and} \quad \lim_{x \to a} g(x) = g(a)$$

so

$$\lim_{x \to a} (f + g)(x) = f(a) + g(a) = (f + g)(a)$$

which means that the function $f + g$ is also continuous at a. Similarly, the functions $f - g$, $f \cdot g$, and f/g (if $g(a) \neq 0$) are each continuous at a provided that both f and g are continuous at a.

EXAMPLE 4.4-6

Verify that the function $f(x) = [(2x^2) - (1/|x|)] \cdot (\sqrt{x})$ is continuous at each number $a > 0$.

SOLUTION

This function is of the form

$$f(x) = [h(x) - g(x)]s(x)$$

where $h(x) = 2x^2$, $g(x) = 1/|x|$, and $s(x) = \sqrt{x}$. Now h is a polynomial function and is thus continuous at each a, and s is continuous at each $a > 0$, as seen from Fig. 4.4-2. The function g is of the form $g(x) = p(x)/q(x)$, where $p(x) = 1$ and $q(x) = |x|$; we have seen that $q(x) = |x|$ is continuous (see Fig. 4.4-3), and of course, $p(x)$ is a constant function, so $g(x)$ is also continuous at each a for which $q(a) \neq 0$, i.e., for all $a \neq 0$. Hence $h(x) - g(x)$ is continuous, and finally, $[h(x) - g(x)] \cdot s(x) = f(x)$ is continuous at each a for which h, g, and s are all continuous, i.e., for each $a > 0$.

Exercises 4.4

Note that every polynomial function p is also a rational function, since $p(x) = p(x)/q(x)$, where q is the constant (polynomial) function $q(x) = 1$. But a rational function r may not be a polynomial function, as is illustrated by $r(x) = 1/x$.

In Exercises 1 through 30, state whether the given function f is a polynomial or a rational function (or both) and find $\lim_{x \to a} f(x)$ at the given number a.

1. $f(x) = x^2 + 2x - 1$, $a = 0$

2. $f(x) = -3x^2 + 2x$, $a = 1$

3. $f(x) = x^3 + \frac{1}{2}x + 1$, $a = 2$

4. $f(x) = -x^3 + 2x - \frac{1}{2}$, $a = -1$

5. $f(x) = \frac{1}{4}x^3 - \frac{1}{2}x - 2$, $a = 2$

6. $f(x) = -\frac{1}{2}x^3 + x - 1$, $a = -2$

7. $f(x) = x^4/4 + \sqrt{2}\,x - \frac{1}{2}$, $a = 0$

8. $f(x) = \dfrac{-x^4}{2} - \dfrac{x^3}{2} + 4$, $a = 2$

9. $f(x) = \dfrac{\sqrt{2}\,x^3}{3} - \dfrac{x^2}{2} + 1$, $a = 1$

10. $f(x) = \dfrac{x^6}{4} - \dfrac{3x^4}{2} + \dfrac{x^3}{2} + \dfrac{3}{2}$, $a = 2$

11. $f(x) = \dfrac{x^2 - x + 1}{x}$, $a = 1$

12. $f(x) = \dfrac{x^2 + x - 2}{-x - 1}$, $a = 1$

13. $f(x) = \dfrac{2}{x^2 + x - 1}$, $a = -1$

14. $f(x) = \dfrac{-x^3 + 2x - \frac{1}{2}}{4}$, $a = 2$

15. $f(x) = \dfrac{\sqrt{2}\,x^3 - \frac{3}{2}x}{\sqrt{3}}$, $a = -1$

16. $f(x) = \dfrac{-\sqrt{7}\,x^4 + \pi x^2 - 2}{\sqrt{3}\,x^2 - 1}$, $a = 0$

17. $f(x) = \dfrac{\frac{1}{3}x^5 - 2x^3 + \sqrt{3}}{-x^3 + 2}$, $a = -1$

18. $f(x) = \dfrac{3x^3 - 2x}{\frac{x}{2} + 1}$, $a = \frac{1}{2}$

19. $f(x) = 2\sqrt{2}\,x^2 - 4x + \frac{1}{4}$, $a = -\sqrt{2}$

20. $f(x) = \dfrac{\sqrt{3}\,x^3 - 2x}{\sqrt{2}\,x - 1}$, $a = \sqrt{3}$

21. $f(x) = \dfrac{-2x^2 - x}{x - 2}$, $a = \frac{1}{4}$

22. $f(x) = \dfrac{x^3}{4} - \dfrac{2x^2}{3} + \dfrac{1}{2}$, $a = -\frac{1}{2}$

23. $f(x) = \dfrac{\sqrt{7}\,x^3 - \sqrt{5}\,x + 2}{\sqrt{3}\,x}$, $a = \frac{1}{4}$

24. $f(x) = \dfrac{3x^2}{4} - \dfrac{2x}{3} + 1$, $a = -\frac{1}{8}$

25. $f(x) = \dfrac{\dfrac{3x^3}{4} - 3x}{\dfrac{\sqrt{3}\,x}{2} + 1}$, $a = \frac{1}{2}$

26. $f(x) = \dfrac{x+1}{x} + \dfrac{\sqrt{2}}{x-2}$, $a = \frac{3}{2}$

27. $f(x) = \dfrac{x^2 + 2x - 3}{x} - \dfrac{\sqrt{7}}{x^2}$, $a = -\frac{1}{5}$

28. $f(x) = \dfrac{2x^2/\sqrt{2} - 1}{x+1} + x^3$, $a = -\frac{1}{3}$

29. $f(x) = \dfrac{\sqrt{2}\,x^2 - 3}{2x - 1} + 10x$, $a = -4$

30. $f(x) = \dfrac{\sqrt{3}}{x} - \dfrac{2}{x+1} + \dfrac{7}{x^2}$, $a = 3$

31. Do you think that the sum, difference, and product of two polynomial functions is again a polynomial function?

32. Do you think that the sum, difference, product, and quotient of two rational functions is again a rational function?

In Exercises 33 through 40, the graph of a function f is shown. From the graph, determine whether $\lim_{x \to a} f(x)$ exists and whether f is continuous at a.

33.

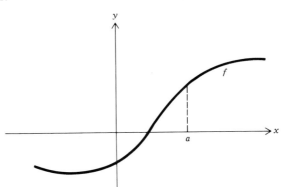

34.

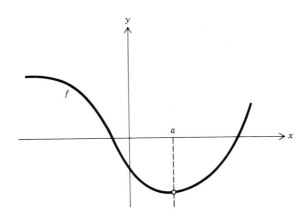

35.

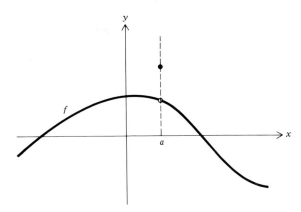

36.

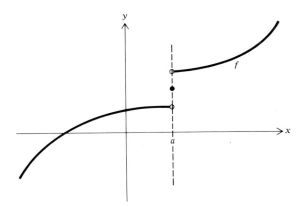

37.

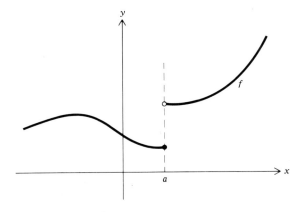

38.

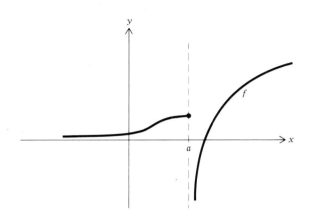

39.

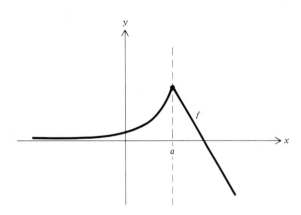

40.

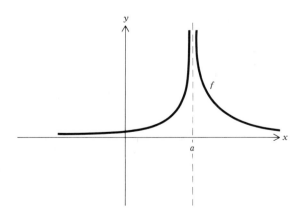

In Exercises 41 through 55, sketch the graph of the given function f and decide from the graph whether the function is continuous. Find $\lim_{x \to a} f(x)$ for the given number a if the limit exists.

Recall that a continuous function is continuous at each point of its domain.

41. $f(x) = \sqrt{x + 2}$, $a = -1$

42. $f(x) = -\sqrt{x^2}$, $a = 0$

43. $f(x) = |x| + 1$, $a = -1$

44. $f(x) = |x - 4|$, $a = \frac{1}{3}$

45. $f(x) = 2 - |x|$, $a = -\frac{1}{4}$

46. $f(x) = |4 - x^2|$, $a = 2$

47. $f(x) = |x^3 - 1|$, $a = -1$

48. $f(x) = |x^5 - 1|$, $a = -1$

49.

$$f(x) = \begin{cases} -1 & \text{if } x < 0 \\ 0 & \text{if } x = 0 \\ 1 & \text{if } x > 0 \end{cases} \qquad a = 0$$

50.

$$f(x) = \begin{cases} 1 & \text{if } x > 0 \\ -1 & \text{if } x \leqslant 0 \end{cases} \qquad a = 0$$

51.

$$f(x) = \begin{cases} 1 & \text{if } x \geqslant 1 \\ x & \text{if } -1 < x < 1 \\ -1 & \text{if } x \leqslant -1 \end{cases} \qquad a = 1$$

52.

$$f(x) = \begin{cases} 2x + 1 & \text{if } x > 0 \\ 1 & \text{if } x \leqslant 0 \end{cases} \qquad a = 0$$

53.

$$f(x) = \begin{cases} \dfrac{|x|}{x} & \text{if } x \neq 0 \\ 0 & \text{if } x = 0 \end{cases} \qquad a = 0$$

54.

$$f(x) = \begin{cases} \dfrac{1}{x} & \text{if } x \neq 0 \\ 0 & \text{if } x = 0 \end{cases} \qquad a = 0$$

55.

$$f(x) = \begin{cases} \dfrac{1}{x^2} & \text{if } x \neq 0 \\ 2 & \text{if } x = 0 \end{cases} \qquad a = 0$$

In Exercises 56 through 65, a function f is given that is not continuous at the given number a. Find a function g that *is* continuous at a and that has the same graph as that of f except at a (see Fig. 4.4-5). Sketch the graphs of f and g, and use the fact that $\lim_{x \to a} f(x) = \lim_{x \to a} g(x)$ to evaluate $\lim_{x \to a} f(x)$.

56. $f(x) = \dfrac{2x^2 - 3x}{x}$, $a = 0$

57. $f(x) = \dfrac{-3x^2 + 2x}{x}$, $a = 0$

58. $f(x) = \dfrac{x^2 - 4}{x + 2}$, $a = -2$

59. $f(x) = \dfrac{x^2 - 9}{x - 3}$, $a = 3$

60. $f(x) = \dfrac{x^2 - 16}{x + 4}$, $a = -4$

61. $f(x) = \dfrac{x^2 - 2x + 1}{x - 1}$, $a = 1$

62. $f(x) = \dfrac{x^2 + x - 2}{x + 2}$, $a = -2$

63. $f(x) = \dfrac{2x^2 - x - 1}{2x + 1}$, $a = -\frac{1}{2}$

64. $f(x) = \dfrac{-3x^3 + 2x^2 + x}{x}$, $a = 0$

65. $f(x) = \dfrac{-x^3/2 + 3x^2/4 + x}{x}$, $a = 0$

A continuous function f is "predictable" in the sense that the value of f at a point a is determined by the values of f at numbers x close to a, because $\lim_{x \to a} f(x) = f(a)$. In Exercises 66 through 69, a situation is described that gives rise to a continuous function f. Estimate the value of f at the indicated point from the values of f at points nearby.

66. The seawater temperature f at a location t mi up the coast from Los Angeles is given by the following table.

t	50	40	30	20	10
$f(t)$	64°	65°		67°	68°

What would you estimate the seawater temperature to be at a point 30 mi up the coast?

67. The speed f (in miles per hour) of an airplane t sec after takeoff is given by the following table.

t	10	20	30	40	50	60
$f(t)$	120	140	160		200	220

What would you estimate the speed of the airplane to be 40 sec after takeoff?

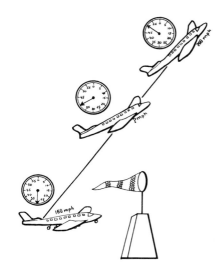

68. The population of a certain culture of bacteria is measured each hour after the sample arrives at the laboratory. The following table gives some values of the population f (in millions) at times t (hours after arrival in the laboratory).

t	1	2	3	4	5	6
$f(t)$	2	4	8		32	64

What would you estimate the population to be 4 hr after arrival in the laboratory?

69. Wildcat Wells, Inc., estimates that the number of barrels f (in millions) of crude oil it should produce per month is given by $f(x) = (x - 6)/(\sqrt{x + 3} - 3)$, where x is the selling price per barrel in dollars. Note that the function f is not defined when $x = 6$, although 6 is certainly a reasonable value for x. What value would you assign to $f(6)$, i.e., how many barrels should the company produce per month when the selling price is \$6 per barrel? (*Hint*: Let

$$g(x) = (x - 6)(\sqrt{x + 3} + 3)/(\sqrt{x + 3} - 3)(\sqrt{x + 3} + 3)$$

$$= \sqrt{x + 3} + 3$$

Then $g(x) = f(x)$ for $x \neq 6$, so $\lim_{x \to 6} g(x) = \lim_{x \to 6} f(x)$.)

70. The monthly service charge f (in dollars) for a checking account at a certain bank is given by the function

$$f(x) = \begin{cases} 0.5 & \text{if } 1 \leqslant x < 5 \\ 0.75 & \text{if } 5 < x \leqslant 15 \\ 1 & \text{if } 15 < x \leqslant 30 \end{cases}$$

where x is the number of checks written during the month. Sketch the graph of f and find those points where f is not continuous.

71. Air Freight, Inc. charges $\$f(x)$ to ship x lb of merchandise from Miami to Chicago, where

$$f(x) = \begin{cases} 0.5x & \text{if } 0 < x \leqslant 100 \\ 0.4x & \text{if } 100 < x \leqslant 300 \\ 0.3x & \text{if } 300 < x \end{cases}$$

Sketch the graph of f and find those points where f is not continuous.

72. Use rule (1) of Section 4.3 to show that if f and g are each continuous at a, then $f - g$ is also continuous at a.

73. Use rule (2) of Section 4.3 to show that if f and g are each continuous at a, then $f \cdot g$ is also continuous at a.

74. Use rule (3) of Section 4.3 to show that if f and g are each continuous at a and $g(a) \neq 0$, then f/g is also continuous at a.

4.5. The Exponential and Logarithmic Functions

Two important functions that appear frequently in applications of mathematics are the exponential function and the logarithmic function. These functions constantly arise in the fields of medicine, biology, business, ecology, psychology, and other disciplines. They are used both as computational tools and as building blocks for constructing functions which give mathematical descriptions of experimental observations. Our purpose here is to define these functions, examine some of their properties, and discuss the relationship between them.

We begin by introducing an interesting and important irrational number, called e, which arises from the compound interest law. Suppose for the moment that money were in such short supply that banks were paying 100% annual interest on savings accounts. If the interest is compounded annually (computed and credited to an account once a year), then $1 deposited for 1 year would earn $1 in interest, so at the end of a year there would be $1 (deposit) + $1 (interest) = $2 in the account. If the interest is compounded twice a year, then six months

The formula for computing interest is $I = Prt$, where I is interest earned, P is the amount of principle, r is the annual rate of interest, and t is the time (in years) the principle is on deposit. Thus the interest for $P = \$1$, $r = 100\% = 1.00$, and $t = 1$ year is $I = (1)(1.00)(1) = \$1$.

after the dollar is deposited, interest of $(\$1)(1.00)\frac{1}{2} = \$\frac{1}{2}$ is added to the account (see margin); then interest is paid on the total of $1 + \frac{1}{2}$ dollars in the account over the next six-month period, this amounting to

$$\left(1 + \tfrac{1}{2}\right)(1.00)\tfrac{1}{2} = \left(1 + \tfrac{1}{2}\right)\tfrac{1}{2}$$

Thus at the end of a year there would be $\left(1 + \frac{1}{2}\right)$ (at midyear) $+ \left(1 + \frac{1}{2}\right)$ $(1.00)\frac{1}{2}$ (interest for second six months), or

$$\left(1 + \tfrac{1}{2}\right) + \left(1 + \tfrac{1}{2}\right)\tfrac{1}{2} = \left(1 + \tfrac{1}{2}\right)\left(1 + \tfrac{1}{2}\right) = \$\left(1 + \tfrac{1}{2}\right)^2$$

in the account, which is \$2.25. Similar computations show that if the interest is compounded three times a year, then there would be $\$(1 + \frac{1}{3})^3$ in the account after 1 year, which is \$2.37. In general, if the interest is compounded n times a year, the value of the account (in which \$1 is initially deposited) after 1 year would be $\$(1 + 1/n)^n$. Table 4.5-1 shows the value of the account after 1 year for several different values of n. Clearly, the more often interest is compounded, the greater the amount in the account after 1 year. How large can the amount in the account at year's end become if we allow the interest to be compounded more and more often, for example, every day, every minute, or every second? Intuition may lead us to expect a fortune, but in truth the amount will never exceed \$2.72 no matter how often the interest is compounded. This is because the amount in the account at the end of a year is $(1 + 1/n)^n$ when interest is compounded n times a year, and it turns out that the limit of $(1 + 1/n)^n$ as n grows larger and larger is $2.71828182845904523536 \cdots$. The letter e is used to designate this number, i.e., $\lim(1 + 1/n)^n = e$; the value of e is customarily approximated as 2.718. Unlikely as it may seem, the number e, like the irrational number π, appears frequently in applications. As we shall see, it is this number that is used to define the exponential function.

Again, $I = Prt$, but now $t = \frac{1}{2}$ year.

Note that $x + xy = x(1 + y)$, so if $x = 1 + \frac{1}{2}$ and $y = \frac{1}{2}$, we have $\left(1 + \frac{1}{2}\right) + \left(1 + \frac{1}{2}\right)\left(\frac{1}{2}\right) = \left(1 + \frac{1}{2}\right)\left(1 + \frac{1}{2}\right)$.

Definition of the number e

The number e is an irrational number, so its decimal representation is infinite and nonrepeating.

TABLE 4.5-1

Interest compounded	n	Value of account after one year	
Annually	1	$(1 + 1)^1$	$= \$2.00$
Semiannually	2	$(1 + \frac{1}{2})^2$	$= \$2.25$
Every four months	3	$(1 + \frac{1}{3})^3$	$= \$2.37$
Quarterly	4	$(1 + \frac{1}{4})^4$	$= \$2.44$
Monthly	12	$(1 + \frac{1}{12})^{12}$	$= \$2.61$
Daily	365	$(1 + \frac{1}{365})^{365}$	$= \$2.67$
Hourly	8760	$(1 + \frac{1}{8760})^{8760}$	$= \$2.71$

From the approximation $e = 2.718$, we can compute approximations for $e^2, e^3, e^4, \cdots$, as well as $1/e = e^{-1}$, $1/e^2 = e^{-2}, \cdots$. For ex-

For any real number $a \neq 0$ and any positive integer n, $1/a^n$ is denoted by a^{-n}.

We will write $e = 2.718$, $e^2 = 7.388$, etc., but it should be kept in mind that these decimal numbers are merely adequate approximations.

For any real number $a \neq 0$, a^0 is defined to be 1, so $e^0 = 1$.

The definition of the exponential function $\exp(x) = e^x$

Since $\exp(x)$ is a continuous function, $\lim_{x \to a} e^x = e^a$ for any real number a.

FIGURE 4.5-1

When constructing the graph of $\exp(x) = e^x$ shown in Fig. 4.5-1, considerably more points of the form (x, e^x) than those given by Table 4.5-2 should be plotted before joining them with a smooth curve, to improve the accuracy of the graph. Table 4.5-3 contains more values of x and gives the more detailed graph of Fig. 4.5-2.

ample, $e^2 = (2.718)(2.718) = 7.388$, and $e^{-2} = 1/e^2 = 1/7.388 = 0.135$. The values of e^x for several different values of x are shown in Table 4.5-2.

TABLE 4.5-2

x	-2	-1	0	1	2	3
e^x	0.135	0.368	1	2.718	7.388	20.079

If we plot the points (x, e^x) from this table in a Cartesian plane and connect the points with a smooth curve, we have the graph of a continuous function whose domain consists of all real numbers (see Fig. 4.5-1). This function is called the *exponential function*, denoted exp; the value of the function exp at x is given by the rule

$$\exp(x) = e^x$$

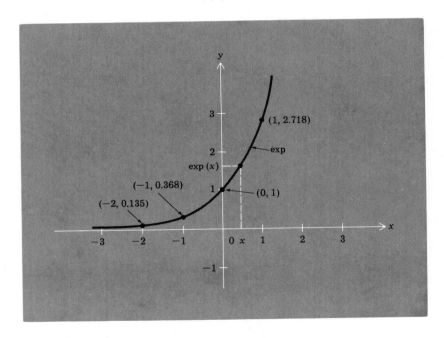

Actually, a detailed study of the meaning of e^x for an arbitrary real number x (such as $x = \sqrt{2}$) is quite intricate, so we will not pursue the question here. However, from the graph of exp in Fig. 4.5-2 we can estimate the value of e^x for any number x by locating x on the x axis, moving vertically to the graph, and then moving horizontally to e^x on the y axis. Thus for $x = \sqrt{2}$ (which is approximately 1.41), we can estimate from Fig. 4.5-2 that $\exp(\sqrt{2}) = e^{\sqrt{2}} = 4.1$. Alternatively, from Table A in the back of the book we find that $\exp(\sqrt{2}) = 4.096$.

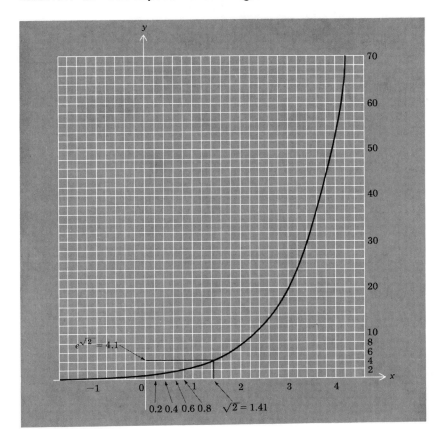

FIGURE 4.5-2   Some sample values
of exp(x) are:

x	exp(x)
− 2	0.14
− 1	0.37
0	1
0.1	1.11
0.2	1.22
0.5	1.65
1.0	2.72
1.2	3.32
1.4	4.06
1.6	4.95
1.8	6.05
2.0	7.39
2.2	9.03
2.4	11.02
2.6	13.46
2.8	16.45
3.0	20.08
4.0	54.60

We introduced $\exp(x) = e^x$ by plotting the points from Table 4.5-2 in a Cartesian plane and connecting these points with a smooth curve to obtain the graph of exp. If we interchange the top and bottom rows of this table (see Table 4.5-3), plot the points from the new table in a Cartesian plane, and connect them with a smooth curve, we produce the graph of a continuous function whose importance rivals that of exp. This new function is called the *natural logarithm function* and is denoted by ln. The value of ln at x, $\ln(x)$, is called the natural logarithm of x; from Table 4.5-3 we see that $\ln(1) = 0$, $\ln(e) = 1$, $\ln(0.135) = -2$, etc.

TABLE 4.5-3

x	$e^{-2} = 0.135$	$e^{-1} = 0.368$	$e^0 = 1$	$e^1 = 2.718$	$e^2 = 7.388$	$e^3 = 20.079$
$\ln(x)$	-2	-1	0	1	2	3

The graph of the function ln is shown in Fig. 4.5-3; the domain of ln consists of the positive real numbers, since only numbers of the form e^x (for some x) are in the domain of ln and e^x is always positive.

FIGURE 4.5-3

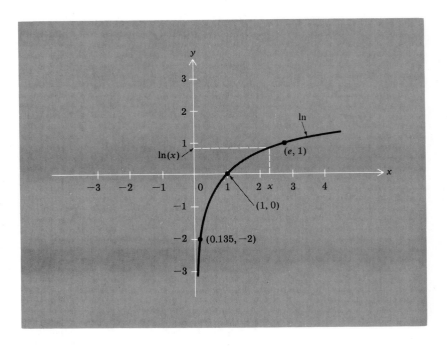

If the graphs of these two functions, exp and ln, are drawn on the same coordinate system as shown in Fig. 4.5-4 and if the plane is folded along the dotted line (the line $y = x$), then the two graphs will coincide. Such graphs are said to be *reflections* of each other through the line $y = x$.

FIGURE 4.5-4

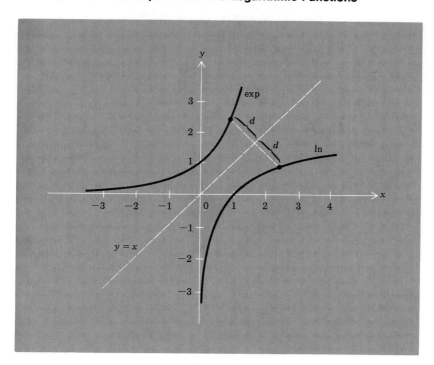

To compute the value of ln(x) for a given number x, we can locate x on the x axis in Fig. 4.5-5, move vertically to the graph of ln, and then move horizontally to the y axis for an approximate value of ln(x). Thus for $x = 2.5$, we might estimate ln(2.5) = 0.9. For more accurate values of ln(x), we use Table A in the back of this book, where we find that ln(2.5) = 0.9163.

Note: since ln(x) is a continuous function, $\lim_{x \to a} \ln(x) = \ln(a)$ for any positive real number a.

The key relationships between the functions exp and ln are indicated by Tables 4.5-2 and 4.5-3. From these tables we see that

$$\ln\left[\exp(0)\right] = \ln(1) = 0$$

$$\ln\left[\exp(2)\right] = \ln(7.388) = 2$$

$$\exp\left[\ln(1)\right] = \exp(0) = 1$$

$$\exp\left[\ln(0.368)\right] = \exp(-1) = 0.368$$

and

$$\exp\left[\ln(20.079)\right] = \exp(3) = 20.079.$$

These computations suggest that the composition functions exp ∘ ln and ln ∘ exp are given by the rules

$$(\ln \circ \exp)(x) = \ln\left[\exp(x)\right] = x$$

Recall that, given functions f and g, the composition function $f \circ g$ is defined by $(f \circ g)(x) = f\left[g(x)\right]$ for each x for which $g(x)$ is in the domain of f (see Section 2.5).

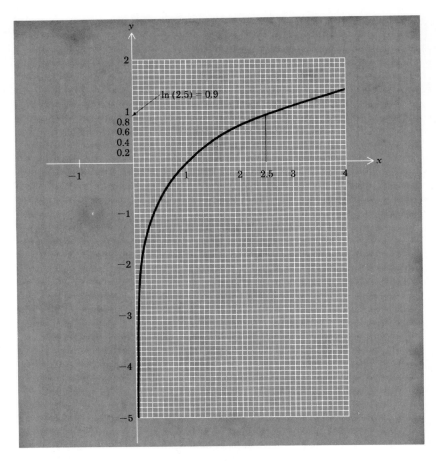

FIGURE 4.5-5 Some sample values of ln(x) are:

x	ln(x)
0.01	− 4.61
0.05	− 3.00
0.10	− 2.30
0.20	− 1.61
0.30	− 1.20
0.40	− 0.92
0.50	− 0.69
0.60	− 0.51
0.80	− 0.22
1.00	0.00
1.20	0.18
1.40	0.34
1.60	0.47
1.80	0.59
2.00	0.69
2.50	0.92
3.00	1.10
4.00	1.39

for any real number x, and

$$(\exp \circ \ln)(x) = \exp\left[\ln(x)\right] = x$$

for all real numbers $x > 0$; this is indeed the case. These rules are also expressed by the formulas

$$\ln(e^x) = x \quad \text{and} \quad e^{\ln x} = x \quad (x > 0)$$

For any x, $\ln(e^x) = x$, and for any $x > 0$, $e^{\ln x} = x$.

Many properties of the functions exp and ln are used as computational tools. We will not verify all these properties here, but in Table 4.5-4 we list some of the important ones that we shall have occasion to use in our later work.

TABLE 4.5-4

Properties of exp	Properties of ln
$\exp(0) = 1$	$\ln(1) = 0$
$\exp(1) = e$	$\ln(e) = 1$
$\exp(x + y) = \exp(x) \cdot \exp(y)$	$\ln(xy) = \ln(x) + \ln(y)$
$\exp(-x) = 1/\exp(x)$	$\ln(x^n) = n \ln(x)$, for any integer n

Note that the formula $\exp(x + y) = \exp(x) \cdot \exp(y)$ can also be written as $e^{x+y} = e^x \cdot e^y$; similarly, $\exp(-x) = 1/\exp(x)$ can be written $e^{-x} = 1/e^x$.

EXAMPLE 4.5-1

Given that $\ln(2) = 0.693$ and $\ln(5) = 1.609$, find $\ln(10)$, $\ln(100)$, and $\ln(25)$.

SOLUTION

Since $10 = (2)(5)$, we have

$$\ln(10) = \ln\left[(2)(5)\right] = \ln(2) + \ln(5)$$

$$= 0.693 + 1.609 = 2.302.$$

Since $100 = 10^2$, we have

$$\ln(100) = \ln(10^2) = 2 \ln(10)$$

$$= 2(2.302) = 4.604.$$

Finally, since $25 = 5^2$, we have

$$\ln(25) = \ln(5^2) = 2 \ln(5) = 2(1.609) = 3.218$$

EXAMPLE 4.5-2

Without using tables to evaluate $\ln(7)$ and $\ln(3)$, find $\exp[\ln(7) + 4 \ln(3)]$.

SOLUTION

Since

$$4 \ln(3) = \ln(3^4) = \ln(81)$$

we have

$$\ln(7) + 4 \ln(3) = \ln(7) + \ln(81) = \ln\big[(7)(81)\big] = \ln(567)$$

Thus

Recall that for any $x > 0$, $\exp[\ln(x)] = x$.

$$\exp\big[\ln(7) + 4 \ln(3)\big] = \exp\big[\ln(567)\big] = 567$$

EXAMPLE 4.5-3

Continuous compound interest

If $\$P$ is placed in an account that pays $r\%$ annual interest compounded continuously, it can be shown that the formula $F = Pe^{rt}$ gives the amount F in the account after t years. If $\$100$ is placed in an account that pays 5% annual interest compounded continuously, find the amount F in the account after 3 years.

SOLUTION

We must convert 5% to a number before using it in the formula, and $5\% = 5/100 = 0.05$.

We use the formula $F = Pe^{rt}$ with $P = 100$, $r = 5\% = 0.05$, and $t = 3$. Hence

$$F = 100e^{(0.05)(3)} = 100e^{0.15}$$

From Table A in the back of this book, $e^{0.15} = 1.162$, so

$$F = 100e^{0.15} = (100)(1.162) = \$116.20$$

Exercises 4.5

In Exercises 1 through 8, use the graph of $\exp(x) = e^x$ in Fig. 4.5-2 to approximate the value of exp at the given value of x. Then check your approximation by referring to Table A in the back of the book.

1. $x = 1.5$ 2. $x = -1.5$
3. $x = 2.5$ 4. $x = 3.5$
5. $x = 1.9$ 6. $x = 2.1$
7. $x = -0.4$ 8. $x = 0.8$

In Exercises 9 through 16, use the graph of ln in Fig. 4.5-5 to approximate the value $\ln(x)$ at the given value of x. Check your approximation by referring to Table A in the back of the book.

9. $x = 1.5$ 10. $x = 2.2$
11. $x = 0.7$ 12. $x = 3.5$
13. $x = 1.3$ 14. $x = 2.9$
15. $x = 1.9$ 16. $x = 0.9$

Table A gives values of $\exp(x)$ only for $-10 \leqslant x \leqslant 10$. In Exercises 17 through 24, write the given number x as a sum $x = a + b$ where a and b are numbers between -10 and 10. Then use Table A and the formula $\exp(x) = \exp(a + b) = \exp(a) \cdot \exp(b)$ to evaluate $\exp(x)$.

17. $x = 12$ 18. $x = 16.2$
19. $x = -14$ 20. $x = -18.1$
21. $x = -13.86$ 22. $x = 14.25$
23. $x = 10.01$ 24. $x = -10.01$

25. Write the number 24.21 as a sum of numbers a, b, and c, each between -10 and 10. Then by repeated use of the formula $\exp(x + y) = \exp(x) \cdot \exp(y)$ and of Table A, evaluate $\exp(24.21)$.

26. Write the number -27.02 as a sum of numbers a, b, and c, each between -10 and 10. Then by repeated use of the formula $\exp(x + y) = \exp(x) \cdot \exp(y)$ and of Table A, evaluate $\exp(-27.02)$.

Table A gives values of $\ln(x)$ only for $0 < x \leqslant 10$. In Exercises 27 through 34, write the given number x as a product $x = ab$ where a and b are numbers between 0 and 10. Then use Table A and the formula $\ln(ab) = \ln(a) + \ln(b)$ to evaluate $\ln(x)$.

27. $x = 36$ 28. $x = 42$
29. $x = 54.6$ 30. $x = 48.6$
31. $x = 76.5$ 32. $x = 15.33$
33. $x = 38$ 34. $x = 90.25$

35. Write the number 142 as a product of three numbers a, b, and c, each between 0 and 10 (possibly equal to 10). Then by repeated use of the formula $\ln(xy) = \ln(x) + \ln(y)$ and of Table A, evaluate $\ln(142)$.

36. Write the number 954 as a product of three numbers a, b, and c, each between 0 and 10 (possibly equal to 10). Then by repeated use of the formula $\ln(xy) = \ln(x) + \ln(y)$ and of Table A, evaluate $\ln(954)$.

In Exercises 37 through 48, given that $\exp(3) = 20.086$, $\exp(5) = 148.41$, $\ln(3) = 1.099$, and $\ln(5) = 1.609$, evaluate the given function at the given number using only the properties of exp and ln given in Table 4.5-4.

37. $\exp(-3)$ 38. $\ln(3^2)$
39. $\exp(8)$ 40. $\ln(15)$
41. $\exp(-8)$ 42. $\ln(27)$
43. $\exp(13)$ 44. $\ln(5^{12})$
45. $\exp(-13)$ 46. $\ln(75)$
47. $\exp(4)$ 48. $\ln(15e)$

In Exercises 49 through 60, evaluate the given expressions using only the properties of exp and ln in Table 4.5-4 and the facts that $\exp[\ln(x)] = x$ and $\ln[\exp(x)] = x$.

49. $\exp[\ln(5)]$ 50. $\ln[\exp(-3)]$

51. $\exp[\ln(2) + \ln(10)]$ 52. $\ln[\exp(2) \cdot \exp(10)]$

53. $\ln[1/\exp(12)]$ 54. $\exp[4 \ln(2)]$

55. $\ln\left(\dfrac{\exp(7)}{\exp(-3)} \right)$ 56. $\exp[\ln(0.5) + 2 \ln(4)]$

57. $\ln\left(\dfrac{e}{\exp(2)} \right)$ 58. $\exp[1 + 3 \ln(2)]$

59. $\exp[3 \ln(2) + 2 \ln(3)]$ 60. $\ln\left(\dfrac{\exp(7) \cdot \exp(-3)}{\exp(2)} \right)$

Note that $\exp(x - y) = \exp(x)/\exp(y)$ can also be written as $e^{x-y} = e^x/e^y$.

Recall that $1/y = y^{-1}$.

61. Using the properties of exp given in Table 4.5-4, verify that $\exp(x - y) = \exp(x)/\exp(y)$.

62. Using the properties of ln given in Table 4.5-4, verify that for any two positive numbers x and y, $\ln(x/y) = \ln(x) - \ln(y)$.

63. If $2000 is placed in an account that pays 6% annual interest compounded continuously, find the amount in the account after 5 years (see Example 4.5-3).

64. If $10,000 is placed in an account that pays 8% annual interest compounded continuously, find the amount in the account after 10 years (see Example 4.5-3).

65. If $P is invested in an account that pays 5% annual interest compounded continuously, how many years will it take for the amount of money in the account to double? (*Hint*: we want to find t such that $2P = Pe^{0.05t}$, i.e., $2 = e^{0.05t}$. Solve this equation for t by evaluating ln of both sides of the equation.)

For Exercise 65, use the formula $F = Pe^{rt}$ (see Example 4.5-3).

66. Medical researchers, testing a new drug, find that the maximum concentration $c(0)$ (measured in milligrams per cubic centimeter) of the drug in the bloodstream is reached 30 sec after the drug is injected. They also find that t hr after the maximum is reached, the concentration c of the drug in the bloodstream is given by $c(t) = c(0)e^{-t}$. Find the concentration of the drug in the bloodstream 5 hr after the maximum, if the maximum is $c(0) = 0.1$ mg/cc.

67. A biologist estimates that the population P of a certain type of bacteria is given by the equation $P = P_0e^{0.7t}$, where P_0 is the population at time $t = 0$ and t is measured in hours. If a sample initially contains 4000 bacteria, what is the population of the sample 5 hr later?

68. The Ideal Business College estimates that the typing rate R (in words per minute) of their average student after x weeks in class is given by $R(x) = 40 - 40e^{-x}$, for $0 < x \leqslant 5$. How many words per minute can their average student type after 4 weeks of study?

69. The Toast-Rite Toaster Company has estimated from experience that when a new employee is hired, the number of toasters N the employee will be able to assemble per day is given by $N(x) = 80 - 80e^{-0.2x}$ where x is the number of days worked. How many toasters can an employee be expected to assemble per day after working 6 days?

70. Easy Wrap, Inc., is putting a new plastic wrap on the market. Studies of similar products show that the number of cases N of the product sold per week is given by $N(x) = 8000 - 8000e^{-0.02x}$, where x is the number of weeks the product has been on the market. How many cases should the company sell during the fifth week after the product is marketed?

71. The Shave-Ease Company estimates that its monthly sales s (in dollars) increases logarithmically according to the equation $s(x) = 1500 \ln(x)$, where x is the amount (in dollars) spent on advertising per month. What is the company's monthly sales when it spends $500 a month on advertising?

72. A psychologist conducting an experiment on learning finds that the percent P of nonsense syllables from a list that are retained t min after the end of the learning period is given by $P(t) = 600/[(\ln(t))^2 + 10]$ for $t \geqslant 1$. What percent of the syllables are retained 1 min after the end of the learning period? Thirty minutes after the end of the learning period?

Self Test   Part II

1. Compute the first six terms of the sequence $\{2 + (-1)^n\}$ and construct the line graph.

2. Let g be the function defined by

$$g(x) = \begin{cases} 0 & \text{if } x < 0 \\ 1 & \text{if } x \geq 0 \end{cases}$$

Show that $\lim_{x \to 0} g(x)$ does not exist.

In questions 3 through 13, find the given limit if it exists.

3. $\lim_{x \to 5} e^{2x+1}$

4. $\lim_{n \to \infty} \dfrac{(-1)^n 2}{n^2}$

5. $\lim_{x \to 1} \dfrac{-1}{\sqrt{x-1}}$

6. $\lim_{x \to -3} e^x$

7. $\lim_{x \to 1} (-3x^3 - 2x + 1)$

8. $\lim_{x \to 2} (3\sqrt{x+7} - 1)$

9. $\lim_{x \to \infty} (2 - x^2)$

10. $\lim_{x \to 2} \dfrac{\sqrt{3x+2}}{\sqrt{x}}$

11. $\lim_{x \to 1} \dfrac{2x^2 - x - 1}{1 - x}$

12. $\lim_{x \to 7} \left[(\ln x)^2 \cdot e^x \right]$

13. $\lim_{x \to -1} \dfrac{-2x^3 + 3x}{2x^2 - 3}$

14. Evaluate $\exp[1 + 2\ln(3)]$

15. A piece of industrial machinery, which cost $50,000 new, depreciates each year by an amount equal to $\frac{1}{10}$ of its value at the beginning of the year. Let s be the sequence whose value at each positive integer n is the amount the machinery is worth at the end of the nth year after it is purchased. Find $s(1)$, $s(2)$, $s(3)$, and $s(4)$. Show that the equation $s(n) = 50,000(\frac{9}{10})^n$ gives the value of the nth term of the sequence s for each positive integer n. (*Hint*: the value of the machinery at the end of any year is always $\frac{9}{10}$ its value at the end of the preceding year.)

16. The population P (in hundreds) of a certain colony of bacteria at time t (in hours) is found to be $P(t) = 2t^2 + t$. Sketch the graph of P, and from the graph, find $\lim_{t \to +\infty} P(t)$. Interpret the value of this limit in the context of the problem.

17. An experiment on memory retention reveals that the percent p of specified material remembered by one of the participants after t months is given by $p(t) = 100/(1 + 0.5t)$. Find $\lim_{t \to 4} p(t)$.

18. A company estimates that its weekly production cost C (in dollars) is given by the equation $C(x) = x \ln(x) + (3x/\ln(x))$, where x is the number of units produced per week, $x > 1$. Find the production cost for a week when 400 units are produced.

The Derivative and its Applications

PART III

The Derivative

5.1. Introduction

Now that we have defined limits of functions, we are ready to study the first of the two central topics in calculus, the *derivative* of a function. The derivative will be a powerful addition to our supply of mathematical tools. It will substantially reduce the effort presently required to sketch the graph of a given function accurately, and it will also lead us to solutions of a wide variety of applied problems; for example, it will enable us to solve problems involving the rate at which one quantity changes as a result of changes in another and it will provide us with the means to select optimum methods of procedure. Moreover, many problems that we already know how to solve can be dispatched more efficiently using the derivative.

The two central topics in calculus are the *derivative* and the *integral*. The concept of the integral of a function will be discussed in Chapter 7.

5.2. Tangent Lines and the Derivative

As a means of introducing the derivative, let us first discuss what we mean by the line *T tangent* to the graph of a function *f* at some given point *P* on the graph.

Recall from elementary geometry that a tangent line to a circle at a given point *P* on the circle is defined as the line which meets the circle only at *P* (see Fig. 5.2-1). Unfortunately, this definition is not suitable for other curves. Consider, for example, a function *f* whose graph is as

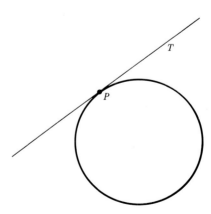

FIGURE 5.2-1 The line *T* tangent to a circle at a point *P* on the circle.

shown in Fig. 5.2-2(a). There are many different lines passing through the point P which intersect the graph only at P, but none of these lines seem to be appropriate candidates for the tangent. However, the graph of f *near* P coincides with the circle shown in Fig. 5.2-2(b), and the tangent line to the *circle* at P *does* seem an appropriate choice for the line T tangent to the graph of f at P, even though T meets the graph of f more than once.

Granting, for the moment, that the line T in Fig. 5.2-2(b) is the line tangent to the graph of f at P, what is the *slope* of T? We know that T passes through P, but the formula for the slope of a line requires that we

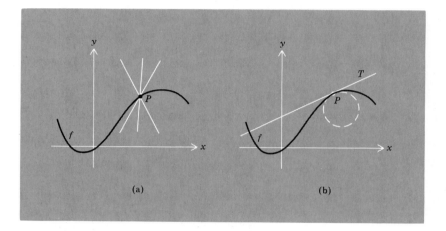

(a)   (b)

FIGURE 5.2-2

Recall from Section 2.3 that the slope m of a line L passing through the points $P_1 = (x_1, y_1)$ and $P_2 = (x_2, y_2)$ is given by the formula $m = (y_2 - y_1)/(x_2 - x_1)$, provided that $x_2 \neq x_1$.

know *two* distinct points through which the line passes; thus we cannot compute the slope of T directly from the formula. Instead we use a limit process to obtain the slope. We begin by recalling that since P is on the graph of f, the coordinates of P are of the form $(x_0, f(x_0))$. Now, let x be a number close to x_0 on the x axis; then the point $Q = (x, f(x))$ will be close to $P = (x_0, f(x_0))$ on the graph of f (see Fig. 5.2-3). If we construct the line L through the points P and Q, it is evident from Fig. 5.2-3 that as x moves closer to x_0, the point $Q = (x, f(x))$ moves closer to the point

Since L passes through the points $P = (x_0, f(x_0))$ and $Q = (x, f(x))$, the slope of L is $(f(x) - f(x_0))/$ $(x - x_0)$.

$P = (x_0, f(x_0))$ on the graph of f and the slant of the line L through P and Q becomes closer and closer to the slant of the tangent line T. This suggests that the slope of the line T should be the limiting value of the slope of the line L through P and Q as Q moves closer and closer to P, i.e., the slope of T should be

$$\lim_{x \to x_0} \frac{f(x) - f(x_0)}{x - x_0}$$

The definition of the tangent line T to the graph of a function f at a point P on the graph

Indeed, this is the definition that mathematicians have found most useful, and we now *define* the line T *tangent* to the graph of a function f

FIGURE 5.2-3

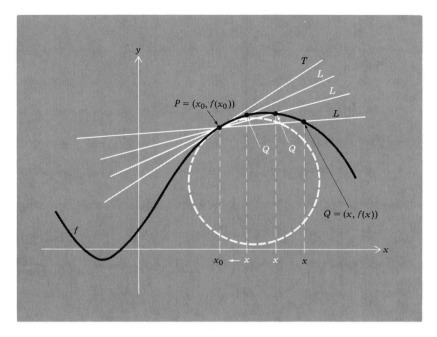

at a point $P = (x_0, f(x_0))$ on the graph to be the line through P having slope

$$\lim_{x \to x_0} \frac{f(x) - f(x_0)}{x - x_0}$$

provided that this limit exists and is finite.

A function f may not have a tangent line in the sense in which we have just defined the tangent at some point P on its graph. For example, the graph of $f(x) = |x|$ does not have a tangent line at $(0, 0)$ (see Fig. 5.2-4). But if the tangent line T exists at $P = (x_0, f(x_0))$ and is not vertical, then the slope of T is given by $\lim_{x \to x_0}(f(x) - f(x_0))/(x - x_0)$.

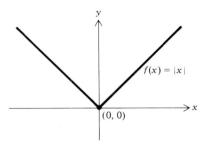

FIGURE 5.2-4   The graph of $f(x) = |x|$. This graph does not have a tangent line at $(0, 0)$. Note that while for $x > 0$ we have $(f(x) - f(0))/(x - 0) = x/x = 1$, for $x < 0$ we have $(f(x) - f(0))/(x - 0) = -x/x = -1$. Thus $\lim_{x \to 0}(f(x) - f(0))/(x - 0)$ does not exist, so there is no tangent line at $(0, 0)$. In general if the graph of a function has a "corner" at some point P, then there is no line tangent to the graph at P.

EXAMPLE 5.2-1

Find the slope of the line tangent to the graph of $f(x) = x^2$ at the point $P = (1, 1)$.

SOLUTION

The slope of this tangent line will be given by

$$\lim_{x \to x_0} \frac{f(x) - f(x_0)}{x - x_0}$$

with $x_0 = 1$ and $f(x) = x^2$. Since $f(x_0) = f(1) = 1^2 = 1$, we have

$$\lim_{x \to x_0} \frac{f(x) - f(x_0)}{x - x_0} = \lim_{x \to 1} \frac{x^2 - 1}{x - 1}$$

$$= \lim_{x \to 1} \frac{(x - 1)(x + 1)}{x - 1}$$

$$= \lim_{x \to 1} (x + 1) = 2$$

Thus the slope of the line tangent to the graph of $f(x) = x^2$ at $P = (1, 1)$ is 2 (see Fig. 5.2-5).

FIGURE 5.2-5

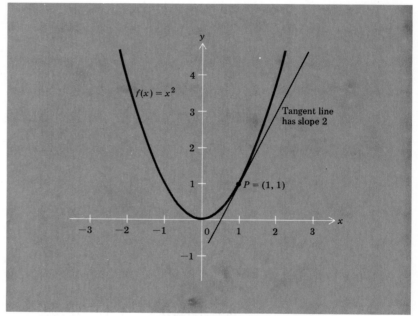

EXAMPLE 5.2-2.

Find the equation of the line tangent to the graph of $f(x) = x^2$ at the point $P = (-2, 4)$.

SOLUTION

The slope of this tangent line is given by

$$\lim_{x \to x_0} \frac{f(x) - f(x_0)}{x - x_0}$$

with $x_0 = -2$ and $f(x) = x^2$. Since $f(x_0) = f(-2) = 4$, the slope is

$$\lim_{x \to -2} \frac{x^2 - 4}{x - (-2)} = \lim_{x \to -2} \frac{(x-2)(x+2)}{x+2}$$

$$= \lim_{x \to -2} (x - 2) = -4$$

Thus the tangent line passes through the point $P = (-2, 4)$ and has slope -4, so its equation is $y - 4 = -4(x - (-2))$ or $y = -4x - 4$ (see Fig. 5.2-6).

Recall from Section 2.3 that the line through the point $P = (x_0, y_0)$ with slope m has the equation $y - y_0 = m(x - x_0)$.

FIGURE 5.2-6

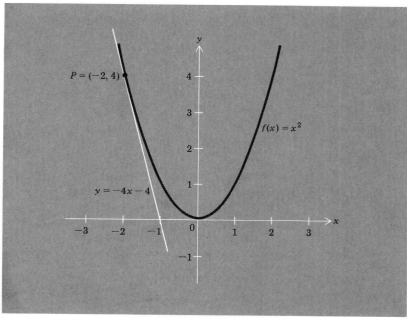

Given a function f with values $f(x)$ and a number x_0 in the domain of f, we will denote the value of the limit $\lim_{x \to x_0}(f(x) - f(x_0))/(x - x_0)$ by $D_x f(x_0)$ if the limit exists and is finite. Thus for each x_0 in the domain of f (for which $D_x f(x_0)$ exists), we produce a number

$$D_x f(x_0) = \lim_{x \to x_0} \frac{f(x) - f(x_0)}{x - x_0}$$

Definition of the derivative of a function

which is the slope of the line tangent to the graph of f at $P = (x_0, f(x_0))$. Thus we have derived a new function from the function f; to each x_0 in the domain of f (for which $D_x f(x_0)$ exists) corresponds the number $D_x f(x_0)$. This new function is called the *derivative* of the function f and is denoted by $D_x f$. (The notation $D_x f$ is usually read "the derivative of f with respect to x.") As we will see, when the function f has a simple equation, it is usually possible to find a simple equation for its derivative $D_x f$ as well.

To say that $D_x f(x_0)$ exists means $\lim_{x \to x_0}(f(x) - f(x_0))/(x - x_0)$ exists and is finite, and this means that the line tangent to the graph of f at $P = (x_0, f(x_0))$ exists and is not vertical. Thus $D_x f(x_0)$ will not exist if the graph of f has either a vertical tangent or a "corner" at $P = (x_0, f(x_0))$ as in Fig. 5.2-4.

EXAMPLE 5.2-3

Find an equation for the derivative of the function $f(x) = x^2$. What is the domain of the derivative function?

SOLUTION

If x_0 is in the domain of a function f and if $D_x f(x_0)$ exists, then $P = (x_0, f(x_0))$ is a point on the graph of f and $D_x f(x_0)$ is the slope of the line tangent to the graph of f at P.

For each x_0 in the domain of f, the value of $D_x f$ at x_0 is, by definition,

$$D_x f(x_0) = \lim_{x \to x_0} \frac{f(x) - f(x_0)}{x - x_0}$$

$$= \lim_{x \to x_0} \frac{x^2 - x_0^2}{x - x_0}$$

$$= \lim_{x \to x_0} \frac{(x - x_0)(x + x_0)}{x - x_0}$$

$$= \lim_{x \to x_0} (x + x_0) = x_0 + x_0 = 2x_0$$

The domain of the derivative function $D_x f$ consists of all numbers x_0 in the domain of f for which $D_x f(x_0)$ exists.

Thus the derivative function $D_x f$ is the function determined by the rule $D_x f(x) = 2x$. This means that at any x_0 in the domain of $f(x) = x^2$, the slope of the line tangent to the curve at $(x_0, f(x_0))$ is $2x_0$. Since $D_x f(x_0)$ exists for each x_0 in the domain of f, the domain of $D_x f$ consists of all numbers in the domain of f, i.e., all real numbers.

Definition of the term *differentiable*

Whenever $D_x f(x_0)$ exists, f is said to be *differentiable* at x_0. A function which is differentiable at each point of its domain is called a *differentiable function*.

EXAMPLE 5.2-4

Find an equation for the derivative of the function $f(x) = 3x + 7$. What is the domain of the derivative function?

SOLUTION

For each x_0 in the domain of f, we have

$$D_x f(x_0) = \lim_{x \to x_0} \frac{f(x) - f(x_0)}{x - x_0}$$

$$= \lim_{x \to x_0} \frac{(3x + 7) - (3x_0 + 7)}{x - x_0}$$

$$= \lim_{x \to x_0} \frac{3(x - x_0)}{x - x_0} = \lim_{x \to x_0} 3 = 3$$

There are several alternative notations commonly used for the derivative function $D_x f$. Among these are f' and df/dx. However, in this book we will always denote the derivative function by $D_x f$.

Thus $D_x f(x_0) = 3$, and the derivative function $D_x f$ is the function determined by the rule $D_x f(x) = 3$. Since $D_x f(x_0)$ exists for each x_0 in the domain of f, the domain of $D_x f$ is identical with the domain of f, namely, all real numbers. The equation $D_x f(x) = 3$ means that the slope of the line tangent to the graph of $f(x) = 3x + 7$ at each point $(x, f(x))$ on the graph is 3. But since $f(x) = 3x + 7$ is a linear function, the graph of f is itself a line with slope 3. This means that the line tangent to the graph of $f(x) = 3x + 7$ at any point on the graph is the graph of f itself (see Fig. 5.2-7).

Recall that the graph of a linear function $g(x) = ax + b$ is a line with slope a (see Section 2.3).

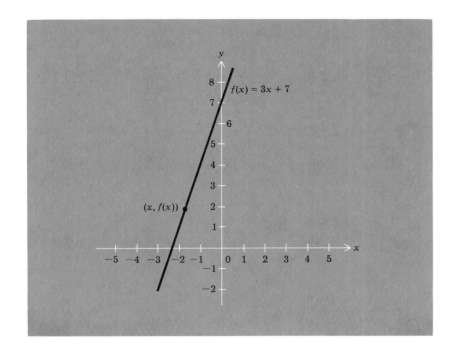

FIGURE 5.2-7   The line tangent to the graph of the linear function $f(x) = 3x + 7$ at any point $(x, f(x))$ on the graph is the graph itself.

EXAMPLE 5.2-5

Find an equation for the derivative of the function $f(x) = \sqrt{x}$.

SOLUTION

The domain of f consists of all numbers greater than or equal to 0, and for each x_0 in the domain, we have

Remember that $(\sqrt{x})^2 = x$ (for $x \geqslant 0$), so $(\sqrt{x} - \sqrt{x_0})$ $(\sqrt{x} + \sqrt{x_0}) = (\sqrt{x})^2 - (\sqrt{x_0})^2$ $= x - x_0$.

$$D_x f(x_0) = \lim_{x \to x_0} \frac{f(x) - f(x_0)}{x - x_0}$$

$$= \lim_{x \to x_0} \frac{\sqrt{x} - \sqrt{x_0}}{x - x_0}$$

$$= \lim_{x \to x_0} \frac{(\sqrt{x} - \sqrt{x_0})(\sqrt{x} + \sqrt{x_0})}{(x - x_0)(\sqrt{x} + \sqrt{x_0})}$$

$$= \lim_{x \to x_0} \frac{(x - x_0)}{(x - x_0)(\sqrt{x} + \sqrt{x_0})}$$

$$= \lim_{x \to x_0} \frac{1}{\sqrt{x} + \sqrt{x_0}}$$

Thus for $x_0 > 0$,

$D_x f(0) = \lim_{x \to 0}[1/(\sqrt{x} + \sqrt{0})]$; since this limit is not finite, $D_x f(0)$ does not exist.

$$D_x f(x_0) = \lim_{x \to x_0} \frac{1}{\sqrt{x} + \sqrt{x_0}} = \frac{1}{\sqrt{x_0} + \sqrt{x_0}} = \frac{1}{2\sqrt{x_0}}$$

so the derivative function $D_x f$ is given by the rule $D_x f(x)$ $= 1/(2\sqrt{x})$. Note that 0 is not in the domain of $D_x f$, since $D_x f(0)$ does not exist. Thus the domain of $D_x f$ consists of all numbers greater than 0. $D_x f(0)$ does not exist because the line most suited to be the tangent line to the graph of $f(x) = \sqrt{x}$ at $(0, 0)$ is vertical and so has no slope (see Fig. 5.2-8).

EXAMPLE 5.2-6

Find the slope of the line tangent to the graph of $f(x) = \sqrt{x}$ at $P = (1, 1)$ and at $P = (2, \sqrt{2})$.

SOLUTION

In Example 5.2-5, we found that $D_x f(x) = 1/(2\sqrt{x})$. Thus the slope of the line tangent to the graph of $f(x) = \sqrt{x}$ at $P = (1, 1)$ is $D_x f(1) = 1/(2\sqrt{1}) = \frac{1}{2}$. Similarly, the slope of the line tangent to the graph at $P = (2, \sqrt{2})$ is $D_x f(2) = 1/(2\sqrt{2})$ (see Fig. 5.2-9).

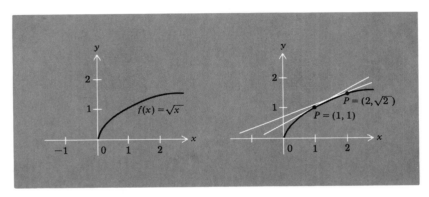

FIGURE 5.2-8   The graph of $f(x)$ $= \sqrt{x}$. The line tangent to the graph at $(0, 0)$ is the y axis, which is vertical and has no slope.

FIGURE 5.2-9

Computations similar to those used to find the derivatives of the functions in Examples 5.2-3 through 5.2-5 yield the rules shown in Table 5.2-1.

TABLE 5.2-1

	$f(x)$	$D_x f(x)$
(1)	$f(x) = ax + b$	$D_x f(x) = a$
(2)	$f(x) = x^n$ (n any integer)	$D_x f(x) = nx^{n-1}$
(3)	$f(x) = \sqrt{x}$	$D_x f(x) = 1/(2\sqrt{x})$

Since any constant function $f(x) = b$ is of the form $f(x) = ax + b$ with $a = 0$, we see from rule (1) of Table 5.2-1 that $D_x f(x) = a = 0$. Thus the derivative of any constant function $f(x) = b$ is the constant function $D_x f(x) = 0$.

The derivative of any constant function $f(x) = b$ is the constant function $D_x f(x) = 0$.

Recall the discussion of polynomial functions in Section 2.4.

For any integer n, $x^{-n} = 1/x^n$.

Rule (2) of Table 5.2-1 is of particular importance, for as we shall see in the next section, repeated applications of this rule will enable us to find the derivative of any polynomial function. Notice that this rule is valid for negative as well as positive integers n. Thus, since x^{-n} is simply a shorthand notation for $1/x^n$, rule (2) allows us to find the derivative of such functions as $h(x) = 1/x^2 = x^{-2}$ and $g(x) = 1/x^{10} = x^{-10}$, in addition to the familiar power functions $f(x) = x^n$, where n is a *positive* integer (see Section 2.4).

EXAMPLE 5.2-7

Find the derivatives of the functions $f(x) = x^3$, $h(x) = x^{-2}$, $s(u) = u^6$, and $g(z) = z^{-10}$.

SOLUTION

If we are given a function s with values $s(u)$, then the derivative of s is with respect to u (instead of x) and thus is denoted $D_u s$. Similarly, for a function g with values $g(z)$, the derivative is $D_z g$, and so on.

Using rule (2) of Table 5.2-1, we see that $D_x f(x) = 3x^2$, $D_x h(x) = -2x^{-3} = -2/x^3$, $D_u s(u) = 6u^5$, and $D_z g(z) = -10z^{-11} = -10/z^{11}$.

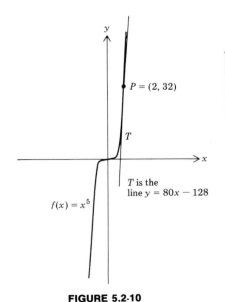

$f(x) = x^5$

T is the line $y = 80x - 128$

$P = (2, 32)$

T

FIGURE 5.2-10

EXAMPLE 5.2-8

Find the equation of the line tangent to the graph of $f(x) = x^5$ at the point $P = (2, 32)$.

SOLUTION

From rule (2) of Table 5.2-1, we see that the derivative of $f(x) = x^5$ is the function $D_x f(x) = 5x^4$. Thus the slope of the line T tangent to the graph of f at $P = (2, 32)$ is $D_x f(2) = 5(2)^4 = 80$. Since T passes through the point $P = (2, 32)$, the equation of T is $y - 32 = 80(x - 2)$, or $y = 80x - 128$ (see Fig. 5.2-10).

EXAMPLE 5.2-9

Find all points P on the graph of $f(x) = x^3$ such that the line tangent to the graph of f at P has slope 12.

SOLUTION

The slope of the line tangent to the graph of $f(x) = x^3$ at a point $P = (x_0, f(x_0))$ on the graph is $D_x f(x_0)$. Thus we wish to find values of x_0 such that $D_x f(x_0) = 12$. From Table 5.2-1 we have $D_x f(x) = 3x^2$, so we want values of x_0 such that $D_x f(x_0) = 3x_0^2 = 12$. Solving this equation, we have $x_0^2 = 4$, which means that $x_0 = 2$ or -2. Thus the points $P = (x_0, f(x_0))$ where the tangent line has slope 12 are the points $(2, f(2)) = (2, 8)$ and $(-2, f(-2)) = (-2, -8)$ (see Fig. 5.2-11).

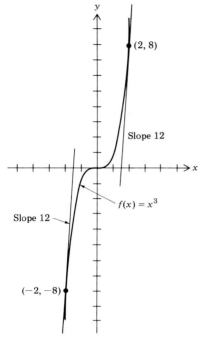

FIGURE 5.2-11

Exercises 5.2

In Exercises 1 through 25, use Table 5.2-1 to find the derivative of the given function. Then use the derivative to find the slope of the line tangent to the graph of the function at the given point P on the graph.

1. $f(x) = 5x + 2$, $P = (0, 2)$

2. $f(x) = 7x - 3$, $P = (1, 4)$

3. $f(x) = -3x + 2$, $P = (-1, 5)$

4. $f(x) = -10x - 7$, $P = (-1, 3)$

5. $f(x) = x$, $P = (4, 4)$

6. $f(x) = x - 1$, $P = (4, 3)$

7. $f(x) = \frac{1}{4}x + \frac{1}{2}$, $P = (2, 1)$

8. $f(x) = -\frac{7}{8}x - \frac{1}{4}$, $P = (-2, \frac{3}{2})$

9. $h(t) = 5t - 12$, $P = (2, -2)$

10. $g(z) = -3z - 14$, $P = (-3, -5)$

11. $w(s) = -\frac{3}{5}s - \frac{1}{5}$, $P = (-5, \frac{14}{5})$

12. $p(v) = (\frac{2}{3})v + \frac{1}{6}$, $P = (2, \frac{3}{2})$

13. $f(x) = x^4$, $P = (2, 16)$

14. $f(x) = x^5$, $P = (-2, -32)$

15. $h(t) = t^3$, $P = (\frac{1}{2}, \frac{1}{8})$

16. $s(w) = w^9$, $P = (-1, -1)$

17. $w(s) = s^2$, $P = (-\frac{1}{3}, \frac{1}{9})$

18. $z(t) = t^6$, $P = (-\frac{1}{2}, \frac{1}{64})$

19. $f(x) = x^{-3}$, $P = (2, \frac{1}{8})$

20. $f(x) = x^{-5}$, $P = (-1, -1)$

21. $g(w) = w^{-4}$, $P = (2, \frac{1}{16})$

22. $s(t) = t^{-7}$, $P = (1, 1)$

23. $p(v) = v^{-1}$, $P = (-3, -\frac{1}{9})$

24. $h(s) = \sqrt{s}$, $P = (16, 4)$

25. $g(z) = \sqrt{z}$, $P = (9, 3)$

In Exercises 26 through 38, find the equation of the line T tangent to the graph of the given function at the given point P on the graph. Sketch the graph of the function, locate the point P on the graph, and then sketch the graph of the line T on the same coordinate system.

26. $f(x) = x^3$, $P = (1, 1)$

27. $f(x) = \sqrt{x}$, $P = (4, 2)$

28. $f(x) = x^4$, $P = (-1, 1)$

29. $f(x) = x^2$, $P = (-2, 4)$

30. $f(x) = 3x - 1$, $P = (2, 5)$

31. $f(x) = -4x + 2$, $P = (-1, 6)$

32. $h(x) = \frac{1}{2}x + 2$, $P = (2, 3)$

33. $p(v) = -\frac{1}{2}v - 4$, $P = (-2, -3)$

34. $s(t) = \sqrt{t}$, $P = (9, 3)$

35. $f(x) = x^{-1}$, $P = (1, 1)$

36. $f(x) = x^{-2}$, $P = (2, \frac{1}{4})$

37. $w(z) = z^{-5}$, $P = (-1, -1)$

38. $z(t) = t^{-4}$, $P = (-1, 1)$

39. Find the point P on the graph of $f(x) = x^2$ where the slope of the tangent line is 4.

40. At what point P on the graph of $f(x) = \sqrt{x}$ is the slope of the tangent line $\frac{1}{4}$?

41. Find all points P on the graph of $f(x) = x^3$ where the tangent line has slope 27.

42. Let $f(x) = x^4$. Find the point P on the graph of f where the tangent line is parallel to the line L defined by the equation $g(x) = -4x + 10$.

Recall that parallel lines have the same slope.

43. Find the point P on the graph of $f(x) = \sqrt{x}$ where the tangent line is parallel to the line L defined by the equation $g(x) = \frac{1}{8}x - 2$.

In each of Exercises 44 through 49, the graph of a function f is shown. Determine from the graph (a) whether the function is continuous at x_0 and (b) whether the function is differentiable at x_0 (see Fig. 5.2-4 and Section 4.4).

Remember that a function f is not differentiable at x_0 (i.e., $D_x f(x_0)$ does not exist) if the graph of f has either a vertical tangent or a "corner" at the point $(x_0, f(x_0))$.

44.

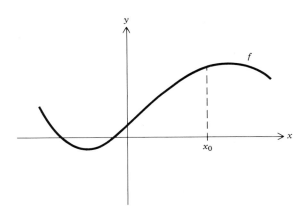

45.

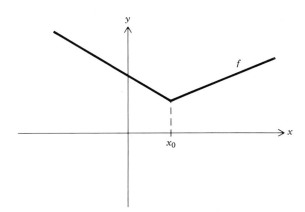

46.

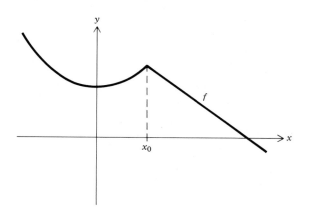

47.

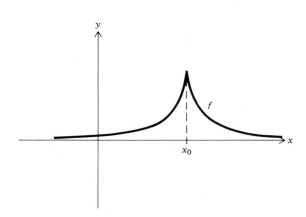

48.

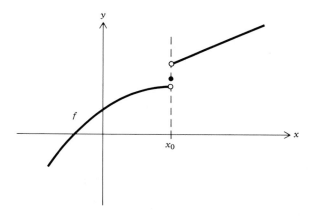

49.

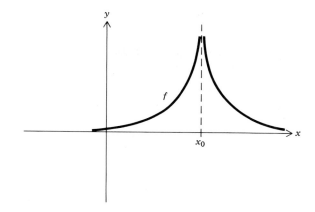

In Exercises 50 through 54, sketch the graph of the function f, and find the values x_0 for which $D_x f(x_0)$ does not exist.

50. $f(x) = \begin{cases} 2x + 1 & \text{if } x \leqslant 1 \\ 3x & \text{if } x > 1 \end{cases}$

51. $f(x) = \begin{cases} \sqrt{x} & \text{if } x \geqslant 0 \\ -x & \text{if } x < 0 \end{cases}$

52. $f(x) = \begin{cases} x^3 & \text{if } x \geqslant -1 \\ -3x - 4 & \text{if } x < -1 \end{cases}$

53. $f(x) = \begin{cases} x^2 & \text{if } x \leqslant 2 \\ 4 & \text{if } x > 2 \end{cases}$

54. $f(x) = \begin{cases} \sqrt{x} & \text{if } x \geqslant 1 \\ -1 & \text{if } x < 1 \end{cases}$

5.3.   Rate of Change

As we mentioned earlier, the derivative enables us to solve problems involving *rate of change*. Indeed, we shall soon discover that the technique of interpreting the derivative as a rate of change is of utmost importance in applications. Let us begin by discussing the concept of rate of change; then we will investigate its connection with the derivative.

Change is something we encounter consistently in our daily lives: Our mood, our ideals, and our salaries change, as does the weather, the temperature, the population of the world, and the time of day. Virtually everything is subject to change. In Table 5.3-1 we have recorded the daily weight of a man, whom we will call Sam, over a two-week period. Sam's weight is a function w of time t (measured in days), and the table gives the value of w at t for $t = 1$ through $t = 14$. The change in Sam's

TABLE 5.3-1

t (days)	1	2	3	4	5	6	7	8	9	10	11	12	13	14
$w(t)$ (pounds)	164	164	165	165	167	169	170	171	172	175	175	179	180	182

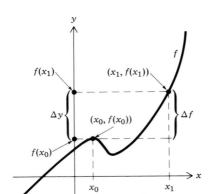

FIGURE 5.3-1   Note that Δf = $f(x_1) - f(x_0)$ geometrically corresponds to distance on the y axis. For this reason Δf is sometimes denoted Δy.

weight from the first day to the fifth day is $w(5) - w(1) = 167 - 164$ = 3 lbs, and from the third day to the ninth day it is $w(9) - w(3)$ = $172 - 165 = 7$ lbs; his total weight change from the first to the fourteenth day is $w(14) - w(1) = 182 - 164 = 18$ lbs. In general, given a function f with values $f(x)$, the *change in f* from $x = x_0$ to $x = x_1$ (where x_0 and x_1 are in the domain of f) is the difference $f(x_1) - f(x_0)$. The difference $x_1 - x_0$, often called the *increment in x*, is denoted by Δx, whereas the difference $f(x_1) - f(x_0)$, called the *increment in f*, is denoted by Δf (see Fig. 5.3-1).

EXAMPLE 5.3-1

Given $f(x) = 2x^2 + 1$, find the increments Δx and Δf for (a) $x_0 = 0$ and $x_1 = 1$ and (b) $x_0 = 1$ and $x_1 = 4$.

SOLUTION

(a) Since $\Delta x = x_1 - x_0$, we have $\Delta x = 1 - 0 = 1$. Furthermore, $\Delta f = f(x_1) - f(x_0) = f(1) - f(0) = (2(1)^2 + 1) - (2(0)^2 + 1) = 3 - 1 = 2$ (see Fig. 5.3-2).

(b) $\Delta x = x_1 - x_0 = 4 - 1 = 3$, and $\Delta f = f(x_1) - f(x_0)$ = $f(4) - f(1) = (2(4)^2 + 1) - (2(1)^2 + 1) = 33 - 3 = 30$ (see Fig. 5.3-3).

Although change is an important phenomenon in our daily lives, it should be evident that the *rate* at which things change is equally important. Our friend Sam might not mind gaining 18 lb over a period of 10 or 15 years, but he must find it alarming to gain 18 lb in two weeks! Similarly, we all expect the world population to double eventually, but it would be altogether disquieting to learn that it would double by tomorrow.

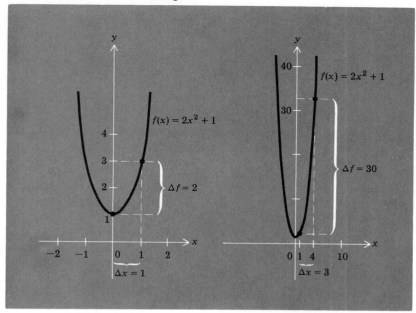

FIGURE 5.3-2   **FIGURE 5.3-3**

To compute the *average* rate of change of Sam's weight from the first to the fifth day (see Table 5.3-1), we divide his change in weight over this period by the number of days during which this change occurred; thus we obtain $[w(5) - w(1)]/(5 - 1) = (167 - 164)/4 = \frac{3}{4}$ lb per day. Similarly, the average rate of change of Sam's weight from the third day to the ninth day is $[w(9) - w(3)]/(9 - 3) = (172 - 165)/6 = \frac{7}{6}$ lb per day, and from the first to the fourteenth day it is $[w(14) - w(1)]/(14 - 1) = (182 - 164)/13 = \frac{18}{13}$ lb per day. In general, given a function f with values $f(x)$, the *average rate of change of f with respect to x*, as x changes from x_0 to x_1, is the quotient of the two differences $f(x_1) - f(x_0)$ and $x_1 - x_0$, that is,

$$\frac{f(x_1) - f(x_0)}{x_1 - x_0}$$

This quotient is usually referred to as the *difference quotient.*

Average rate of change

The number $[f(x_1) - f(x_0)]/(x_1 - x_0)$ is the average rate of change of f with respect to x as x changes from x_0 to x_1.

Since Δy is sometimes used to denote $\Delta f = f(x_1) - f(x_0)$ and since $\Delta x = x_1 - x_0$, alternative notations for the difference quotient $[f(x_1) - f(x_0)]/(x_1 - x_0)$ are $\Delta y/\Delta x$ and $\Delta f/\Delta x$.

EXAMPLE 5.3-2

Given $f(x) = 2x^2 + 1$, find the average rate of change of f with respect to x as x changes from (a) $x_0 = 0$ to $x_1 = 1$ and (b) $x_0 = 1$ to $x_1 = 4$.

SOLUTION

(a) For $x_0 = 0$ and $x_1 = 1$, we have $x_1 - x_0 = 1$ and $f(x_1) - f(x_0) = 2$ (as found in Example 5.3-1(a)). Thus the average rate of change of f with respect to x as x changes from 0 to 1 is $[f(x_1) - f(x_0)]/(x_1 - x_0) = \frac{2}{1} = 2$.

(b) For $x_0 = 1$ and $x_1 = 4$, we have $f(x_1) - f(x_0) = 30$ (as found in Example 5.3-1(b)) and $x_1 - x_0 = 3$. Thus $[f(x_1) - f(x_0)]/(x_1 - x_0) = \frac{30}{3} = 10$, and this is the average rate of change of f with respect to x as x changes from 1 to 4.

The average rate of change of production cost with respect to the number x of units produced is called the *average marginal cost*.

EXAMPLE 5.3-3

The Futura Furniture Manufacturing Company estimates that the cost C (in dollars) of manufacturing x units of its executive model office desk is given by $C(x) = 100x + (200/x)$. Find the average rate of change of C with respect to x as x changes from $x_0 = 10$ to $x_1 = 20$.

SOLUTION

Since $x_0 = 10$ and $x_1 = 20$, we have $x_1 - x_0 = 20 - 10 = 10$. Moreover, $C(x_0) = C(10) = 100(10) + \frac{200}{10} = 1020$ and $C(x_1) = C(20) = 100(20) + 200/20 = 2010$, so $C(x_1) - C(x_0) = 2010 - 1020 = 990$. Thus $[C(x_1) - C(x_0)]/(x_1 - x_0) = 990/10 = 99$, which means that the average rate of change of C with respect to x as x changes from $x_0 = 10$ to $x_1 = 20$ is $99 per unit (desk) manufactured.

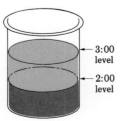

— 3:00 level

— 2:00 level

Now we are ready to investigate the relationship of the derivative to the concept of rate of change. Suppose it is 2:00 PM on a rainy day and we wish to determine in some way the rate at which the rain is falling. If there is a pan of water outside and we find that the level of the water in the pan has risen 2 in. by 3:00 PM, we might be tempted to say that the rate of rainfall is 2 in. per hour. However, this is not necessarily the rate of rainfall *at* 2:00 PM, it is only the *average* rate of rainfall between 2:00 PM and 3:00 PM; it may well have happened that $1\frac{1}{2}$ in. of water collected in the pan between 2:00 and 2:15 and the remaining $\frac{1}{2}$ in. of water resulted from a 45-min drizzle. A more reliable estimate of the rate of rainfall *at* 2:00 would be obtained by checking the water level at

2:30 and finding the average rainfall rate between 2:00 and 2:30; in this way we would exclude all information about the rainfall between 2:30 and 3:00, which has nothing to do with the rate of rainfall at 2:00. The average rate of rainfall between 2:00 and 2:15 would be a still more reliable estimate.

We can easily formalize this line of thought by letting $l(x)$ be the level of the water in the pan at time x. Then the average rainfall between 2:00 and 3:00 is $(l(3) - l(2))/(3 - 2)$, and between 2:00 and 2:30 it is $(l(2.5) - l(2))/(2.5 - 2)$. Thus, in general, the average rainfall between 2:00 and any time x is $(l(x) - l(2))/(x - 2)$; we will get better and better estimates of the rate of rainfall *at* 2:00 by choosing values of x (times of the day) closer and closer to 2:00. This suggests that the best way to determine the rate of rainfall exactly *at* 2:00 (the *instantaneous* rate of rainfall at 2:00) is to calculate

$$\lim_{x \to 2} \frac{l(x) - l(2)}{x - 2}$$

which is the derivative with respect to x of the function $l(x)$ at $x = 2$, i.e., $D_x l(2)$.

Considerations of this kind lead us to make the following definition. Given a function f with values $f(x)$, the *instantaneous rate of change* of f with respect to x at $x = x_0$ is the value of the derivative (with respect to x) of f at x_0, i.e., $D_x f(x_0)$.

> The rate of rainfall *at* 2:00 might be called the *instantaneous* rate. This could be quite different from the *average* rate between 2:00 and 3:00.

> The phrase "instantaneous rate of change" is often shortened to "rate of change." Following this convention, we agree that "rate of change" will always mean instantaneous rate of change unless it is specifically said to mean average rate of change.

EXAMPLE 5.3-4

Given $f(x) = x^5$, find the instantaneous rate of change of f with respect to x at $x_0 = 2$ and $x_0 = -3$.

SOLUTION

Using rule (2) of Table 5.2-1, we see that $D_x f(x) = 5x^4$. Thus the instantaneous rate of change of f with respect to x at $x_0 = 2$ is $D_x f(2) = 5(2)^4 = 5(16) = 80$; at $x_0 = -3$ it is $D_x f(-3) = 5(-3)^4 = 5(81) = 405$.

> $D_x f(x)$ gives the rate of change of $f(x)$ per unit of x. Thus if $f(x)$ is given in pounds and x in days, then $D_x f(x)$ is given in pounds per day. If $f(x)$ is in grams and x in weeks, then $D_x f(x)$ is given in grams per week.

EXAMPLE 5.3-5

Suppose that the function $f(x) = 5x$ gives the distance f (in miles) that a bee has flown from a flower back toward its hive x hr after leaving the flower; the distance the bee has traveled

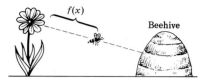

$f(x)$

Beehive

since leaving the flower is thus given as a function of the time elapsed since it left the flower. Find the speed (instantaneous rate of change in distance traveled with respect to time) of the bee $\frac{1}{2}$ hr after it leaves the flower.

SOLUTION

The speed of the bee will be the instantaneous rate of change of f (distance traveled) with respect to x (time) at $x = 0.5$, which is given by $D_x f(0.5)$. From rule (1) of Table 5.2-1, $D_x f$ is the constant function $D_x f(x) = 5$, so for $x = 0.5$, we have $D_x f(0.5) = 5$ mph.

In economic theory, the instantaneous rate of change of the production cost C with respect to the number x of items produced, $D_x C$, is called the *marginal cost*. Marginal cost approximates the additional cost necessary to produce an extra item.

Actually, the domain of the function C consists only of integers. However, in order to apply the techniques of differentiation, we assume that C is defined for nonintegral values of x as well. Occasionally, this leads to results which must be interpreted rather than taken literally.

EXAMPLE 5.3-6

The Crinkle-Care Cosmetics Company estimates that the cost C(in dollars)of producing x bottles of its moisturizing lotion is given by $C(x) = 2x + 10$. Find the instantaneous rate of change of C with respect to x (the *marginal cost*) at $x = 40$.

SOLUTION

Since $C(x) = 2x + 10$, we see from rule (1) of Table 5.2-1 that $D_x C$ is the constant function $D_x C(x) = 2$. Thus the instantaneous rate of change of C with respect to x at $x = 40$ is $D_x C(40) = \$2$ per bottle. This means that if the company were to produce 41 bottles of lotion instead of 40 bottles, the production cost C would increase by $2.

EXAMPLE 5.3-7

After a flu epidemic in the city of Pleasantville, public health officials determined that the percent P of the city's population infected by the disease t days after the first reported case was closely approximated by the equation $P(t) = t^2$ for $0 \leqslant t \leqslant 7$ days. Find the instantaneous rate of change of P with respect to t at $t = 3$ days and at $t = 6$ days.

SOLUTION

From rule (2) of Table 5.2-1, we see that $D_t P(t) = 2t$. Thus the instantaneous rate of change of P with respect to t at $t = 3$ is $D_t P(3) = 2(3) = 6\%$ of the population per day; at $t = 6$ it is $D_t P(6) = 2(6) = 12\%$ of the population per day.

The fact that the rate of change of P with respect to t at $t = 3$ is 6% per day, means that on the third day of the epidemic, the disease is spreading at the rate of 6% of the population per day.

EXAMPLE 5.3-8

It is estimated that the total quantity Q (in hundreds of gallons) of liquid chemical waste that will have been discharged into a lake by a newly constructed chemical manufacturing plant during the first t days of plant operation will be closely approximated by the equation $Q(t) = \sqrt{t}$ for $0 \leqslant t \leqslant 100$ days. Find the rate at which the amount of chemical waste in the lake will be increasing (the instantaneous rate of change of Q with respect to t) at $t = 25$ days after the plant begins operation and at $t = 100$ days after plant operation begins.

SOLUTION

Since $Q(t) = \sqrt{t}$, we see from rule (3) of Table 5.2-1 that $D_t Q(t) = 1/(2\sqrt{t}\,)$. Thus the instantaneous rate of change of Q with respect to t at $t = 25$ is $D_t Q(25) = 1/(2\sqrt{25}\,) = \frac{1}{10}$ hundreds of gallons or 10 gal per day. At $t = 100$ days, the rate at which the amount of chemical in the lake will be increasing will be $D_t Q(100) = 1/(2\sqrt{100}\,) = 1/(2(10)) = \frac{1}{20}$ hundreds of gallons or 5 gal per day.

Exercises 5.3

In Exercises 1 through 10, find the increments $\Delta x = x_1 - x_0$ and $\Delta f = f(x_1) - f(x_0)$ for the given function f and the given values x_0 and x_1.

1. $f(x) = 2x + 3$, $x_0 = 2$, $x_1 = 5$

2. $f(x) = -3x + 4$, $x_0 = 0$, $x_1 = 6$

3. $f(x) = 2x^2 - 4$, $x_0 = -2$, $x_1 = 4$

4. $f(x) = 3x^2 + 2$, $x_0 = -2$, $x_1 = 2$

5. $f(x) = -x^3 - 2$, $x_0 = -2$, $x_1 = 1$

6. $f(x) = \sqrt{x + 1}$, $x_0 = 3$, $x_1 = 8$

7. $f(x) = \sqrt{4 - x^2}$, $x_0 = 0$, $x_1 = 2$

8. $f(x) = |2x - 6|$, $x_0 = 1$, $x_1 = 4$

9. $f(x) = x^{-2} + 2$, $x_0 = 1$, $x_1 = 3$

10. $f(x) = (x + 4)^{-2}$, $x_0 = -3$, $x_1 = 2$

Recall that the difference quotient $\Delta f/\Delta x = [f(x_1) - f(x_0)]/(x_1 - x_0)$ gives the average rate of change of f with respect to x as x changes from x_0 to x_1.

In Exercises 11 through 22, find the average rate of change of the given function f with respect to x as x changes from the given value x_0 to the given value x_1.

11. $f(x) = 4x - 4$, $x_0 = 0$, $x_1 = 4$

12. $f(x) = -x + 6$, $x_0 = 2$, $x_1 = 4$

13. $f(x) = 2x^2 - 3$, $x_0 = 1$, $x_1 = 3$

14. $f(x) = -3x^2 + 2$, $x_0 = -1$, $x_1 = 4$

15. $f(x) = x^2 + 2x + 4$, $x_0 = 0$, $x_1 = 5$

16. $f(x) = 3x^2 - x + 1$, $x_0 = -2$, $x_1 = 2$

17. $f(x) = |x - 3|$, $x_0 = \frac{1}{2}$, $x_1 = 2$

18. $f(x) = |-4x + 1|$, $x_0 = -1$, $x_1 = \frac{3}{2}$

19. $f(x) = x^{-2} + 4$, $x_0 = 1$, $x_1 = 4$

20. $f(x) = x^{-4} - 6$, $x_0 = 1$, $x_1 = 2$

21. $f(x) = (x + 1)^{-3}$, $x_0 = -3$, $x_1 = -2$

22. $f(x) = (-2x + 4)^{-2}$, $x_0 = -2$, $x_1 = 0$

Given a function g with values $g(t)$, the average rate of change of g with respect to t as t changes from t_0 to t_1 is $\Delta g/\Delta t = [g(t_1) - g(t_0)]/(t_1 - t_0)$.

In Exercises 23 through 30, find the average rate of change of the given function.

23. $g(t) = t^2 + 2t$, $t_0 = 1$, $t_1 = 4$

24. $s(z) = -2z + 3$, $z_0 = -2$, $z_1 = 2$

25. $p(v) = |v^2 - v|$, $v_0 = -2$, $v_1 = 3$

26. $w(t) = |2t - t^2|$, $t_0 = -1$, $t_1 = 2$

27. $r(z) = 4z^{-2} - 3$, $z_0 = 2$, $z_1 = 4$

28. $h(t) = -2t^{-1} + 1$, $t_0 = \frac{1}{2}$, $t_1 = 3$

29. $v(s) = (2 - s)^{-2}$, $s_0 = -\frac{3}{2}$, $s_1 = 1$

30. $p(t) = |4t^2 - 10|^{-2}$, $t_0 = -\frac{1}{2}$, $t_1 = \frac{3}{2}$

In Exercises 31 through 45, find the instantaneous rate of change of the given function at the given value of its variable.

31. $f(x) = x^3, x_0 = 2$

32. $f(x) = x^{-2}, x_0 = 4$

33. $f(x) = -4x + 1, x_0 = -1$

34. $f(x) = 10x + 4, x_0 = \frac{1}{2}$

35. $f(x) = -7, x_0 = \frac{1}{4}$

36. $f(x) = \sqrt{x}, x_0 = 4$

37. $f(x) = \sqrt{x}, x_0 = \frac{1}{4}$

38. $g(t) = t^5, t_0 = -1$

39. $h(z) = z^{-4}, z_0 = 2$

40. $s(w) = -10, w = \frac{1}{16}$

41. $p(t) = -\frac{1}{2}t - \frac{1}{4}, t_0 = \frac{1}{8}$

42. $v(s) = \frac{1}{5}s, s_0 = -5$

43. $w(t) = t^{-3}, t_0 = -2$

44. $g(v) = v^4, v_0 = -\frac{1}{2}$

45. $r(w) = w^{-5}, w_0 = \frac{1}{2}$

Given a function g with values $g(t)$, the instantaneous rate of change of g with respect to t at $t = t_0$ is $D_t g(t_0)$.

46. A large mass of polluted air is observed spreading westward from an industrial city according to the equation $d(t) = \sqrt{t}$, where d is the distance in miles from the center of the city to the west edge of the air mass and t is time measured in hours. The equation is considered valid for values of t between 25 and 144 inclusive. Find the average rate of change of d with respect to t as t changes from $t_0 = 25$ to $t_1 = 64$ (this is called the *average speed* of the air mass). What is the speed (instantaneous rate of change of d with respect to t) of the air mass at time $t = 81$?

47. The Majestic Mirror Manufacturing Company estimates that the cost C (in dollars) of producing x units of its deluxe hand mirror is given by $C(x) = 3x + 80$. Find the average rate of change of C with respect to x (the average marginal cost) as x changes from $x_0 = 50$ to $x_1 = 100$. What is the marginal cost (instantaneous rate of change of C with respect to x) at $x = 80$?

48. Medical authorities estimate that t days after a community is exposed to a certain communicable disease, the percent p of the population infected by the disease is approximated by the equation $P(t) = t^3$ for $0 \leqslant t \leqslant 4$ days. Find (a) the average rate of change of P with respect to t as t changes from $t_0 = 1$ to $t_1 = 3$ and (b) the instantaneous rate of change of P with respect to t at $t = 2$.

The average rate of change of demand with respect to the selling price x of a product is called the *average marginal demand*. The instantaneous rate of change of demand d with respect to selling price x, i.e., $D_x d$, is called the *marginal demand*.

49. During a recent partial breakdown of its sewage treatment plant, the city of Euphoria discharged raw sewage into a nearby lake. It was estimated that the total quantity Q of sewage discharged into the lake (in thousands of gallons) during the t days immediately following the breakdown was approximated by $Q(t) = t^2$, $0 \leqslant t \leqslant 30$ days. Find (a) the average rate of change of Q with respect to t as t changes from $t_0 = 10$ to $t_1 = 20$ and (b) the instantaneous rate of change of Q with respect to t at $t = 10$.

50. The population P (in thousands) of a certain bacteria culture is given by the equation $P(t) = t^4$ where t is measured in hours and $1 \leqslant t \leqslant 5$. Find (a) the average rate of change of P with respect to t as t changes from $t_0 = 1$ to $t_1 = 3$ and (b) the instantaneous rate of change of P with respect to t at $t = 3$.

51. Automatic Sharpener, Inc., estimates that the demand function for its best electric pencil sharpener is $d(x) = 40 - 2x$, where d is the number of thousands of sharpeners that can be sold per year and x is the selling price of the sharpener in dollars. The equation is considered valid for values of x between \$8 and \$15 inclusive. Find (a) the average rate of change of d with respect to x (called the *average marginal demand*) as x changes from $x_0 = 10$ to $x_1 = 14$ and (b) the instantaneous rate of change of d with respect to x (called the *marginal demand*) at $x = 10$.

52. The Purity Pet Food Company estimates that its total sales revenue S, in dollars per week, is given by $S(x) = 2500 + 100x$, where x is the number of 30-sec television advertisements for its product shown per week. The equation is considered valid for values of x between 20 and 100 inclusive. Find (a) the average rate of change of S with respect to x as x changes from $x_0 = 20$ to $x_1 = 40$ and (b) the instantaneous rate of change of S with respect to x at $x = 40$.

5.4.   Differentiation Techniques and Higher-Order Derivatives

The two interpretations of the derivative discussed thus far are (1) the slope of the tangent line and (2) rate of change.

We now have defined the derivative $D_x f$ of a function f and discussed two of its interpretations, but as yet we have a rather meager supply of functions whose derivatives we can compute (see Table 5.2-1). However, as we shall soon see, there are rules governing the derivative of the sum, difference, product, and quotient of two functions that provide efficient means for finding the derivative of a wide variety of functions.

Suppose, for example, that a swimming pool is being drained and that the number G of gallons of water remaining in the pool t min after draining begins is given by the equation $G(t) = 2500 - 100t + t^2$. Then in order to find the rate of change of G with respect to t (the *drainage rate* at time t) we need to find $D_t G(t)$, but this derivative cannot be computed from Table 5.2-1. However, $G(t) = f(t) + g(t)$, where $f(t) = 2500 - 100t$ and $g(t) = t^2$, and we already know how to compute $D_t f(t)$ and $D_t g(t)$. Thus, if we knew that the derivative of a sum of two functions was the sum of the derivatives of each, we could write $D_t G(t) = D_t[f(t) + g(t)] = D_t f(t) + D_t g(t) = -100 + 2t$, using Table 5.2-1 to compute $D_t f(t) = -100$ and $D_t g(t) = 2t$. Fortunately, it is indeed the case that the derivative of the sum of two functions is the sum of their derivatives; moreover, other rules exist for the derivatives of differences, products, and quotients of functions. These rules are given in Tables 5.4-1 and 5.4-2, along with two additional frequently used rules of differentiation. Proofs of rules (3), (5), and (6) are given in Examples 5.4-12, 5.4-13, and 5.4-14.

TABLE 5.4-1

(1) For any constant function $f(x) = c$, $D_x f(x) = 0$
(2) For any real number c, $D_x[cf(x)] = cD_x f(x)$
(3) $D_x[f(x) + g(x)] = D_x f(x) + D_x g(x)$
(4) $D_x[f(x) - g(x)] = D_x f(x) - D_x g(x)$

EXAMPLE 5.4-1

Given $f(x) = 3x^2 - \sqrt{x}$, find $D_x f(x)$.

SOLUTION

$$D_x f(x) = D_x(3x^2 - \sqrt{x})$$

$$= D_x(3x^2) - D_x(\sqrt{x}) \qquad \text{(rule (4))}$$

$$= 3D_x(x^2) - D_x(\sqrt{x}) \qquad \text{(rule (2))}$$

$$= 3(2x) - 1/(2\sqrt{x}) = 6x - 1/(2\sqrt{x})$$

When f is defined by an equation such as $f(x) = 3x^2 - \sqrt{x}$, we write $D_x(3x^2 - \sqrt{x})$ for $D_x f(x)$.

Recall from Table 5.2-1 that $D_x(x^2) = 2x$ and $D_x(\sqrt{x}) = 1/(2\sqrt{x})$.

EXAMPLE 5.4-2

Given $f(x) = 5x^{-3} - \frac{1}{4}x^2 + 2$, find $D_x f(x)$.

SOLUTION

$$D_x f(x) = D_x \left(5x^{-3} - \tfrac{1}{4}x^2 + 2 \right)$$

$$= D_x \left[\left(5x^{-3} - \tfrac{1}{4}x^2 \right) + 2 \right]$$

$$= D_x \left(5x^{-3} - \tfrac{1}{4}x^2 \right) + D_x(2) \qquad \text{(rule (3))}$$

$$= D_x \left(5x^{-3} \right) - D_x \left(\tfrac{1}{4}x^2 \right) + D_x(2) \qquad \text{(rule (4))}$$

$$= 5 D_x \left(x^{-3} \right) - \tfrac{1}{4} D_x \left(x^2 \right) + 0 \qquad \text{(rules (1) and (2))}$$

$$= 5(-3x^{-4}) - \tfrac{1}{4}(2x)$$

$$= -15x^{-4} - \tfrac{1}{2}x$$

Notice that rule (3) of Table 5.4-1 may be used repeatedly to show that the derivative of the sum of any finite number of functions is the sum of the derivatives of the functions. For example, $D_x[f(x) + g(x) + h(x)] = D_x[(f(x) + g(x)) + h(x)] = D_x[f(x) + g(x)] + D_x h(x)$ (rule (3)) $= D_x f(x) + D_x g(x) + D_x h(x)$ (rule (3)). This extended version of rule (3), together with other rules from Tables 5.2-1 and 5.4-1, enables us to find the derivative of any polynomial function p. For if

Recall from Section 2.4 that a polynomial function p is any function defined by an equation of the form $p(x) = a_n x^n + a_{n-1}x^{n-1} + \cdots + a_1 x + a_0$, where $a_n, a_{n-1}, \ldots, a_0$ are fixed real numbers with $a_n \neq 0$.

$$p(x) = a_n x^n + a_{n-1}x^{n-1} + \cdots + a_1 x + a_0$$

then

$$D_x p(x) = D_x \left(a_n x^n + a_{n-1}x^{n-1} + \cdots + a_1 x + a_0 \right)$$

$$= D_x \left(a_n x^n \right) + D_x \left(a_{n-1}x^{n-1} \right) + \cdots + D_x(a_1 x) + D_x(a_0)$$

$$= a_n D_x \left(x^n \right) + a_{n-1} D_x \left(x^{n-1} \right) + \cdots + a_1 D_x(x) + D_x(a_0)$$

From Table 5.2-1, $D_x(x) = D_x(1 \cdot x + 0) = 1$.

$$= a_n n x^{n-1} + a_{n-1}(n-1)x^{n-2} + \cdots + a_1$$

EXAMPLE 5.4-3

Given $p(x) = 2x^5 - 5x^4 + x^2 + 3x + 1$, find $D_x p(x)$.

SOLUTION

$$D_x p(x) = D_x(2x^5 - 5x^4 + x^2 + 3x + 1)$$

$$= D_x(2x^5) + D_x(-5x^4) + D_x(x^2) + D_x(3x) + D_x(1)$$

$$= 2D_x(x^5) + (-5)D_x(x^4) + D_x(x^2) + 3D_x(x) + D_x(1)$$

$$= 2(5x^4) - 5(4x^3) + 2x + 3(1)$$

$$= 10x^4 - 20x^3 + 2x + 3.$$

EXAMPLE 5.4-4

Given $f(t) = 7t^8 - 4t^3 + 3\sqrt{t} + 1$, find $D_t f(t)$.

SOLUTION

Note that f is not a polynomial function, because of the presence of the term $3\sqrt{t}$. However, we can still use the extended version of rule (3) of Table 5.4-1 to find the derivative of f.

$$D_t f(t) = D_t(7t^8 - 4t^3 + 3\sqrt{t} + 1)$$

$$= D_t(7t^8) + D_t(-4t^3) + D_t(3\sqrt{t}) + D_t(1)$$

$$= 7D_t(t^8) + (-4)D_t(t^3) + 3D_t(\sqrt{t}) + D_t(1)$$

$$= 7(8t^7) + (-4)(3t^2) + 3\left(\frac{1}{2\sqrt{t}}\right) + 0$$

$$= 56t^7 - 12t^2 + \frac{3}{2\sqrt{t}}$$

TABLE 5.4-2

(5) $D_x[f(x)g(x)] = f(x)D_x g(x) + g(x)D_x f(x)$
(6) $D_x[f(x)/g(x)] = [g(x)D_x f(x) - f(x)D_x g(x)]/[g(x)]^2, \ (g(x) \neq 0)$

Notice in Table 5.4-2 that the derivative of the product (or quotient) of two functions is *not* the product (or quotient) of the derivatives of the functions.

EXAMPLE 5.4-5

Given $h(t) = t^3/\sqrt{t}$, find $D_t h(t)$.

SOLUTION

$$D_t h(t) = D_t \frac{t^3}{\sqrt{t}}$$

$$= \frac{\sqrt{t}\, D_t(t^3) - t^3 D_t \sqrt{t}}{(\sqrt{t})^2} \qquad \text{(rule (6))}$$

$$= \frac{\sqrt{t}\,(3t^2) - t^3\left(1/(2\sqrt{t})\right)}{(\sqrt{t})^2}$$

$$= \frac{3t^2\sqrt{t} - (t^3/2\sqrt{t})}{t} = 3t\sqrt{t} - \frac{t^2}{2\sqrt{t}}$$

EXAMPLE 5.4-6

Given $g(z) = \sqrt{z}\,(3z + 1)$, find $D_z g(z)$.

SOLUTION

$$D_z g(z) = D_z\left(\sqrt{z}\,(3z+1)\right)$$

$$= \sqrt{z}\, D_z(3z+1) + (3z+1)D_z(\sqrt{z}) \qquad \text{(rule (5))}$$

From Table 5.2-1, $D_z(3z+1)$ = 3.

$$= \sqrt{z}\,(3) + (3z+1)\frac{1}{2\sqrt{z}} = \frac{6(\sqrt{z})^2 + 3z + 1}{2\sqrt{z}}$$

$$= \frac{9z+1}{2\sqrt{z}}$$

EXAMPLE 5.4-7

A group of medical researchers finds that an injection of x cc of an experimental drug causes a change in temperature of $R°$ Fahrenheit in the human body, where R is closely approximated by the equation $R(x) = x^2(2 - (x/3))$ for $0 \leqslant x \leqslant 2$. Find the rate of change of R with respect to x at $x = 1.5$.

In Example 5.4-7, the function R is called the *reaction* to the injection of x cc of the drug, and the rate of change of R with respect to x, $D_x R$, is called the *sensitivity* of the drug.

SOLUTION

The rate of change of R with respect to x at $x = 1.5$ is the value of $D_x R(x)$ at $x = 1.5$. Using rule (5) of Table 5.4-2, we have

$$D_x R(x) = D_x \left[x^2 \left(2 - \frac{x}{3} \right) \right]$$

$$= x^2 D_x \left(2 - \frac{x}{3} \right) + \left(2 - \frac{x}{3} \right) D_x(x^2)$$

$$= x^2 \left[D_x(2) - D_x \left(\frac{x}{3} \right) \right] + \left(2 - \frac{x}{3} \right) D_x(x^2)$$

$$= x^2 \left(0 - \frac{1}{3} \right) + \left(2 - \frac{x}{3} \right)(2x)$$

$$= -\frac{x^2}{3} + 4x - \frac{2x^2}{3}$$

$$= 4x - x^2$$

Thus the rate of change of R with respect to x at $x = 1.5$ is $D_x R(1.5) = 4(1.5) - (1.5)^2 = 3.75°$ per cubic centimeter.

$D_x R(1.5) = 3.75°$ per cubic centimeter can be interpreted to mean that if 1 cc of the drug is administered immediately after a dose of 1.5 cc is injected, there will be a rise in temperature of 3.75° *in addition to the temperature change resulting from the 1.5-cc injection.*

Although the rules given in Tables 5.2-1, 5.4-1, and 5.4-2 enable us to compute the derivatives of many functions, they do not provide us with the means to find the derivatives of the exponential and natural logarithm functions introduced in Section 4.5. The rules for finding the derivatives of these two functions are given in Table 5.4-3.

TABLE 5.4-3

(1)	$D_x \exp(x) = \exp(x)$
(2)	$D_x \ln(x) = 1/x$

Since $\exp(x) = e^x$, rule (1) of Table 5.4-3 can also be written as $D_x e^x = e^x$.

EXAMPLE 5.4-8

Given $f(x) = x^2/\exp(x)$, find $D_x f(x)$.

SOLUTION

Using rule (6) of Table 5.4-2 and rule (1) of Table 5.4-3, we have

$$D_x f(x) = D_x \left(\frac{x^2}{\exp(x)} \right)$$

$$= \frac{\exp(x) D_x(x^2) - x^2 D_x \exp(x)}{(\exp(x))^2}$$

$$= \frac{2x \exp(x) - x^2 \exp(x)}{(\exp(x))^2}$$

$$= \frac{\exp(x)(2x - x^2)}{(\exp(x))^2}$$

$$= \frac{2x - x^2}{\exp(x)}$$

EXAMPLE 5.4-9

Given $h(t) = t^2 \ln(t)$, find $D_t h(t)$.

SOLUTION

Using rule (5) of Table 5.4.2 and rule (2) of Table 5.4.3, we have $D_t h(t) = D_t[t^2 \ln(t)] = t^2 D_t \ln(t) + \ln(t) D_t(t^2) = t^2(1/t) + (\ln(t))2t = t + 2t \ln(t)$.

There are frequent occasions when it is profitable to compute not only the derivative of a function but also the derivative of the derivative. This gives rise to what are termed *higher-order derivatives* of a function. No extension of technique is required here, but we must introduce some new notation. Given a function f with values $f(x)$, the derivative (first-order) of f is denoted $D_x f$, the second-order derivative (the derivative of $D_x f$) is denoted $D_x^2 f$, the third-order derivative (the derivative of $D_x^2 f$) is denoted $D_x^3 f$, and in general, the nth-order derivative (obtained by starting with f and taking n derivatives) is denoted $D_x^n f$. We illustrate the process of finding higher-order derivatives in the following examples.

Alternative notations for $D_x^2 f$ are f'' and $d^2 f / dx^2$; for $D_x^3 f$, f''' and $d^3 f / dx^3$; and for $D_x^n f$, $f^{(n)}$ and $d^n f / dx^n$.

EXAMPLE 5.4-10

Find the first-order, second-order, third-order, and fourth-order derivatives of the function f defined by $f(x) = 2x^3 - 5x^2 + 7$.

SOLUTION

The (first-order) derivative is $D_x f(x) = D_x(2x^3 - 5x^2 + 7)$ $= 2D_x(x^3) - 5D_x(x^2) + D_x(7) = 6x^2 - 10x$. Now, the second-order derivative $D_x^2 f$ is the derivative of $D_x f$, i.e., $D_x^2 f(x)$ $= D_x[D_x f(x)] = D_x(6x^2 - 10x) = 6D_x(x^2) - 10D_x(x) = 12x$ $- 10$. To obtain the third-order derivative $D_x^3 f$, we simply find the derivative of $D_x^2 f$, i.e., $D_x^3 f(x) = D_x[D_x^2 f(x)] = D_x(12x -$ $10) = 12D_x(x) - D_x(10) = 12$. Finally, $D_x^4 f(x) = D_x[D_x^3 f(x)]$ $= D_x(12) = 0$. It is perhaps worth noting that $D_x^5 f(x)$ $= D_x[D_x^4 f(x)] = D_x(0)$ is also 0, as is $D_x^6 f(x)$; indeed, every $D_x^n f(x)$ for $n \geqslant 4$ is 0. These results are summarized in Table 5.4-4.

TABLE 5.4-4

Name of function	Notation	Defining equation
The function f	f	$f(x) = 2x^3 - 5x^2 + 7$
First-order derivative of f	$D_x f$	$D_x f(x) = 6x^2 - 10x$
Second-order derivative of f	$D_x^2 f$	$D_x^2 f(x) = 12x - 10$
Third-order derivative of f	$D_x^3 f$	$D_x^3 f(x) = 12$
Fourth-order derivative of f	$D_x^4 f$	$D_x^4 f(x) = 0$
nth-order derivative of f $(n > 4)$	$D_x^n f$	$D_x^n f(x) = 0$

EXAMPLE 5.4-11

Find the first-order, second-order, and third-order derivatives of the function h defined by $h(z) = \ln(z) + z^3 - \exp(z)$.

SOLUTION

The first-order derivative is

$$D_z h(z) = D_z\left[\ln(z) + z^3 - \exp(z)\right]$$

$$= D_z \ln(z) + D_z(z^3) - D_z \exp(z)$$

$$= (1/z) + 3z^2 - \exp(z)$$

$$= z^{-1} + 3z^2 - \exp(z)$$

To find the second-order derivative $D_z^2 h$, we compute the derivative of $D_z h$, i.e.,

$$D_z^2 h(z) = D_z\left[D_z h(z)\right]$$

$$= D_z\left[z^{-1} + 3z^2 - \exp(z)\right]$$

$$= D_z(z^{-1}) + 3D_z(z^2) - D_z \exp(z)$$

$$= -z^{-2} + 6z - \exp(z)$$

Finally, $D_z^3 h(z) = D_z[D_z^2 h(z)]$, so we have

$$D_z^3 h(z) = D_z\left[-z^{-2} + 6z - \exp(z)\right]$$

$$= D_z(-z^{-2}) + 6D_z(z) - D_z \exp(z)$$

$$= 2z^{-3} + 6 - \exp(z)$$

Table 5.4-5 summarizes these results.

TABLE 5.4-5

Name of function	Notation	Defining equation
The function h	h	$h(z) = \ln(z) + z^3 - \exp(z)$
First-order derivative of h	$D_z h$	$D_z h(z) = z^{-1} + 3z^2 - \exp(z)$
Second-order derivative of h	$D_z^2 h$	$D_z^2 h(z) = -z^{-2} + 6z - \exp(z)$
Third-order derivative of h	$D_z^3 h$	$D_z^3 h(z) = 2z^{-3} + 6 - \exp(z)$

EXAMPLE 5.4-12

Use the definition of the derivative and the fact that the limit of a sum is the sum of the limits to prove the corresponding fact (rule (3) of Table 5.4-1) about derivatives.

SOLUTION

Let $h(x) = f(x) + g(x)$, and suppose that x_0 is a real number for which $D_x f(x_0)$ and $D_x g(x_0)$ both exist. Then

$$D_x h(x_0) = \lim_{x \to x_0} \frac{h(x) - h(x_0)}{x - x_0}$$

$$= \lim_{x \to x_0} \frac{[f(x) + g(x)] - [f(x_0) + g(x_0)]}{x - x_0}$$

$$= \lim_{x \to x_0} \frac{[f(x) - f(x_0)] + [g(x) - g(x_0)]}{x - x_0}$$

$$= \lim_{x \to x_0} \left[\frac{f(x) - f(x_0)}{x - x_0} + \frac{g(x) - g(x_0)}{x - x_0} \right]$$

$$= \lim_{x \to x_0} \frac{f(x) - f(x_0)}{x - x_0} + \lim_{x \to x_0} \frac{g(x) - g(x_0)}{x - x_0}$$

$$= D_x f(x_0) + D_x g(x_0)$$

Thus $D_x h(x_0) = D_x f(x_0) + D_x g(x_0)$ for all such x_0; or $D_x h(x) = D_x [f(x) + g(x)] = D_x f(x) + D_x g(x)$.

EXAMPLE 5.4-13

Use the definition of the derivative and the rules governing limits to establish rule (5) of Table 5.4-2.

SOLUTION

Let $h(x) = f(x) g(x)$, and suppose that x_0 is a real number for which $D_x f(x_0)$ and $D_x g(x_0)$ both exist. We wish to show that

$$D_x [f(x) g(x)] = f(x) D_x g(x) + g(x) D_x f(x)$$

By definition,

$$D_x h(x_0) = \lim_{x \to x_0} \frac{h(x) - h(x_0)}{x - x_0} = \lim_{x \to x_0} \frac{f(x)\,g(x) - f(x_0)\,g(x_0)}{x - x_0}$$

For reasons that will soon become clear, we adjust the form of the numerator by adding $-f(x_0)\,g(x) + f(x_0)\,g(x)$ which equals 0 and so does not change the value of the numerator. Thus

$$f(x)\,g(x) - f(x_0)\,g(x_0) = f(x)\,g(x) + 0 - f(x_0)\,g(x_0)$$

$$= f(x)\,g(x) - f(x_0)\,g(x) + f(x_0)\,g(x) - f(x_0)\,g(x_0)$$

$$= [\,f(x) - f(x_0)\,]\,g(x) + [\,g(x) - g(x_0)\,]\,f(x_0)$$

Consequently,

$$D_x h(x_0) = \lim_{x \to x_0} \frac{f(x)\,g(x) - f(x_0)\,g(x_0)}{x - x_0}$$

$$= \lim_{x \to x_0} \frac{[\,f(x) - f(x_0)\,]\,g(x) + [\,g(x) - g(x_0)\,]\,f(x_0)}{x - x_0}$$

$$= \lim_{x \to x_0} \left(\frac{[\,f(x) - f(x_0)\,]\,g(x)}{x - x_0} + \frac{[\,g(x) - g(x_0)\,]\,f(x_0)}{x - x_0} \right)$$

$$= \lim_{x \to x_0} \frac{[\,f(x) - f(x_0)\,]\,g(x)}{x - x_0} + \lim_{x \to x_0} \frac{[\,g(x) - g(x_0)\,]\,f(x_0)}{x - x_0}$$

$$= \left(\lim_{x \to x_0} g(x) \right) \lim_{x \to x_0} \frac{f(x) - f(x_0)}{x - x_0}$$

$$+ f(x_0) \left(\lim_{x \to x_0} \frac{g(x) - g(x_0)}{x - x_0} \right)$$

$$= \left(\lim_{x \to x_0} g(x) \right) D_x f(x_0) + f(x_0) D_x g(x_0)$$

If g is continuous at x_0, then we have $\lim_{x \to x_0} g(x) = g(x_0)$ and thus

$$D_x h(x_0) = g(x_0) D_x f(x_0) + f(x_0) D_x g(x_0)$$

for all x_0 at which g is continuous; this means that

$$D_x h(x) = D_x[\,f(x)\,g(x)\,] = g(x) D_x f(x) + f(x) D_x g(x)$$

It is a fact, proved in more advanced texts, that if $D_x g(x_0)$ exists, then g is continuous at x_0; thus rule (5) is established, since we are assuming that $D_x g(x_0)$ exists.

EXAMPLE 5.4-14

Use the definition of the derivative and the rules governing limits to establish rule (6) of Table 5.4-2.

SOLUTION

Let $h(x) = 1/g(x)$, and suppose that x_0 is any real number for which $D_x g(x_0)$ exists and $g(x_0) \neq 0$. Then

$$D_x h(x_0) = \lim_{x \to x_0} \frac{h(x) - h(x_0)}{x - x_0} = \lim_{x \to x_0} \frac{1/g(x) - 1/g(x_0)}{x - x_0}$$

We can rewrite

$$\frac{1}{g(x)} - \frac{1}{g(x_0)}$$

as $(g(x_0) - g(x))/(g(x) g(x_0))$, then

$$D_x h(x_0) = \lim_{x \to x_0} \frac{1/g(x) - 1/g(x_0)}{x - x_0}$$

$$= \lim_{x \to x_0} \frac{\left[(g(x_0) - g(x))/(g(x) g(x_0)) \right]}{x - x_0}$$

$$= \lim_{x \to x_0} \frac{g(x_0) - g(x)}{g(x) g(x_0)(x - x_0)}$$

$$= \left[\lim_{x \to x_0} \frac{g(x_0) - g(x)}{x - x_0} \right]\left[\lim_{x \to x_0} \frac{1}{g(x) g(x_0)} \right]$$

$$= \left[- \lim_{x \to x_0} \frac{g(x) - g(x_0)}{x - x_0} \right]\left[\lim_{x \to x_0} \frac{1}{g(x) g(x_0)} \right]$$

$$= - D_x g(x_0) \lim_{x \to x_0} \frac{1}{g(x) g(x_0)}$$

$$= - D_x g(x_0) \frac{1}{g(x_0)} \lim_{x \to x_0} \frac{1}{g(x)}$$

Again if g is continuous at x_0, then since $g(x_0) \neq 0$, we have

$$\lim_{x \to x_0} \frac{1}{g(x)} = \frac{1}{\lim_{x \to x_0} g(x)} = \frac{1}{g(x_0)}$$

and consequently, $D_x h(x_0) = - D_x g(x_0) \cdot 1/g(x_0) \cdot 1/g(x_0) = - D_x g(x_0)/[g(x_0)]^2$ for all such x_0. So we have shown that

$D_x[1/g(x)] = -D_x g(x)/[g(x)]^2$. Then by rule (5) of Table 5.4-2 we get

$$D_x\left(\frac{f(x)}{g(x)}\right) = D_x\left(f(x)\,\frac{1}{g(x)}\right)$$

$$= \frac{1}{g(x)}\,D_x f(x) + f(x) D_x\left(\frac{1}{g(x)}\right)$$

$$= \frac{D_x f(x)}{g(x)} - \frac{f(x) D_x g(x)}{[g(x)]^2}$$

$$= \frac{g(x) D_x f(x)}{[g(x)]^2} - \frac{f(x) D_x g(x)}{[g(x)]^2}$$

$$= \frac{g(x) D_x f(x) - f(x) D_x g(x)}{[g(x)]^2}$$

As noted in the previous example, it is not necessary to assume that g is continuous at x_0, because if $D_x g(x_0)$ exists, then g is necessarily continuous at x_0.

Exercises 5.4

In Exercises 1 through 40, find the derivative of the given function.

1. $f(x) = 4$

2. $f(x) = 7x^3$

3. $f(x) = 2x^{-4}$

4. $f(x) = -10\sqrt{x}$

5. $f(x) = 2x^3 + \sqrt{x}$

6. $f(x) = 3x^2 - 2\sqrt{x}$

7. $f(x) = 4x^3 - 2x - 1$

8. $f(x) = -2x^3 - 6x + \frac{1}{2}$

9. $f(x) = 5x^7 - 3x^5 + 3x^2 - 1$

10. $f(x) = -7x^{11} + 5x^4 - 2x + 3$

11. $f(x) = -6x^{-4} + \frac{1}{2}x - \sqrt{x}$

12. $f(x) = \frac{1}{8}\sqrt{x} - 3x^{-3} + 1$

13. $f(x) = \exp(x) - 4\ln(x)$

14. $f(x) = 6\exp(x) - 2x^{-3}$

15. $f(x) = 2x^3(\sqrt{x} + 1)$

16. $f(x) = -3\sqrt{x}\,(2x^2 - x)$

17. $f(x) = (x^2 + \sqrt{x})\exp(x)$

18. $f(x) = (\sqrt{x} - 5)\ln(x)$

19. $f(x) = \sqrt{x}\,\exp(x) + 2x^{-2}$

20. $f(x) = 3x^2\ln(x) - 3x^4$

21. $f(x) = \dfrac{x^5}{\sqrt{x}}$

22. $f(x) = \dfrac{\sqrt{x}}{2x^4}$

23. $f(x) = \dfrac{4 \ln(x)}{\sqrt{x}}$

24. $f(x) = \dfrac{-8\sqrt{x}}{\exp(x)}$

25. $f(x) = \dfrac{7x^5 - 3x^2 + x}{2x^3 - 1}$

26. $f(x) = \dfrac{8x^7 + 6x^3 - 2}{3x^2 - 1}$

27. $f(x) = \dfrac{4x^{-3} - \sqrt{x}}{\exp(x) + x^2}$

28. $f(x) = \dfrac{2\exp(x) - \ln(x)}{2x^3 + x^2}$

29. $f(x) = -\frac{1}{5}x^3 + \sqrt{x} - \exp(x) \cdot \ln(x)$

30. $f(x) = (\frac{1}{2}\exp(x) - \frac{1}{5}\ln(x))^2$

31. $g(t) = -3t^3\sqrt{t} + 2\ln(t)$

32. $h(z) = 4z^2 \exp(z) - \dfrac{2}{\ln(z)}$

33. $p(v) = (3v^2 - 6v^{-1})(3\ln(v) + 6)$

34. $s(w) = -3w\exp(w) + 5\sqrt{w}$

35. $r(s) = \dfrac{-\frac{1}{2}s^7 - \frac{2}{3}\exp(s)}{\exp(s)}$

36. $u(z) = \dfrac{z^2}{2} - \dfrac{z\ln(z)}{3}$

37. $p(t) = \sqrt{t}\exp(t) - t\ln(t) + \dfrac{3}{\exp(t)}$

38. $r(x) = x^{-3}\ln(x) + 2x^2\exp(x) - 6x^{-4}$

39. $w(r) = \dfrac{3\exp(r) + \sqrt{r}}{r\ln(r) + r^3}$

40. $t(z) = \dfrac{z\ln(z)}{z\exp(z) + z^2}$

In Exercises 41 through 50, find the first-order, second-order, and third-order derivatives of the given function.

41. $f(x) = 4x^3 + 2x^2 - 3x$

42. $f(x) = 6x^5 - 7x^2 - 3x + 1$

43. $f(x) = \frac{1}{2}x^6 - \frac{1}{3}x^3 + 3x^{-3}$

44. $f(x) = -2x^{-8} + \frac{1}{5}x^5 - \frac{1}{2}x^4 + 2x$

45. $f(x) = 3\ln(x) + 4\exp(x) - 6$

46. $g(z) = 2z^3 - 6\ln(z) + 2z^{-2}$

47. $h(t) = t^4 \ln(t)$

48. $p(v) = v^{-2} \exp(v) + \frac{1}{2} v^3$

49. $s(w) = w \ln(w) - \dfrac{\exp(w)}{w}$

50. $u(v) = 2v^2 \exp(v) - \dfrac{\ln(v)}{v}$

In Exercises 51 through 56, find the equation of the line tangent to the graph of the given function at the given point P on the graph.

Recall that the equation of the line L through a given point $P = (x_0, y_0)$ with slope m is $y - y_0 = m(x - x_0)$.

51. $f(x) = x^3 - 4x - 1$, $P = (2, -1)$

52. $f(x) = -2x^3 + 4x + 2$, $P = (-1, 0)$

53. $f(x) = 16x^{-2} + 4\sqrt{x}$, $P = (4, 9)$

54. $f(x) = 4 \ln(x) + 2x^2$, $P = (1, 2)$

55. $f(x) = (x \exp(x) - 2x^2)(3x + 1)$, $P = (0, 0)$

56. $f(x) = \dfrac{x \ln(x)}{\sqrt{x}}$, $P = (1, 0)$

57. Find the point P on the graph of $f(x) = 2x^2 - 4x + 1$ where the slope of the tangent line is 4.

58. Find all points P on the graph of the function $f(x) = x - (9/x)$ where the slope of the tangent line is 2.

59. At what point P on the graph of $f(x) = (\exp(x))(x^2 + 1)$ is the tangent line parallel to the x axis?

60. At what point P on the graph of $f(x) = \ln(x)/x$ is the tangent line horizontal? (*Hint:* Recall from Section 4.5 that $\exp[\ln(x)] = x$ and $\exp(1) = e^1 = e$, where e is approximately equal to 2.718.)

61. Use rule (5) of Table 5.4-2 to verify that the derivative of the product of three functions f, g, and h is given by the formula
$D_x[f(x) \cdot g(x) \cdot h(x)] = f(x) \cdot g(x) \cdot D_x h(x) + h(x) \cdot f(x) \cdot D_x g(x) + h(x) \cdot g(x) \cdot D_x f(x)$.

62. Medical researchers, in testing the reaction of the body to an experimental drug, find that an injection of x cc of the drug causes an increase of R beats per minute in the pulse rate, where R is given by the equation $R(x) = 4x^2 - x^3/3$, for $0 \le x \le 3$. Find the rate of change of R with respect to x at $x = 1$ and $x = 2$.

Recall that the rate of change of R with respect to x, $D_x R$, is called the *sensitivity* of the drug.

63. Delta Electronics, Inc., estimates that its cost C (in dollars) to produce x units of its economy model clock radio is given by the equation $C(x) = -x^2/400 + 8x + 200$ for $500 \le x \le 1500$. Find the average rate of change of C with respect to x (average marginal cost) as x changes from $x_0 = 1000$ to $x_1 = 1200$. What is the marginal cost (instantaneous rate of change of C with respect to x) at a production output of $x = 1000$ units?

64. The Amazon Alarm Company finds that its total revenue R (in dollars) from the production of x units of its home burglar alarm system is given by $R(x) = (x^3/100) - (x^2/40) - 2x$ for $45 \leqslant x \leqslant 55$. The value of the derivative $D_x R$ at $x = x_0$ is called the *marginal revenue* at $x = x_0$. Find the marginal revenue at $x = 50$.

The instantaneous rate of change of revenue R with respect to the number x of units produced, $D_x R$, is called the *marginal revenue*.

65. The Public Health Department of the city of Euphoria estimates that t days after the detection of type A influenza in the community, the percent P of the city's population infected by the disease will be given by $P(t) = (60t^2 - t^3)/640$ for $0 \leqslant t \leqslant 50$. Find (a) the average rate of change of P with respect to t as t changes from $t_0 = 20$ to $t_1 = 40$, (b) the rate of change of P with respect to t at $t = 20$, and (c) the rate of change of P with respect to t at $t = 40$.

66. A circus performer is shot from a cannon on the ground to a trapeze suspended 70 ft above ground level. The height d (in feet) of the performer above the ground t sec after the cannon is fired is given by the equation $d(t) = -16t^2 + 64t + 6$, for $0 \leqslant t \leqslant 2$. Find the speed of the performer (instantaneous rate of change of distance with respect to time) at $t = 1$ sec after the cannon is fired and at $t = 2$ sec after the cannon is fired.

67. A group of psychologists conducting an experiment on learning prepared a special set of materials to be studied and learned by participants in the experiment during a specified learning period. The participants were then tested at different times after the learning period to see how much of the material they retained. It was found that the percent P of the material retained t min after the end of the learning period was closely approximated by the equation $P(t) = 900/[\ln(t) + 15]$ for $t \geqslant 1$. Find the rate of change of P with respect to t at $t = 1$ and at $t = 10$. (Use Table A in the back of the book to find the value of $\ln(10)$.) Of what significance is it that these rates of change are negative?

68. A biologist finds that the population P of a certain culture of bacteria is approximated by the equation $P(t) = P_0 \exp(t) + 1000t$, where P_0 is the population at time $t = 0$ and t is measured in hours. If a population initially contains 2000 bacteria, what is the rate of change of P with respect to t at $t = 4$ hr and at $t = 10$ hr? (Use Table A in the back of the book to find the values of $\exp(4)$ and $\exp(10)$.)

69. Grantham Industries manufactures portable utility buildings. The company estimates that its monthly production cost C (in dollars) is given by the equation $C(x) = 300x - (10x/\ln(x))$ where x is the number of units manufactured during the month and $100 \leqslant x \leqslant 140$. Find the marginal cost at a production output of 100 units per month and at 120 units per month. (Use Table A in the back of the book to find the values of $\ln(100)$ and $\ln(120)$.)

70. Permalite Products, Inc. manufactures rechargeable flashlights.

The company estimates that the demand function for its product is $d(x) = 200 - 50 \ln(x)$ where d is the number of flashlights that can be sold per month and x is the wholesale selling price of the flashlight in dollars. The equation is considered valid for values of x between 1 and 3 inclusive. Find the marginal demand (rate of change of d with respect to x) at $x = 1$, $x = 2$, and $x = 3$.

In Exercise 71, the rate of change of R with respect to x is not the marginal revenue, since x is the selling price of the product rather than the number of units produced. Marginal revenue is the rate of change of revenue with respect to the number of units produced.

71. In Exercise 70, justify to yourself that the total monthly revenue R of the Permalite Company is given by the equation $R(x) = xd(x) = 200x - 50x \ln(x)$. Find the rate of change of R with respect to x at $x = 1$ and $x = 2$. (Use Table A in the back of the book to find the value of $\ln(2)$.)

72. A company determines that its monthly production cost C (in dollars) is given by the equation $C(x) = 150x - x^2$ and its total monthly revenue R (in dollars) is given by $R(x) = x^3 - 7x^2 - 5x$, where x is the number of hundreds of units of its product produced per month and $20 \leqslant x \leqslant 25$. Then the monthly profit P of the company is given by $P(x) = R(x) - C(x)$. Find $P(x)$ and $D_x P(x)$. What is the rate of change of P with respect to x when $x = 20$ and when $x = 24$?

73. The County Pollution Control Board estimates that the quantity Q (in gallons) of liquid waste that will be discharged into a lake during the first t days of operation of a new chemical manufacturing plant is given by the equation $Q(t) = 100(t^2 + 1)\ln(t)$ for $1 \leqslant t \leqslant 30$. Find the rate at which the amount of waste in the lake is increasing 10 days after the plant begins operation and 20 days after the plant begins operation. (Use Table A in the back of the book to find the values of $\ln(10)$ and $\ln(20)$.)

Applications: Maxima and Minima

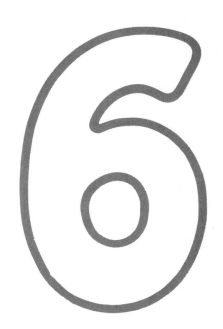

6.1. The Chain Rule

Just as there are rules for finding derivatives of sums, differences, products, and quotients of functions, there is also a rule, called the *chain rule*, for finding the derivative of a composite function $g \circ f$. Aside from providing us with a rather spectacular labor-saving method for computing derivatives, the chain rule is frequently an indispensible differentiation technique, as will become evident.

The chain rule can be described by analyzing the rate of change of a composite function $g \circ f$ in terms of the rates of change of the functions f and g. To assist us in this analysis, consider the following simple situation. Suppose that 1 oz of pure lemon juice produces 2 oz of lemonade concentrate and that 1 oz of concentrate produces 3 pt of lemonade. Then how many pints of lemonade are produced per ounce of lemon juice? We easily answer this question by multiplying the *rate* at which lemonade is produced per ounce of concentrate by the *rate* at which concentrate is produced per ounce of pure lemon juice, i.e., we have 3 (pints of lemonade per ounce of concentrate) · 2 (ounces of concentrate per ounce of lemon juice) = 6 (pints of lemonade per ounce of lemon juice). Now let us see how this illustration enables us to find the derivative of a composite function $g \circ f$.

Let $f(x)$ be the number of ounces of concentrate produced from x oz of lemon juice, and let $g(u)$ be the number of pints of lemonade produced from u oz of concentrate. Then $(g \circ f)(x) = g[f(x)]$ is the

Recall from Section 2.5 that given two functions f and g, the composite function $g \circ f$ is defined by the rule $(g \circ f)(x) = g[f(x)]$.

Note that for x oz of lemon juice, we get $f(x)$ oz of concentrate, so $f(x)$ is in the domain of g, i.e., $g(u)$ exists at $u = f(x)$.

To check that $D_x(g \circ f) = D_u g \cdot D_x f$ in the lemonade illustration, note that $f(x) = 2x$, $g(u) = 3u$ and $(g \circ f)(x) = g[f(x)] = 6x$. Thus $D_x f = 2$, $D_u g = 3$ and $D_x(g \circ f) = 6$.

Since several alternative notations are available for the derivative, the chain rule can be expressed in different ways. For example, $[g \circ f]'(x) = g'[f(x)]f'(x)$ or $[d(g \circ f)/dx] = (dg/df) \cdot (df/dx)$ are alternative ways of writing the chain rule.

For a review of how to express a given function h as a composition of two other functions f and g, see Section 2.5.

number of pints of lemonade produced from x oz of lemon juice. Thus the *rate* at which concentrate is produced per ounce of lemon juice is $D_x f$, the *rate* at which lemonade is produced per ounce of concentrate is $D_u g$, and, as we saw above, the product of these rates is the rate at which lemonade is produced per ounce of lemon juice, i.e., $D_x(g \circ f)$. This is precisely the *chain rule* for finding derivatives: If g is a function of u and $u = f(x)$ is a function of x, then the derivative of $g \circ f$ with respect to x is given by

$$D_x(g \circ f)(x) = D_u g[f(x)]D_x f(x)$$

EXAMPLE 6.1-1

Use the chain rule to find the derivative of $h(x) = (x^2 + 2x + 3)^4$.

SOLUTION

We first find functions f and g such that $h = g \circ f$. To do this, we let $g(u) = u^4$ and $u = f(x) = x^2 + 2x + 2$; then $(g \circ f)(x) = g[f(x)] = g(x^2 + 2x + 3) = (x^2 + 2x + 3)^4 = h(x)$, so indeed $h = g \circ f$. Now we use the chain rule to find the derivative of $g \circ f$. Since $g(u) = u^4$, we see that $D_u g(u) = 4u^3$, so at $u = f(x)$, $D_u g[f(x)] = 4[f(x)]^3$. Furthermore, since $f(x) = x^2 + 2x + 3$, we have $D_x f(x) = 2x + 2$. Thus

$$D_x h(x) = D_x(g \circ f)(x)$$

$$= D_u g[f(x)]D_x f(x)$$

$$= 4[f(x)]^3(2x + 2)$$

$$= 4(x^2 + 2x + 2)^3(2x + 2)$$

EXAMPLE 6.1-2

Use the chain rule to find the derivative of $h(x) = 1/(\sqrt{x} + 1)$.

SOLUTION

Letting $g(u) = u^{-1} = 1/u$ and $u = f(x) = \sqrt{x} + 1$, we have $(g \circ f)(x) = g[f(x)] = g[\sqrt{x} + 1] = 1/(\sqrt{x} + 1) = h(x)$, so $h = g \circ f$. Using the chain rule to find the derivative of $g \circ f$, we see that $D_u g(u) = -u^{-2}$, so at $u = f(x)$, $D_u g[f(x)] = -[f(x)]^{-2}$. Since $f(x) = \sqrt{x} + 1$, we have $D_x f(x) = 1/(2\sqrt{x})$. Thus

$$D_x h(x) = D_x(g \circ f)(x)$$

$$= D_u g[f(x)]D_x f(x)$$

$$= -[f(x)]^{-2} \frac{1}{2\sqrt{x}}$$

$$= \frac{-1}{[f(x)]^2} \frac{1}{2\sqrt{x}}$$

$$= \frac{-1}{(\sqrt{x} + 1)^2} \frac{1}{2\sqrt{x}}$$

$$= \frac{-1}{2\sqrt{x}(\sqrt{x} + 1)^2}$$

EXAMPLE 6.1-3

Use the chain rule to find the derivative of $p(t) = \sqrt{t^3 - 2t}$.

SOLUTION

Letting $g(u) = \sqrt{u}$ and $u = f(t) = t^3 - 2t$, we have $(g \circ f)(t) = g[f(t)] = g(t^3 - 2t) = \sqrt{t^3 - 2t} = p(t)$, so $p = g \circ f$. Thus $D_t p = D_t(g \circ f)$, and from the chain rule, $D_t(g \circ f)(t) = D_u g[f(t)]D_t f(t)$ (note that t has replaced x). But $D_t f(t) = 3t^2 - 2$ and $D_u g(u) = 1/(2\sqrt{u})$, so at $u = f(t)$, $D_u g[f(t)] = 1/(2\sqrt{f(t)})$. Thus we have

$$D_t p(t) = D_t(g \circ f)(t)$$

Replacing x by t in the formula $D_x(g \circ f)(x) = D_u g[f(x)]D_x f(x)$ yields the formula $D_t(g \circ f)(t) = D_u g[f(t)]D_t f(t)$.

$$= D_u g[f(t)] D_t f(t)$$

$$= \frac{1}{2\sqrt{f(t)}} (3t^2 - 2)$$

$$= \frac{1}{2\sqrt{t^3 - 2t}} (3t^2 - 2)$$

$$= \frac{3t^2 - 2}{2\sqrt{t^3 - 2t}}$$

EXAMPLE 6.1-4

Use the chain rule to find the derivative of $f(z) = \ln[3z^2 + \exp(z)]$.

SOLUTION

Letting $g(u) = \ln(u)$ and $u = h(z) = 3z^2 + \exp(z)$, we have $(g \circ h)(z) = g[h(z)] = g[3z^2 + \exp(z)] = \ln[3z^2 + \exp(z)] = f(z)$, so $f = g \circ h$. Thus $D_z f = D_z(g \circ h)$, and from the chain rule, $D_z(g \circ h)(z) = D_u g[h(z)] D_z h(z)$ (note that z has replaced x and h has replaced f). But $D_z h(z) = 6z + \exp(z)$ and $D_u g(u) = 1/u$, so at $u = h(z)$, $D_u g[h(z)] = 1/h(z)$. Thus we have

Replacing x by z and f by h in the formula $D_x(g \circ f)(x) = D_u g[f(x)] D_x f(x)$ yields the formula $D_z(g \circ h)(z) = D_u g[h(z)] D_z h(z)$.

$$D_z f(z) = D_z(g \circ h)(z)$$

$$= D_z g[h(z)] D_z h(z)$$

$$= \frac{1}{h(z)} [6z + \exp(z)]$$

$$= \frac{1}{3z^2 + \exp(z)} [6z + \exp(z)]$$

$$= \frac{6z + \exp(z)}{3z^2 + \exp(z)}$$

As the process of identifying a given function as the composition of two other functions becomes familiar, it is customary to identify the two functions mentally rather than writing them explicitly as was done in the earlier examples. Often one function is thought of as the "inside" function and the other as the "outside" function. In $g \circ f$, the formula $(g \circ f)(x) = g[f(x)]$ suggests that f be called the "inside" and g the "outside" function. The chain rule is then somewhat cryptically stated as "The derivative of the composition of two functions is the derivative of the outside function times the derivative of the inside function."

EXAMPLE 6.1-5

Use the chain rule to find the derivative of $f(y) = (2y + 1)^{27}$.

SOLUTION

Identifying $2y + 1$ as the "inside" function and u^{27} as the "outside" function, we find

$$D_y f(y) = D_y (2y + 1)^{27}$$

$$= 27(2y + 1)^{27-1} D_y (2y + 1)$$

$$= 27(2y + 1)^{26}(2) = 54(2y + 1)^{26}$$

EXAMPLE 6.1-6

Use the chain rule to find the derivative of $h(z) = \ln(z^2 - z)$.

SOLUTION

Identifying $z^2 - z$ as the "inside" function and $\ln u$ as the "outside" function, we find

$$D_z h(z) = D_z \ln(z^2 - z) = \frac{1}{z^2 - z} D_z (z^2 - z)$$

$$= \frac{1}{z^2 - z} (2z - 1) = \frac{2z - 1}{z^2 - z}$$

Situations where the "outside" functions are $g(x) = x^n$, $g(x) = \ln x$, or $g(x) = e^x$ result in the differentiation rules given in Table 6.1-1.

TABLE 6.1-1

$$(1) \quad D_x[f(x)]^m = m[f(x)]^{m-1}D_x f(x)$$

$$(2) \quad D_x \ln(f(x)) = \frac{1}{f(x)} D_x f(x)$$

$$(3) \quad D_x e^{f(x)} = e^{f(x)} D_x f(x)$$

EXAMPLE 6.1-7

The Planning Commission of Riverview estimates that the city's population P during the next 10 yr will be approximated by the function $P(t) = 800t^2 + 20{,}000$, where t is time in years with $t = 0$ corresponding to the present year. The Pollution Control Board also estimates that the average daily pollution index I (in parts per million) is a function of the population P of the city as given by the equation $I(P) = 35 + \sqrt{P}/10$. Find the rate of change of the pollution index I with respect to t at $t = 5$ yr from now.

SOLUTION

Replacing x by t, g by I, and f by P in the formula $D_x(g \circ f)(x)$ $= D_u g[f(x)]D_x f(x)$ yields the formula $D_t(I \circ P)(t) = D_u I[P(t)]$ $D_t P(t)$.

Let $u = P(t) = 800t^2 + 20{,}000$; then $I(u) = 35 + \sqrt{u}/10$. The equation $(I \circ P)(t) = I[P(t)] = I(800t^2 + 20{,}000)$ $= 35 + \sqrt{800t^2 + 20{,}000}/10$ gives the pollution index as a function of t. Thus we need to find $D_t(I \circ P)(t)$ and to evaluate this derivative at $t = 5$. By the chain rule, $D_t(I \circ P)(t)$ $= D_u I[P(t)] \cdot D_t P(t)$. Now $D_u I(u) = \frac{1}{10}(1/(2\sqrt{u}))$, so at $u = P(t)$ we have $D_u I[P(t)] = 1/(20\sqrt{P(t)})$. For $t = 5$, $P(5)$ $= 800(5)^2 + 20{,}000 = 800(25) + 20{,}000 = 40{,}000$, so $D_u I[P(5)] = 1/(20\sqrt{40{,}000}) = 1/((20)(200))$. Also, $D_t P(t)$ $= 1600t$, so $D_t P(5) = 1600(5)$. Finally, $D_t(I \circ P)(5)$ $= D_u I[P(5)]D_t P(5) = 1600(5)/(20)(200) = 2$; this means that 5 yr from now, the average daily pollution index will be increasing at the rate of $D_t(I \circ P)(5) = 2$ ppm per year.

Exercises 6.1

In Exercises 1 through 52, use the chain rule to find the derivative of the given function.

1. $h(x) = (2x + 3)^6$

2. $h(x) = (-3x - 4)^8$

3. $h(x) = (x^2 - 6x + 1)^4$

4. $h(x) = (-2x^2 + 7x - 4)^5$

5. $h(x) = \sqrt{3x + 1}$

6. $h(x) = \sqrt{1 - 4x}$

7. $h(x) = \dfrac{6}{\sqrt{x}}$

8. $h(x) = \dfrac{-2}{\sqrt{x} + x^2}$

9. $h(x) = \dfrac{1}{2x^4 - 6x}$

10. $h(x) = \dfrac{8}{2x^3 + x^2 - 7}$

11. $h(x) = \dfrac{-3}{(2x^3 + 4x)^3}$

12. $h(x) = \dfrac{5}{(2 - 4x^3)^6}$

13. $h(x) = (3x^7 + 2x)^{-7}$

14. $h(x) = (4x^5 - 8x)^{-10}$

15. $h(x) = \sqrt{x - 7x^2 + x^3}$

16. $h(x) = \sqrt{2x^5 - 3x^2 + 7}$

17. $h(x) = \exp(-x)$

18. $h(x) = \ln(5x)$

19. $h(x) = \exp(2x^2 + 1)$

20. $h(x) = \ln(x^3 - 2x)$

21. $h(x) = \ln(4x^5 + 6x)$

22. $h(x) = \exp(1 - 3x^3)$

23. $h(x) = \exp[2 \ln(x) - 6x]$

24. $h(x) = \ln[\exp(x) - 2x^2]$

25. $p(t) = \sqrt{3t^3 + 2t}$

26. $f(x) = (\sqrt{x} + x^3)^7$

27. $f(z) = [\ln(z) + z^2]^{-2}$

28. $q(t) = [\exp(t) - 2t]^{-3}$

29. $f(x) = \exp(-6x^3)$

30. $s(t) = \ln(\sqrt{t} + 1)$

31. $p(v) = \exp(2\sqrt{v} + v^2)$

32. $r(w) = \exp[\ln(w) + \sqrt{w}]$

33. $v(s) = \sqrt{2 \ln(s) + s^2}$

34. $r(s) = \sqrt{\exp(s) + 2s^2}$

35. $u(v) = [\exp(v)]^2 + 2v^3$

36. $f(r) = [\ln(r)]^4 - 5r^2$

37. $s(t) = (t^3 - 2t)^5 - \sqrt{t^2 + 1}$

38. $p(v) = \sqrt{2v - 1} + (v^2 - 1)^5$

39. $f(x) = \dfrac{7}{\sqrt{x} + 2} - \sqrt{x^2 + 1}$

40. $g(x) = (2x^2 - 6)^{-8} + \sqrt{x^4 + 2}$

41. $h(x) = (4x^3 + 2x - 1)^8 \sqrt{x^2 + 1}$

42. $f(t) = (8t^3 - 2t)^{-6} \sqrt{t^4 + 6}$

Hint: In Exercises 35 through 52, use an appropriate rule from Table 5.4-1 or Table 5.4-2 before applying the chain rule.

43. $g(u) = \sqrt{u^2 + 1}\, \exp(2u + 3)$

44. $s(t) = (2t^3 + 2)^{-3}\ln(t^2 + 1)$

45. $p(v) = (v^5 + v^2 - 1)^{-2}(2v^3 - 1)^2$

46. $g(t) = (2t^{-2} + 4t)^{-5}(t^5 - t)$

47. $f(x) = \dfrac{(x^3 - 6x + 2)^6}{(x^4 + 1)^3}$

48. $h(t) = \dfrac{\sqrt{6t^3 + 1}}{(t - 2t^3)^4}$

49. $g(x) = \dfrac{\left[x^6 + \exp(x)\right]^4}{\ln(x^5)}$

50. $f(x) = \dfrac{\ln(x^4 + 1)}{\exp(x^4)}$

51. $p(t) = \ln(2t^2 + 1)\exp(3t^4)$

52. $h(t) = \dfrac{\ln\left[2t^2 + \exp(t)\right]}{\exp\left[4t^3 - \ln(t)\right]}$

In Exercises 53 through 60, find the derivative of the given function by repeated applications of the chain rule.

53. $h(x) = (\sqrt{x^2 + 1}\,)^{10}$

54. $f(x) = (\sqrt{x^4 + 4x^3}\,)^{-6}$

55. $g(t) = \ln(\sqrt{t^2 + 2}\,)$

56. $f(t) = \exp(\sqrt{2t + 1}\,)$

57. $h(t) = \dfrac{-2}{\sqrt{2t - 1}}$

58. $f(x) = \dfrac{6}{\sqrt{2x^3 + 1}}$

59. $p(t) = [\exp(2t^2 - 1)]^{-3}$

60. $v(u) = \ln[(u^3 + 2)^6]$

Recall that the equation of the line L through a given point $P = (x_0, y_0)$ with slope m is $y - y_0 = m(x - x_0)$.

In Exercises 61 through 66, find the equation of the line tangent to the graph of the given function at the given point P on the graph.

61. $h(x) = (x^2 + 2x - 1)^5$, $\quad P = (0, -1)$

62. $f(x) = \sqrt{x^3 - 2x}$, $\quad P = (2, 2)$

63. $g(t) = \ln(t^2 - 3)$, $\quad P = (2, 0)$

64. $h(t) = \exp(2t^3 - 2)$, $\quad P = (1, 1)$

65. $f(t) = (2t^3 + t - 1)^4\sqrt{t^2 + 4}$, $\quad P = (0, 2)$

66. $f(x) = \dfrac{(4x^3 - 2x)^3}{\sqrt{x^2 + 3}}$, $\quad P = (1, 4)$

67. The Wellmont Manufacturing Company estimates that its monthly production cost C (in dollars) is given by the equation $C(x) = 200(x - \sqrt{x - 50})$, where x is the number of units manufactured during the month, $100 \leqslant x \leqslant 180$. Find the marginal cost at a production output of 150 units per month and of 171 units per month.

68. Century Components Corporation estimates that the demand function for its premium quality stereo receiver is $d(x) = 8000 - 32\sqrt{(x^2/100) + 124}$, where d is the number of receivers that can be sold per year and x is the selling price of the receiver in dollars. The equation is considered valid for values of x between \$250 and \$450 inclusive. Find the marginal demand (rate of change of d with respect to x) when the selling price is \$300 per unit.

69. The population P of a certain sample of bacteria is given by the equation $P = P_0 \exp(0.8t)$, where P_0 is the population at time $t = 0$ and t is measured in hours. If a sample initially contains 5000 bacteria, at what rate is the population increasing 5 hr later? (Use Table A in the back of the book to find the value of $\exp(4)$.)

70. The typing rate R (in words per minute) of an average student at the Universal School of Business is estimated to be $R(x) = 50 - 45 \exp(-x)$, where x is the number of weeks spent in typing class and $0 < x \leqslant 5$. At what rate is R increasing after 2 wk in class? After 3 wk in class? (Use Table A in the back of the book to evaluate $\exp(-2)$ and $\exp(-3)$.)

71. Cosmos Cosmetics, Inc. estimates that its total monthly sales revenue R (in dollars) is approximated by the equation $R(x) = 1000 \ln(2x + 100)$, where x is the amount (in dollars) spent on advertising per month. At what rate is the company's revenue increasing when it spends \$450 a month on advertising? When it spends \$950 a month on advertising? When it spends \$1950 a month on advertising?

72. The Crispy Cracker Corporation is putting a new wine-flavored cracker on the market. The company estimates that the number N of cases of their new product that they will be able to sell per week in $N(x) = 6000 - 6000 \exp(-0.08x)$, where x is the number of weeks elapsed after the product is first introduced on the market. At what rate will N be increasing 10 wk after the product is introduced on the market? 20 wk after it is introduced on the market? (Use Table A in the back of the book to evaluate $\exp(-0.8)$ and $\exp(-1.6)$.)

The area A of a circle with radius r is $A = \pi r^2$.

Recall that e is approximately equal to 2.72.

73. Medical researchers find that 1 min after a new experimental drug is injected, the maximum concentration c_0 (measured in milligrams per cubic centimeter) of the drug in the bloodstream is reached. They also find that t hr after the maximum is reached, the concentration c of the drug in the bloodstream is given by $c(t) = c_0\exp(-t)$. Find the rate at which the concentration c is decreasing 3 hr after the maximum is reached, if the maximum is $c_0 = 0.4$ mg/cc. (Use Table A in the back of the book to evaluate $\exp(-3)$.)

74. The chef at Pete's Pancake Palace claims that after he pours $\frac{1}{2}$ cup of his pancake batter onto the griddle, it spreads in a circle with radius r (in inches) given by $r(t) = 2\ln(t + 1)$, where t is the time elapsed (in seconds) after the batter is poured. He considers this equation to be valid for values of t between 0 and 4 sec. Find the rate of change of the area A of a pancake with respect to t at $t = e - 1$ sec after the batter is poured. (*Hint:* First express the area A as a function of t.)

6.2. Maximum and Minimum Values

The maximum or minimum value of a given quantity is frequently of interest. Indeed, such terms as maximum profit, minimum cost, maximum weight, minimum size, and maximum price are commonplace in our everyday lives. In this section we show how the derivative can be used to find the maximum and minimum values of a given function f.

Since these values (maxima and minima) correspond to the "high" and "low" points on the graph of f, we shall begin our discussion by investigating the relationship between the derivative of f and the general "shape" of the graph of f. To do this, we need to refer to certain types of *intervals* on the x axis. An *interval* is the set of points on a line between two fixed points (called endpoints), say a and b. If the endpoints are included in the interval, we call the interval *closed* and denote it $[a, b]$. If we exclude the endpoints, we call the interval *open* and denote it (a, b). Thus $[a, b]$ consists of all numbers x for which $a \leqslant x \leqslant b$, and (a, b) consists of all x for which $a < x < b$. The notation (a, ∞) stands for all numbers x such that $x > a$, and the notation $(-\infty, a)$ stands for all numbers x such that $x < a$. It is often convenient to call both (a, ∞) and $(-\infty, a)$ intervals, though this is not strictly correct.

Consider the graph of the function f shown in Fig. 6.2-1(a); the graph of f is rising from left to right over the interval (a, b). To describe this situation more carefully, we say that if x_1 and x_2 are any two points in (a, b) such that x_1 is to the left of x_2 ($x_1 < x_2$), then $(x_1, f(x_1))$ is lower

Definition of an interval on a line

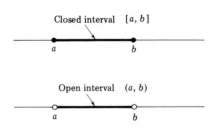

Closed interval $[a, b]$

Open interval (a, b)

Although the notation (a, b) for an open interval is the same as that for an ordered pair, context will prevent confusion between the two meanings.

FIGURE 6.2-1

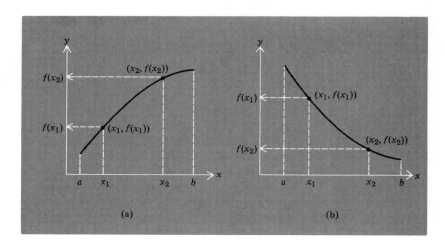

(a)   (b)

FIGURE 6.2-1

than $(x_2, f(x_2))$, i.e., $f(x_1) < f(x_2)$. On the other hand, the graph shown in Fig. 6.2-1(b) is falling from left to right over the interval (a, b). This means that if x_1 and x_2 are any two points in (a, b) such that x_1 is to the left of x_2 $(x_1 < x_2)$, then $(x_1, f(x_1))$ is higher than $(x_2, f(x_2))$ (i.e., $f(x_1) > f(x_2)$).

Now, consider the graph of the function f shown in Fig. 6.2-2. Note that the graph is rising from left to right over the open intervals (x_0, x_1) and (x_2, x_3) on the x axis and falling from left to right over the interval (x_1, x_2). Moreover, for each x in (x_0, x_1) and (x_2, x_3) the line tangent to the graph at $(x, f(x))$ is rising, and for each x in (x_1, x_2) the line tangent to the graph at $(x, f(x))$ is falling. Consequently, for x in (x_0, x_1) or in (x_2, x_3) the slope of the line tangent to the graph at $(x, f(x))$ is positive, and for x in (x_1, x_2) the slope of the line tangent to the graph at $(x, f(x))$ is negative. Since $D_x f(x)$ is the slope of the line tangent to the graph of f

Recall that a line that rises as we move from left to right has positive slope and a line that falls as we move from left to right has negative slope (see Section 2.3).

FIGURE 6.2-2

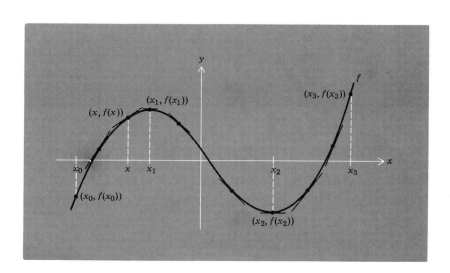

at the point $(x, f(x))$, we see that $D_x f(x)$ is positive if x is in (x_0, x_1) or in (x_2, x_3) and negative if x is in (x_1, x_2). These observations are typical in the following sense. If a function f has a positive derivative at each x in an interval (a, b), then the graph of f rises over the interval (a, b); if a function f has a negative derivative at each x in an interval (a, b), then the graph of f falls over the interval (a, b). These statements are also applicable to the "infinite" intervals $(-\infty, b)$ and (a, ∞). If the graph of a function f rises over an interval (a, b), then f is said to be *increasing* over (a, b); similarly, if the graph of f falls over (a, b), then f is said to be *decreasing* over (a, b). So the function f whose graph is shown in Fig. 6.2-2 is increasing over the intervals (x_0, x_1) and (x_2, x_3) and decreasing over the interval (x_1, x_2).

A function f is *increasing* over (a, b) provided that whenever $x_1 < x_2$ in (a, b), $f(x_1) < f(x_2)$. Similarly, f is *decreasing* over (a, b) provided that whenever $x_1 < x_2$ in (a, b), $f(x_1) > f(x_2)$.

A point x is called a *critical point* of f if either (1) the graph of f has a horizontal tangent line at $(x, f(x))$, or (2) the graph of f has no tangent line at $(x, f(x))$. So if x is a critical point of f at which $D_x f(x)$ exists, we must have $D_x f(x) = 0$. When $D_x f(x)$ exists for each x in the domain of f, we say that f has a *smooth* graph.

Definition of a critical point of a function

In Fig. 6.2-2, the critical points of the function f are at $x = x_1$ and $x = x_2$. A point on the graph of a function f is called a *relative maximum* if it is higher than any nearby point on the graph, and a *relative minimum* if it is lower than any nearby point. In Fig. 6.2-2, the point $(x_1, f(x_1))$ is a relative maximum, and the point $(x_2, f(x_2))$ is a relative minimum; thus we say that f has a relative maximum value at x_1, namely $f(x_1)$, and a relative minimum value at x_2, namely $f(x_2)$.

Definition of relative maximum ·

Definition of relative minimum

FIGURE 6.2-3

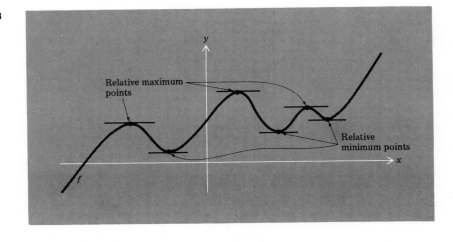

For a smooth graph, it seems evident that at any relative maximum or minimum point on the graph, the tangent line must be horizontal (see Fig. 6.2-3). This is indeed the case for graphs of functions whose domains consist of either all real numbers or of one or more open intervals; thus for such functions f, the relative maximum or minimum values can occur only at critical points of f, that is, values of x where

Our discussion here concerns functions encountered in this text, most of which have smooth graphs. Functions whose graphs have "corners," such as $f(x) = |x|$, are exceptions; for these functions we rely on a more detailed examination of their graphs (or defining equations) to find their relative maximum and minimum values. For example, it is clear from Fig. 6.2-4 that the graph of $f(x) = |x|$ has a relative minimum at $(0, 0)$.

$D_xf(x) = 0$. It is also evident that the graph must be rising just to the left of a relative maximum point and falling just to the right; similarly, the graph must be falling just to the left of a relative minimum and rising just to the right (see Fig. 6.2-3). These considerations give us the following three-step procedure for finding the relative maximum and minimum values of a function f that has a smooth graph and a domain consisting of all real numbers or of open intervals.

Step (1). Find D_xf. (Since the graph of f is smooth, D_xf exists.)

Step (2). Find the critical points of f by solving the equation $D_xf(x) = 0$ to obtain those values of x for which the line tangent to the graph at $(x, f(x))$ will be horizontal.

Step (3). For each critical point c of f (each c such that $D_xf(c) = 0$), evaluate D_xf at a point slightly to the left of c and at a point slightly to the right of c. If $D_xf(x)$ changes from negative to positive as x moves from the left to the right of c, then f has a relative minimum value at c. If $D_xf(x)$ changes from positive to negative as x moves from the left to the right of c, then f has a relative maximum value at c. Finally, if $D_xf(x)$ does not change sign as x moves from the left to the right of c, then f has neither a relative maximum nor a relative minimum value at c (see Fig. 6.2-5).

A function does not necessarily have a relative maximum or minimum value at a critical point. See Fig. 6.2-5(c) and (d).

Finding the relative maximum and relative minimum values of a function f

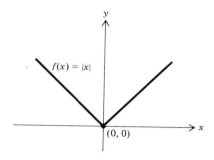

FIGURE 6.2-4 The graph of $f(x) = |x|$ has a relative minimum at $(0, 0)$.

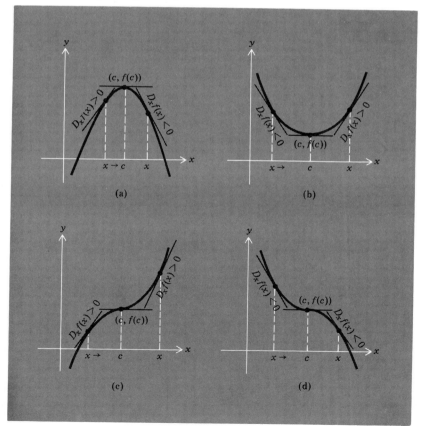

FIGURE 6.2-5 (a) Relative maximum at c. (b) Relative minimum at c. (c) Neither maximum nor minimum at c. (d) Neither maximum nor minimum at c.

Recall that f is increasing over an interval (a, b) if the graph of f rises from left to right over the interval; f is decreasing if the graph of f falls from left to right over the interval.

EXAMPLE 6.2-1

Find the relative maximum and minimum values of the function $f(x) = 2x^3 - 3x^2 - 12x$; determine the intervals over which f is increasing and those over which f is decreasing.

SOLUTION

Step (1) of the procedure outlined above is to find $D_x f$, and, since $f(x) = 2x^3 - 3x^2 - 12x$, we have $D_x f(x) = 6x^2 - 6x - 12$. Next, according to step (2), we set $D_x f(x) = 0$ and solve for x. We have

$$D_x f(x) = 6x^2 - 6x - 12$$

$$= 6(x^2 - x - 2)$$

$$= 6(x - 2)(x + 1) = 0$$

from which we conclude that $x = 2$ or $x = -1$; these are the critical points of f. Now we perform step (3) at each of the critical points, -1 and 2. If x is to the left of -1, then $x < -1$, so $x + 1 < 0$ and $x - 2 < -3$; thus $x + 1$ and $x - 2$ are both negative, and hence $D_x f(x) = 6(x - 2)(x + 1)$ is positive. If x is to the right of -1 and to the left of 2, then $x > -1$ and $x < 2$, so that $x + 1 > 0$ and $x - 2 < 0$; hence $D_x f(x) = 6(x - 2)(x + 1)$ is negative. Finally, if x is to the right of 2, then $x > 2$, so $x - 2 > 0$ and $x + 1 > 3$; thus $D_x f(x) = 6(x - 2)(x + 1)$ is positive. We see then that $D_x f(x)$ changes from positive to negative as x moves from the left to the right of -1, and from negative to positive as x moves from the left to the right of 2. This means that f has a relative maximum value at -1, namely $f(-1) = 2(-1)^3 - 3(-1)^2 - 12(-1) = 7$, and a relative minimum value at 2, namely $f(2) = 2(2)^3 - 3(2)^2 - 12(2) = -20$. Also, since we have found that $D_x f(x)$ is positive for $x < -1$ and for $x > 2$, the function f is increasing over the intervals $(-\infty, -1)$ and $(2, \infty)$. Similarly, since $D_x f(x)$ is negative for $-1 < x < 2$, f is decreasing over the interval $(-1, 2)$. The graph of f is shown in Fig. 6.2-6, and Table 6.2-1 summarizes our results.

TABLE 6.2-1

Critical point	Interval	$D_x f(x)$	$f(x) = 2x^3 - 3x^2 - 12x$	Results
	$(-\infty, -1)$	Positive	Increasing	Relative maximum value at -1
$x = -1$		0	$f(-1) = 7$	
	$(-1, 2)$	Negative	Decreasing	Relative minimum value at 2
$x = 2$		0	$f(2) = -20$	
	$(2, \infty)$	Positive	Increasing	

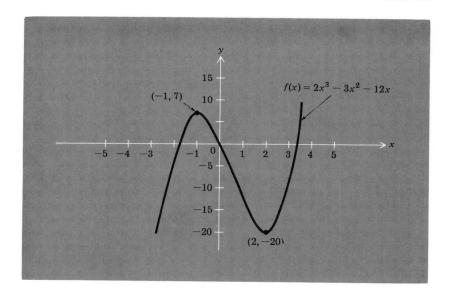

FIGURE 6.2-6

As the solution for Example 6.2-1 illustrates, step (3) of our procedure for finding the relative maximum and minimum values of a function f is usually the most difficult of the three steps. However, if the derivative $D_x f$ is continuous at each real number x, then the only points where the derivative can change sign (positive to negative or negative to positive) are points where the derivative is 0, the critical points of f. So if c_1 and c_2 are adjacent critical points of f (if there is no other critical point of f between c_1 and c_2), then either $D_x f(x)$ is positive for all values of x between c_1 and c_2 or $D_x f(x)$ is negative for all values of x between c_1 and c_2. (Otherwise there would have to be a critical point of f between c_1 and c_2, so c_1 and c_2 would not be adjacent critical points.) We can use this observation to reduce the effort required by step (3) for functions f whose derivatives $D_x f$ are continuous at every real number x.

Remember that if a function g is continuous at each real number x, then the graph of g is all in one piece. Thus the only points where the graph of g can move from one side of the x axis to the other (where g can change sign) are points where the graph crosses the x axis, that is, points x where $g(x) = 0$ (see Fig. 6.2-7).

FIGURE 6.2-7   The function g
changes sign only at x_1, x_3, and x_4.

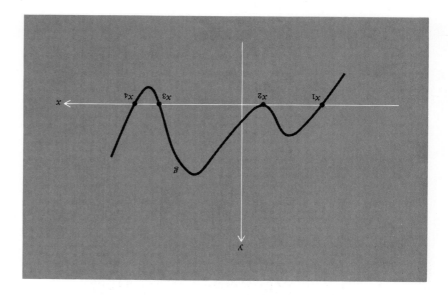

A continuous function does not
have to change sign at a point
where the value of the function is
0. For example, $h(x) = x^2$ is positive everywhere except at $x = 0$,
where $h(0) = 0$. Also, the function
g whose graph is shown in Fig.
6.2-7 does not change sign at x_2,
even though $g(x_2) = 0$.

For example, consider again the function $f(x) = 2x^3 - 3x^2 - 12x$
from Example 6.2-1. Since the derivative $D_x f(x) = 6x^2 - 6x - 12$ is a
polynomial, it is continuous at each real number x; from this we know
that the only points where $D_x f$ can change sign are the points x where
$D_x f(x) = 0$, which are the critical points -1 and 2 of f. Now suppose
we pick a number to the left of -1, say -2; since we find that
$D_x f(-2) = 6(-2)^2 - 6(-2) - 12 = 24$, we can conclude that $D_x f(x)$
must be positive at *each* $x < -1$ (remember that $D_x f$ can not change
sign except at -1 and 2). Similarly, since $1 < 0 < 2$ and $D_x f(0)$
$= 6(0)^2 - 6(0) - 12 = -12$, $D_x f(x)$ is negative at *each* x in the interval
$(-1, 2)$, Finally, since $D_x f(3) = 6(3)^2 - 6(3) - 12 = 24$, $D_x f(x)$ is positive at each $x > 2$ (see Fig. 6.2-8). Thus f is increasing over the intervals
$(-\infty, -1)$ and $(2, \infty)$ and decreasing over $(-1, 2)$, so f has a relative
maximum value at -1 and a relative minimum value at 2. This way of
carrying out step (3) is quicker and simpler than our first way of doing it.

We evaluate $D_x f$ at any convenient number to the left of -1.
Thus, instead of choosing -2, we
could equally well have chosen
-3 or -10. Similarly, we
evaluate $D_x f$ at any convenient
number between -1 and 2, and
finally at any convenient number
to the right of 2.

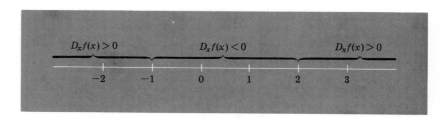

FIGURE 6.2-8

EXAMPLE 6.2-2

Find the relative maximum and minimum values of the function $h(x) = x \exp(x)$; determine the intervals over which h is increasing and those over which h is decreasing.

SOLUTION

First we find $D_x h$. Rule (5) of Table 5.4-2 gives us $D_x h(x) = \exp(x) + x \exp(x)$. Next we perform step (2), setting $D_x h(x) = 0$ and solving for x to find the critical points of h. From the equation

$$D_x h(x) = \exp(x) + x \exp(x) = (1 + x) \exp(x) = 0$$

we conclude that $x = -1$ is the only critical point of h. To implement step (3), notice that since the function $\exp(x)$ is continuous at all x, so are $x \exp(x)$ and $\exp(x) + x \exp(x) = D_x h(x)$. Thus $D_x h$, being continuous, can change sign only at the point -1, the single critical point of h. Since $D_x h(0) = \exp(0) + 0[\exp(0)] = 1$, we conclude that $D_x h(x)$ is positive at each x to the right of -1; since $D_x h(-2) = \exp(-2) - 2\exp(-2) = (1 - 2)\exp(-2) = -\exp(-2) < 0$, we see that $D_x h(x)$ is negative at each x to the left of -1. Thus $D_x h(x)$ changes from negative to positive as x moves from the left to the right of -1. This means that h has a relative minimum value at -1, namely $h(-1) = (-1)\exp(-1) = -\exp(-1) = -1/\exp(1) = -1/e$. Since $D_x h(x)$ is negative at each x to the left of -1, the function $h(x) = x \exp(x)$ is decreasing on $(-\infty, -1)$; similarly, since $D_x h(x)$ is positive at each x to the right of -1, the function h is increasing on $(-1, \infty)$. Table 6.2-2 summarizes our results, and the graph of $h(x) = x \exp(x)$ is shown in Fig. 6.2-10. In sketching the graph, the following simple facts are also used:

(1) $h(x) = xe^x < 0$ for $x < 0$
(2) $h(x) = xe^x = 0$ for $x = 0$
(3) $h(x) = xe^x > 0$ for $x > 0$

Recall that $\exp(x) \neq 0$ for all x, so $(1 + x) \exp(x)$ can be 0 only when $1 + x = 0$. (See Fig. 6.2-9.)

Recall from Section 4.4 that the product of continuous functions is again continuous, as is the sum of continuous functions.

Recall that $\exp(1) = e^1 = e$, where e is approximately equal to 2.718.

TABLE 6.2-2

Critical point	Interval	$D_x h(x)$	$h = x \exp(x)$	Results
	$(-\infty, -1)$	Negative	Decreasing	
$x = -1$		0	$h(-1) = -1/e$	Relative minimum value at -1
	$(-1, \infty)$	Positive	Increasing	

FIGURE 6.2-9   The graph of $\exp(x)$ = e^x.

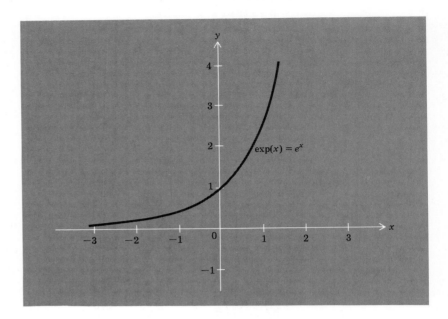

FIGURE 6.2-10

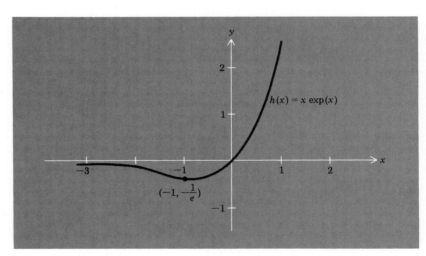

FIGURE 6.2-11

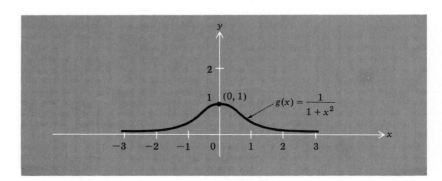

EXAMPLE 6.2-3

Find the relative maximum and minimum values of the function $g(x) = 1/(1 + x^2)$; determine the intervals over which g is increasing and those over which g is decreasing.

SOLUTION

First we find $D_x g$. By the quotient rule we have

$$D_x g(x) = \frac{-2x}{(1 + x^2)^2}$$

Setting $D_x g(x) = 0$, we get $-2x/(1 + x^2)^2 = 0$, or $-2x = 0$ (multiplying both sides by $(1 + x^2)^2$). Thus $x = 0$ is the only critical point. Now the denominator of $D_x g(x)$ is always positive, so the sign of $D_x g(x) = -2x/(1 + x^2)^2$ is the same as the sign of the numerator. But $-2x$ is negative when x is positive and positive when x is negative, so we conclude that $D_x g(x)$ is negative for $x > 0$ and positive for $x < 0$. Hence g is increasing on $(-\infty, 0)$ and decreasing on $(0, \infty)$; thus $g(0) = 1/(1 + 0^2) = 1$ is a relative maximum. There are no relative minimum points. We summarize our results in Table 6.2-3 and sketch the graph of g in Fig. 6.2-11.

TABLE 6.2-3

Critical point	Interval	$D_x g(x)$	$g(x) = 1/(1 + x^2)$	Results
	$(-\infty, 0)$	Positive	Increasing	Relative
$x = 0$		0	$g(0) = 1$	maximum
	$(0, \infty)$	Negative	Decreasing	value at 0

EXAMPLE 6.2-4

Find the relative maximum and minimum values of the function $f(x) = 3x^4 - 4x^3 - 12x^2$; determine the intervals over which f is increasing and those over which f is decreasing.

SOLUTION

The product of the numbers $12x$, $x + 1$, and $x - 2$ can be 0 only if one of the three numbers is 0. This is the same as saying that the product $12x(x + 1)(x - 2)$ can be 0 only if $x = 0$, $x = -1$, or $x = 2$.

First we find $D_x f(x)$, which is $12x^3 - 12x^2 - 24x = 12x(x^2 - x - 2) = 12x(x + 1)(x - 2)$. The solutions of $D_x f(x) = 12x(x + 1)(x - 2) = 0$ are $x = 0$, $x = -1$, and $x = 2$. We arrange these values from smallest to largest, in the order $-1, 0, 2$ (see Figure 6.2-12). Then choose any test point to the left of -1 (say -2), a test point between -1 and 0 (say $-\frac{1}{2}$), a test point between 0 and 2 (say 1), and a test point to the right of 2 (say 3). Evaluating $D_x f(x)$ at these test points will determine whether f is increasing or decreasing in the corresponding interval. Since $D_x f(-2) = 12(-2)(-2 + 1)(-2 - 2) = 12(-2)(-1)(-4) = -96$ is negative, we know $D_x f(x)$ is negative on $(-\infty, -1)$, so f is decreasing on this interval. Since $D_x f(-\frac{1}{2}) = 12(-\frac{1}{2})(-\frac{1}{2} + 1)(-\frac{1}{2} - 2) = 12(-\frac{1}{2})(\frac{1}{2})(-\frac{5}{2}) = \frac{15}{2}$ is positive, we conclude that $D_x f(x)$ is positive on $(-1, 0)$ and that f is increasing on this interval. Since $D_x f(1) = 12(1)(1 + 1)(1 - 2) = (1 + 1)(1 - 2) = 12(1)(2)(-1) = -24$ is negative, we know that $D_x f(x)$ is negative on $(0, 2)$ and that f is decreasing on this interval. Finally, from the fact that $D_x f(3) = 12(3)(3 + 1)(3 - 2) = 12(3)(4)(1) = 144$ is positive, we know $D_x f(x)$ is positive on $(2, \infty)$, so f is increasing on this interval. We summarize our results in Table 6.2-4 and sketch the graph of f in Fig. 6.2-13.

TABLE 6.2-4

Critical point	Interval	$D_x f(x)$	$f(x) = 3x^4 - 4x^3 - 12x^2$	Results
	$(-\infty, -1)$	Negative	Decreasing	Relative minimum value at -1
$x = -1$		0	$f(-1) = -5$	
	$(-1, 0)$	Positive	Increasing	Relative maximum value at 0
$x = 0$		0	$f(0) = 0$	
	$(0, 2)$	Negative	Decreasing	Relative minimum value at 2
$x = 2$		0	$f(2) = -32$	
	$(2, \infty)$	Positive	Increasing	

FIGURE 6.2-12

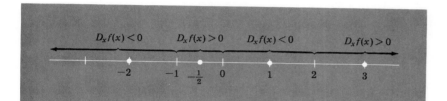

FIGURE 6.2-13

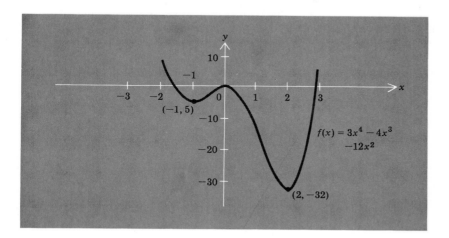

EXAMPLE 6.2-5

A wholesaler has agreed to supply Norm's Novelty Food Store with 10 cans of chocolate-covered grasshoppers per day. It costs the wholesaler $25.60 + $5x to purchase an order of x cans of grasshoppers from the producer, and the average daily cost to store the portion of an order of x cans not yet delivered to Norm's is x cents per day. If the wholesaler has storage space for 300 cans at most, what size order minimizes the average daily cost of supplying Norm's with the grasshoppers?

SOLUTION

Let C be the average daily cost to the wholesaler. Then C is the average purchase cost of the order per day plus the average storage cost per day. Since an order of x cans will last $x/10$ days, the average daily purchase cost in dollars is $(25.60 + 5x)/(x/10) = 256/x + 50$, and we are given that the average storage cost is x cents or $\$x/100$. Thus we see that C is given by the equation

$$C(x) = \frac{x}{100} + \frac{256}{x} + 50$$

The domain of the function C consists only of integers between 10 and 300, since x represents the number of cans in an order. However, in order to apply the techniques of differentiation, we assume that the domain consists of *all* real numbers between 10 and 300, so that the function is defined on an interval. Occasionally, an assumption of this nature leads to results that must be interpreted rather than taken literally.

Since $D_x C$ is continuous at each x in the interval (10, 300), the only point in the interval where it can change sign is at a point x where $D_x C(x) = 0$, namely, at $x = 160$.

C is decreasing over (10, 160) because for each x in the interval (10, 160), $D_x C(x) < 0$. Similarly, since $D_x C(x) > 0$ for each x in the interval (160, 300), C is increasing over the interval (160, 300).

Now we find the number x between 10 and 300 at which C has a relative minimum value; x must be between 10 and 300 because the wholesaler must order at least 10 cans to fill the order from Norm's and at most 300 cans at a time to keep within the available storage space. Since

$$D_x C(x) = \frac{1}{100} - \frac{256}{x^2}$$

the critical points of C are the solutions to the equation $1/100 - 256/x^2 = 0$. Solving this equation, we have

$$\frac{1}{100} = \frac{256}{x^2}$$

which means that $x^2 = 25{,}600$ or $x = \pm 160$. Only the critical point 160 is of interest here, since we are concerned only with values of x between 10 and 300. Choosing the point 100 to the left of 160 and 200 to the right, we find that

$$D_x C(100) = \frac{1}{100} - \frac{256}{10{,}000} = \frac{100}{10{,}000} - \frac{256}{10{,}000} < 0$$

and

$$D_x C(200) = \frac{1}{100} - \frac{256}{40{,}000} = \frac{1}{100} - \frac{64}{10{,}000}$$

$$= \frac{100}{10{,}000} - \frac{64}{10{,}000} > 0$$

Since $D_x C$ is continuous at each x between 10 and 300, this means that $D_x C(x)$ is negative at each x between 10 and 160 and positive at each x between 160 and 300; thus C has a relative minimum value at 160, namely

$$C(160) = \frac{160}{100} + \frac{256}{160} + 50 = 1.6 + 1.6 + 50 = 53.2$$

Since C is decreasing over the entire interval (10, 160) and increasing over the entire interval (160, 300), $C(160) = 53.2$ is not only a relative minimum value of C but is actually *the* minimum value of $C(x)$ for x between 10 and 300 (see Fig. 6.2-14). Thus the wholesaler's smallest possible average daily cost will result from ordering 160 cans at a time from the producer; the daily cost for such an order is $53.20.

FIGURE 6.2-14

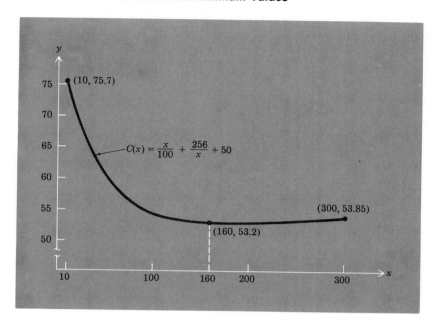

Before closing this section, we mention some important theoretical tools that mathematicians employ when developing a more detailed and precise presentation of the calculus than that undertaken here. One geometrically obvious result that is difficult to justify rigorously is that the graph of a continuous function over a closed interval has a highest point and a lowest point. It is important here that the interval be closed, since the function $f(x) = 1/x$ is continuous at each point of the open interval $(0, 1)$ but the graph of f does not have either a highest or a lowest point over this interval. (see Fig. 6.2-15).

Another frequently used theoretical result is given in what is called the Mean Value Theorem. The word "mean" is used here in the double sense of "average" and "between," as we shall see. Let us first present the exact statement of the theorem, and then examine its geometric interpretation.

The Mean Value Theorem If f is a continuous function on the closed interval $[a, b]$ and if $D_x f(x)$ exists for each x in the open interval (a, b), then there is a number c in (a, b) such that

$$D_x f(c) = \frac{f(b) - f(a)}{b - a}$$

To see the geometric significance of the Mean Value Theorem, consider the graph of the function f shown in Fig. 6.2-16. The number $[f(b) - f(a)]/(b - a)$ is the slope of the line segment L joining the point $(a, f(a))$ to the point $(b, f(b))$, whereas $D_x f(c)$ is the slope of the

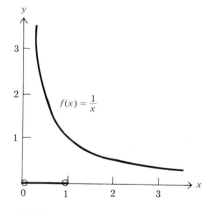

FIGURE 6.2-15 The graph of $f(x) = 1/x$ over $(0, 1)$ does not have a highest point or a lowest point.

line T tangent to the graph of f at the point $(c, f(c))$. Since the slope $[f(b) - f(a)]/(b - a)$ is the average change of f over $[a, b]$, and since $D_x f(c)$ is the instantaneous change of f at c, the Mean Value Theorem asserts that the *average* change of f from a to b is equal to the *instantaneous* change of f at some point c *between* a and b.

FIGURE 6.2-16

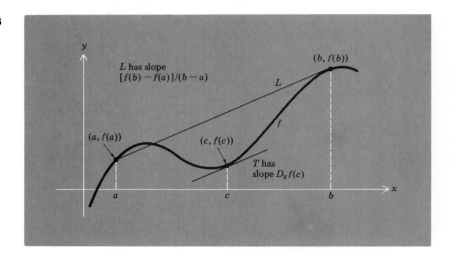

EXAMPLE 6.2-6

Find a value c guaranteed by the Mean Value Theorem for the function $f(x) = x^2 - 2x - 3$ on the interval $[0, 3]$.

SOLUTION

We wish to find a number c in the interval $(0, 3)$ which satisfies the equation $D_x f(c) = [f(b) - f(a)]/(b - a)$, where $f(x) = x^2 - 2x - 3$, $a = 0$, and $b = 3$. Since $f(0) = 0^2 - 2(0) - 3 = -3$ and $f(3) = 3^2 - 2(3) - 3 = 0$, we have $[f(b) - f(a)]/(b - a) = [f(3) - f(0)]/(3 - 0) = [0 - (-3)]/(3 - 0) = \frac{3}{3} = 1$, so we want to find c such that $D_x f(c) = 1$. This means that to find c, we set $D_x f(x) = 2x - 2$ equal to 1 and solve for x. Thus we have $2x - 2 = 1$, which means that $2x = 3$ or $x = \frac{3}{2}$; this is the value of c for which $D_x f(c) = [f(3) - f(0)]/(3 - 0)$ (see Fig. 6.2-17).

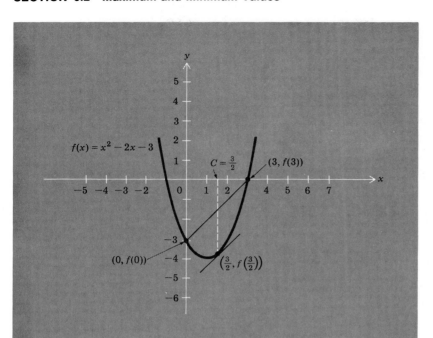

FIGURE 6.2-17

EXAMPLE 6.2-7

Use the Mean Value Theorem to show that if f is continuous on (a, b) and $D_x f(x)$ is positive for each x in (a, b), then f is an increasing function on (a, b).

SOLUTION

Let x_1 and x_2 be two points in (a, b) with $x_1 < x_2$; we wish to show that $f(x_1) < f(x_2)$. Since we know that f is continuous on $[x_1, x_2]$ and differentiable on (x_1, x_2), we may apply the Mean Value Theorem to f and the interval $[x_1, x_2]$ and conclude that there is a number c in (x_1, x_2) such that $D_x f(c)$ $= [f(x_2) - f(x_1)]/(x_2 - x_1)$. Multiplying both sides of this equation by $x_2 - x_1$ yields $f(x_2) - f(x_1) = D_x f(c) \, (x_2 - x_1)$. But since $D_x f(x) > 0$ for all x in (a, b), we have $D_x f(c) > 0$; and $x_1 < x_2$ assures $x_2 - x_1 > 0$, so $f(x_2) - f(x_1) = D_x f(c)$ $(x_2 - x_1) > 0$. From $f(x_2) - f(x_1) > 0$ we obtain $f(x_2)$ $> f(x_1)$ as desired, so f is increasing on (a, b).

Exercises 6.2

In Exercises 1 through 30, find the relative maximum and minimum values of the given function (if any), determine the intervals over which the function is increasing and over which it is decreasing, and sketch the graph.

1. $f(x) = x^2 - 2$

2. $f(x) = 5 - 3x^2$

3. $f(x) = x^4 + 1$

4. $f(x) = 2x^4 + 2$

5. $f(x) = x^3 - 1$

6. $f(x) = 2 - 3x^3$

7. $f(x) = x^2 - 2x + 1$

8. $f(x) = 2x^2 + 4x - 2$

9. $f(x) = -2x^2 - 8x + 1$

10. $f(x) = -3x^2 + 12x - 2$

11. $f(x) = \frac{1}{4}x^2 - \frac{1}{2}x$

12. $f(x) = \frac{-x^2}{2} + 4x - 2$

13. $f(x) = \frac{x^3}{3} + \frac{x^2}{2} - 2x$

14. $f(x) = \frac{x^3}{3} + x^2 - 8x + 1$

15. $f(x) = x^2 \exp(x)$

16. $f(x) = \ln(x) - x$

17. $h(x) = (x^2 - 9)^2$

18. $q(x) = (x - 3)^2(x + 1)^2$

19. $p(t) = (t - 1)^2(t + 2)$

20. $r(s) = \sqrt{s + 1}$

21. $h(t) = \dfrac{1}{t^2 + 1}$

22. $v(u) = \sqrt{u^2 + 4}$

23. $w(t) = t^4 - 2t^2 + 1$

24. $p(v) = (v^2 - 1)^2$

25. $r(v) = \exp(v^2) + 2$

26. $q(w) = [w \exp(w) + 1]^{-1}$

27. $s(t) = 2\ln(t) - t^2$

28. $v(t) = t^2\exp(t) + 3\exp(t)$

29. $r(x) = [\ln(x) - x]^2$

30. $s(t) = \dfrac{1}{\sqrt{t^2 + 2}}$

In Exercises 31 through 38, find a value c guaranteed by the Mean Value Theorem for the given function on the given interval.

31. $f(x) = x^2 + 1$, $[0, 2]$

32. $f(x) = x^2 + 2x + 3$, $[-1, 1]$

33. $h(x) = x^3 + 1$, $[-1, 1]$

34. $g(t) = \sqrt{t + 4}$, $[0, 5]$

35. $v(u) = 3u + 2$, $[-2, 3]$

36. $s(t) = \ln(t)$, $[1, e]$

37. $p(v) = v^{-1}$, $[\frac{1}{4}, 1]$

38. $t(x) = \exp(x) + \exp(-x)$, $[-1, 1]$

39. Show that if n is any even positive integer, the function $f(x) = x^n$ has a relative minimum value at $x = 0$.

40. Show that if n is any odd positive integer, the function $f(x) = x^n$ has no relative maximum or minimum values.

41. For what values of a will the linear function $f(x) = ax + b$ be an increasing function? For what values of a will f be a decreasing function?

42. Use the methods of this section to verify that the vertex of the parabola that is the graph of the quadratic function $f(x) = ax^2 + bx + c$ $(a \neq 0)$ is the point $(p, f(p))$, where $p = -b/2a$ (see Section 2.4).

In Exercises 43 through 59, assume that the maximum and minimum values of the given function over the given interval coincide with the *relative* maximum and minimum values. In general, however, this is not necessarily the case. We shall discuss this topic in the next section.

43. Let x and y be two positive numbers whose sum is 40, i.e., $x + y = 40$. Then $y = 40 - x$, so the product $P = xy$ can be expressed as a function of x alone, $P(x) = x(40 - x)$. Find x and y such that the product P is a maximum.

44. Find two positive numbers x and y whose sum is 450 such that the product of x^2 and y is a maximum.

45. Medical authorities have determined that the percent P of a city's population infected by a communicable disease t days after detection of the disease is given by $P(t) = (30t^2 - t^3)/100$, for values of t between 1 and 30. How many days after detection is the maximum percent of the population infected? What is the maximum percent of the population that will be infected at any given time?

46. Accutype Industries estimates that the monthly sales revenue R (in dollars) received from the production of x units of its economy portable typewriter per month is given by $R(x) = -x^2 + 1600x - 580{,}000$ for $700 \leqslant x \leqslant 1000$. Find (a) the number x_0 of typewriters the company should produce per month to maximize its monthly revenue, (b) the marginal revenue at $x = x_0$, and (c) the maximum monthly revenue the company can receive from the production of the typewriters.

Recall that marginal revenue is the rate of change of revenue with respect to the number of units produced.

47. Patio Products, Inc. estimates that its cost (in dollars) to produce x lounge chairs per month is given by $C(x) = x^2/80 - 45x + 52{,}500$ for $1400 \leqslant x \leqslant 2000$. Find (a) the number x_0 of chairs the company should produce per month to minimize its monthly cost, (b) the marginal cost at $x = x_0$, and (c) the minimum monthly production cost.

48. The pollution index I (in parts per million) on an average day in the city of Euphoria is approximated by the equation $I(t) = -t^2/2 +$

$10t + 25$, where t is time in hours with $t = 0$ corresponding to 8:00 AM and $0 < t < 16$. At what time of day does the pollution index reach a maximum? What is the maximum pollution index?

49. The Supreme Soda Company finds that its monthly profit P from the production of x thousands of bottles of its fruit drink per month is given by $P(x) = 1000(-x^2/2 + 20x - 198)$, where P is in dollars and $19.5 \leqslant x \leqslant 21$. How many thousands of bottles should the company produce per month in order to maximize its profit? What is the maximum monthly profit?

50. A farmer has 100 pigs, each weighing 150 lb. It costs $0.50 a day to keep 1 pig, and the pigs gain weight at 5 lb a day each. The farmer can sell his pigs today for $0.50 a pound, but the price is falling by $0.01 a pound per day. How many days should the farmer wait to sell his pigs in order to maximize his profit? What will his maximum profit be?

51. Mighty-Mite Enterprises, Inc. manufactures minibikes. The company finds that its monthly profit is maximum at a monthly production level of 400 units. If the marginal revenue at the monthly production level of 400 units is $5 per unit, what is the marginal production cost at the same production level? (*Hint:* Recall that if $P(x)$, $R(x)$, and $C(x)$ are, respectively, the profit, revenue, and cost functions, then $P(x) = R(x) - C(x)$.)

52. Medical researchers find that an injection of x cc of an experimental drug causes an increase of R beats per minute in the pulse rate, where R is given by the equation $R(x) = 9[x^2 - x^3/3]$ and $0 \leqslant x \leqslant 2.5$. What amount of the drug causes a maximum increase R in the pulse rate? What is the maximum increase in the pulse rate?

53. Suppose that you are planning to construct a hothouse in your backyard for orchids. You already have a heating unit for the hothouse that will effectively heat 1152 cubic ft of space. You want the hothouse to be rectangular with walls 8 ft high and a flat roof. You estimate that it will cost $2 a square ft to construct the walls and $5 a square ft for the roof. Find the dimensions of the floor that will minimize the construction cost of the hothouse. What is the minimum construction cost? (*Hint:* If x is the length and y the width of the hothouse, then the volume is $8xy$, so you want $8xy = 1152$. Thus $y = 1152/8x = 144/x$. Use this to express the cost C of construction as a function of x alone, and find the value of x that minimizes C.)

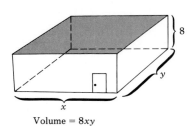

Volume $= 8xy$

54. A peach orchard containing 40 trees per acre yields on the average 300 peaches per tree. For each additional tree planted per acre, the yield decreases by 5 peaches per tree. How many additional trees per acre should be planted in order to maximize the yield per acre? What is the maximum yield per acre?

55. The Super-Swell Soup Company estimates that its monthly cost C (in dollars) to produce x thousands of cans of vegetable soup per month is given by the equation $C(x) = 1000x - 1000 \cdot \ln(x - 19) - 18{,}000$. The equation is considered valid for values of x between 19.5 and 21. Find (a) the number x_0 of thousands of cans of soup the company should produce per month to minimize its monthly cost, (b) the minimum monthly production cost C, and (c) the minimum production cost per can of soup.

56. Standard Stationers, Inc. finds that its monthly profit P (in dollars) is approximated by the function $P(x) = 10{,}000(x - 4)^2 \exp(x - 4) + 1$, where x is the amount (in thousands of dollars) spent on advertising per month. The equation is valid for values of x between 1 and 4. How much should the company spend on advertising per month in order to maximize its monthly profit? What is the maximum monthly profit? (Use Table A in the back of the book to find the value of $\exp(-2)$.)

57. The Triton Company manufactures inflatable rubber rafts. The company finds that its monthly revenue R is given by the equation $R(x) = 10[200x - 50x \ln(x)]$, where x is the selling price of its product, in dollars. The equation is considered valid for values of x between 15 and 25. At what price should the company sell its product in order to maximize its monthly revenue R? (Use Table A in the back of the book to find the value of $\exp(3)$.) What is the maximum monthly revenue?

Recall from Section 4.5 that $\exp[\ln(x)] = x$.

58. A biologist finds that t hr after a culture of 40,000 bacteria is treated with a bactericidal agent, the number N of viable bacteria remaining is given by

$$N(t) = 40{,}000 \left[\frac{\exp(t) + \exp(4 - t)}{1 + e^4} \right]$$

for $0 < t < 5$. How long after treatment is the minimum number of viable bacteria present in the culture? When the population is minimum, how many viable bacteria are present in the culture?

Recall that $e \approx 2.72$ and that $\ln(\exp(x)) = x$.

59. Knifty-Knit Sweaters, Inc. determines that the price S (in dollars) that it should charge per sweater depends upon the number x (in thousands) of sweaters that it manufactures per year according to the equation $S(x) = x^2 - 15x + 72$. This equation is considered valid for values of x between 3 and 8. How many sweaters should the company manufacture per year in order to minimize the selling price of each sweater? What is the minimum selling price of the sweaters?

60. The yearly sales revenue R of Knifty-Knit Sweaters, Inc. (see Exercise 59) is given by $R(x) = 1000x \cdot S(x)$, i.e., $R(x)$ is the number of sweaters produced during the year $(1000x)$ times the selling price $S(x)$ of the sweaters (assuming that all sweaters

produced are sold). Using the equation $S(x) = x^2 - 15x + 72$ from Exercise 59, write the equation that defines $R(x)$. At what production level x does R have a relative maximum value? What is the relative maximum value of R?

61. The yearly profit P of Knifty-Knit Sweaters, Inc. is given by $P(x) = R(x) - C(x)$, where $R(x)$ is the yearly sales revenue, $C(x)$ is the yearly production cost, and x is the number of thousands of sweaters produced per year (see Exercises 59 and 60). If $C(x) = 1000(-1.5x^2 + 12x + 10)$, use the equation for $R(x)$ from Exercise 60 and write the equation that defines $P(x)$. At what production level x does P have a relative maximum value? What is the relative maximum value of P? At what production level does P have a relative minimum value? What is the relative minimum value of P?

6.3.   The Second Derivative Test

Although the procedure given in the previous section for finding the relative maximum and minimum values of a given function f is adequate, it is sometimes difficult to carry out. In this section we develop an alternative procedure, called the *second derivative test*, which frequently provides a much more efficient method. We begin by examining the geometric significance of the second derivative $D_x^2 f$ of a given function f.

Suppose that f is a function whose graph is shown in Fig. 6.3-1. Notice that the graph of f bends in a clockwise direction from left to right (like an upside-down bowl) over the interval (a, b) and in a counterclockwise direction from left to right (like a bowl that is right side up) over the interval (b, c); at the point $(b, f(b))$ on the graph, the direction of bending changes. Such a graph is said to be *concave down* over the interval (a, b) and *concave up* over the interval (b, c); the point $(b, f(b))$

Definition of concavity of a graph

As a memory aid, the phrase "spills water" is frequently associated with a graph that is concave down over an interval, and "holds water" with a graph that is concave up over an interval.

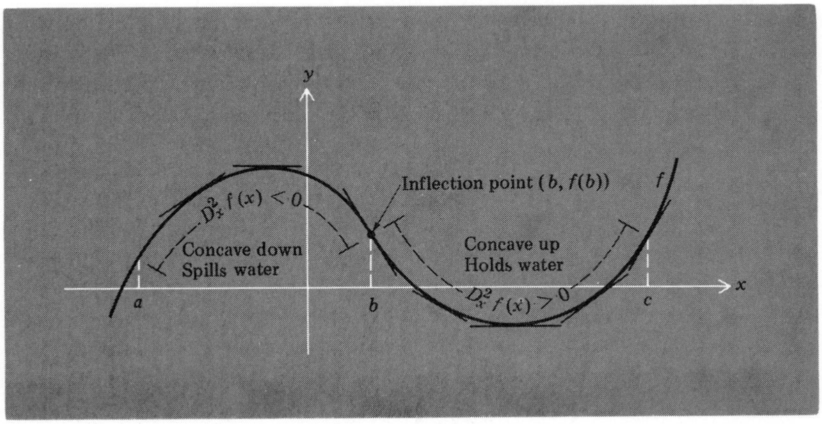

FIGURE 6.3-1   The graph of f is concave down ("spills water") over any interval on which the slopes of the tangents are decreasing from left to right. The graph is concave up ("holds water") over any interval on which the slopes of the tangents are increasing from left to right.

where the graph changes concavity is called an *inflection point* of the graph. Note in Fig. 6.3-1 that when the graph of f is concave down, the tangent lines to the graph lie above the curve, and when the graph is concave up, the tangent lines lie below the curve; at the inflection point $(b, f(b))$ (where the graph changes concavity), the tangent line crosses the curve.

It turns out that the second derivative $D_x^2 f$ of a function f is a reliable sensing device for the concavity of the graph of f. Specifically, if $D_x^2 f(x)$ is negative at each point x in an interval (a, b), then the graph of f is concave down over (a, b); similarly, if $D_x^2 f(x)$ is positive at each x in (b, c), then the graph of f is concave up over (b, c). This connection between the sign of the second derivative and the concavity of f can be discovered by recalling that whenever $D_x^2 f(x)$ is negative at each x in an interval (a, b), then $D_x f$ (whose values are the slopes of the tangent lines to the graph of f) is decreasing over (a, b), which means that the graph of f is concave down over the interval. (Look at Fig. 6.3-1 again and note that the slopes of the lines tangent to f are decreasing from left to right on (a, b).) Similarly, if $D_x^2 f(x)$ is positive at each x in (b, c), then $D_x f$ is increasing over the interval, so the graph of f is concave up over (b, c). (In Fig. 6.3-1 you can see that the slopes of the lines tangent to f are increasing from left to right on (b, c).)

If the slopes of the tangent lines to the graph of a function f decrease as we move along the curve from left to right, then the graph of f is concave down. Similarly, if the slopes of the tangent lines increase, then the graph of f is concave up (see Fig. 6.3-1).

EXAMPLE 6.3-1

Find the intervals over which the graph of the function $f(x) = x^3 + 3x^2 - 9x + 5$ is concave up and those over which the graph is concave down.

SOLUTION

Since $f(x) = x^3 + 3x^2 - 9x + 5$, we see that $D_x f(x) = 3x^2 + 6x - 9$ and $D_x^2 f(x) = 6x + 6 = 6(x + 1)$. Thus for any $x > -1$, we have $x + 1 > 0$ and thus $D_x^2 f(x) = 6(x + 1) > 0$; similarly, for any $x < -1$, we have $x + 1 < 0$ and thus $D_x^2 f(x) < 0$. This means that if x is in $(-1, \infty)$, $D_x^2 f(x) > 0$, so the graph of f is concave up over the interval $(-1, \infty)$. Similarly, for any x in the interval $(-\infty, -1)$, $D_x^2 f(x) < 0$, so the graph of f is concave down over the interval $(-\infty, -1)$. See Fig. 6.3-2.

Since the graph of f is concave up over $(-1, \infty)$ and concave down over $(-\infty, -1)$, the graph changes concavity at the point $(-1, f(-1)) = (-1, 16)$; thus the point $(-1, 16)$ is an inflection point of the graph of f (see Fig. 6.3-2).

FIGURE 6.3-2

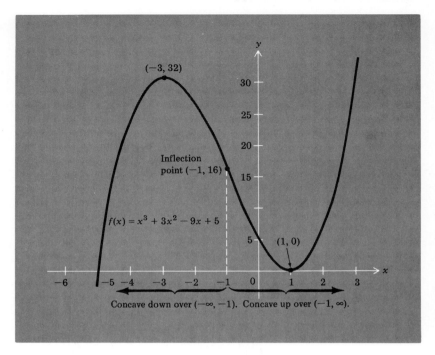

Concave down over $(-\infty, -1)$. Concave up over $(-1, \infty)$.

The relationship between
  (1)   the signs of the first and second derivatives of a function
  (2)   the intervals over which the function is increasing or decreasing
  (3)   the intervals over which the function is concave up or concave down

can be emphasized by comparing the graphs of $D_x f$ and $D_x^2 f$ with the graph of the function $f(x) = x^3 + 3x^2 - 9x + 5$ given in Example 6.3-1. The graph of $D_x f(x) = 3x^2 + 6x - 9$, shown in Fig. 6.3-3(a), is above the x axis ($D_x f(x) > 0$) on $(-\infty, -3)$ and on $(1, \infty)$, which indicates that f is increasing on these intervals; the graph of $D_x f$ is below the x axis ($D_x f(x) < 0$) on $(-3, 1)$, which indicates that f is decreasing on this interval (see Fig. 6.3-2). The graph of $D_x^2 f(x) = 6x + 6$, shown in Fig 6.3-3(b), is above the x axis ($D_x^2 f(x) > 0$) on $(-1, \infty)$, which indicates that f is concave up on $(-1, \infty)$; the graph of $D_x^2 f$ is below the x axis ($D_x^2 f(x) < 0$) on $(-\infty, -1)$, which indicates that f is concave down on $(-\infty, -1)$.

In general, the graph of $D_x f$ will be above the x axis wherever f is increasing and below the x axis wherever f is decreasing; the graph of $D_x^2 f$ will be above the x axis wherever f is concave up and below the x axis wherever f is concave down.

Now suppose that the point c in an interval (a, b) is a critical point of a function f (i.e., $D_x f(c) = 0$). Then we know that the line tangent to

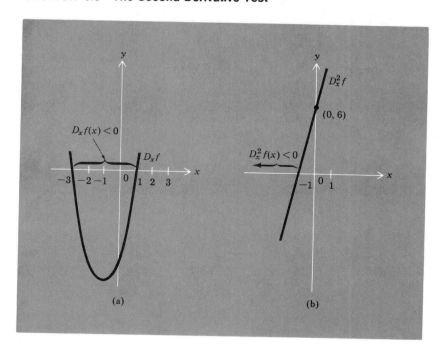

FIGURE 6.3-3 (a) $D_x f(x) = 3x^2 + 6x - 9 = 3(x + 3)(x - 1)$; (b) $D_x^2 f(x) = 6x + 6$.

the graph of f at the point $(c, f(c))$ is horizontal. If, in addition, $D_x^2 f(x)$ is positive at each point x in (a, b), then the graph of f is concave up over (a, b) and f must have a relative minimum value at c (see Fig. 6.3-4(a)). Similarly, if $D_x^2 f(x)$ is negative at each point x in (a, b), then the graph of f is concave down over (a, b) and f must have a relative maximum value at c. (see Fig. 6.3-4(b)). These observations suggest the following procedure, called the *second derivative test*, for finding the relative maximum and minimum values of a function f.

Our discussion here concerns functions f which are defined for each x in (a, b) and whose graphs are smooth over (a, b).

The second derivative test

FIGURE 6.3-4

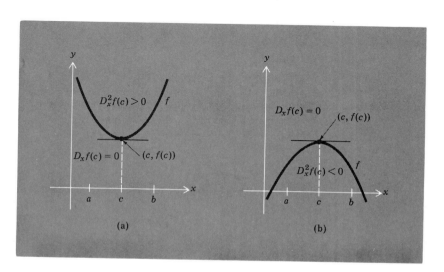

Actually, the second derivative test is valid only for functions f whose second derivative $D_x^2 f$ is continuous in some interval containing the critical point c. Most of the functions in this text have this property.

Step (1)   Find the critical points c of the function f, i.e., those points c such that $D_x f(c) = 0$.

Step (2)   For each critical point c of f, evaluate $D_x^2 f$ at c. If $D_x^2 f(c) > 0$, then f has a relative minimum value at c. If $D_x^2 f(c) < 0$, then f has a relative maximum value at c. If $D_x^2 f(c) = 0$, then no information results; f may have a relative maximum value, a relative minimum value, or neither at c.

If we perform the second derivative test, and find that $D_x^2 f(c) = 0$, then we must resort to the procedure given in the previous section to see whether f has a relative maximum value, a relative minimum value, or neither, at c.

EXAMPLE 6.3-2

Use the second derivative test to find the relative maximum and minimum values of the function $f(x) = x^3 + 3x^2 - 9x + 5$ given in Example 6.3-1.

SOLUTION

By the second derivative test, since $D_x^2 f(1) > 0$ and $D_x^2 f(-3) < 0$, f has a relative maximum value at the critical point $x = -3$ and a relative minimum value at the critical point $x = 1$.

First we find the critical points of f. Since $D_x f(x) = 3x^2 + 6x - 9 = 3(x^2 + 2x - 3) = 3(x + 3)(x - 1)$, we see that $D_x f(x) = 0$ when $x = 1$ and when $x = -3$; these are the critical points of f. Now, since $D_x^2 f(x) = 6x + 6 = 6(x + 1)$, we have $D_x^2 f(1) = 12 > 0$ and $D_x^2 f(-3) = 6(-2) = -12 < 0$. Thus f has a relative maximum value at $x = -3$ (namely, $f(-3) = 32$) and a relative minimum value at $x = 1$ (namely, $f(1) = 0$); see Fig. 6.3-2.

EXAMPLE 6.3-3

A small rectangular grazing pen is to be enclosed by 40 ft of fencing. What should the length and width of the pen be in order to enclose the greatest possible area?

SOLUTION

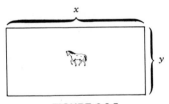

FIGURE 6.3-5

Let x denote the length of the pen and y the width as shown in Fig. 6.3-5. Then, since we have only 40 ft of fencing available, x and y must satisfy $2x + 2y = 40$, so $2y = 40 -$

$2x$ or $y = 20 - x$. Thus the enclosed area A is given by $A = xy = x(20 - x) = 20x - x^2$, and we wish to find the maximum value of the area function $A(x) = 20x - x^2$. Since $D_xA(x) = 20 - 2x = 2(10 - x)$, we see that $D_xA(x) = 0$ when $x = 10$; thus 10 is the critical point of the function A. Also, since $D_x^2A(x) = -2$, we have $D_x^2A(10) = -2 < 0$, so A has a relative maximum value at $x = 10$, namely, $A(10) = 20(10) - 10^2 = 100$ square ft. Since $D_x^2A(x)$ is negative at every value of x $(D_x^2A(x) = -2)$, the graph of A is always concave down; thus the relative maximum value of A determines the highest point on the graph (see Fig. 6.3-6). We conclude that in order to enclose the greatest possible area, the length of the pen should be $x = 10$ ft and the width should be $y = 20 - x = 20 - 10 = 10$ ft, i.e., the pen should be square.

Since $D_x^2A(10) < 0$, the second derivative test tells us that A has a relative maximum value at the critical point $x = 10$.

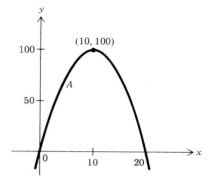

FIGURE 6.3-6   The graph of $A(x)$ $= 20x - x^2$. Recall from Section 2.4 that the graph of A is a parabola opening downward.

EXAMPLE 6.3-4

Use the second derivative test to find the size x of an order that minimizes the average daily cost $C(x) = x/100 + 256/x + 50$ of the wholesaler in Example 6.2-5.

SOLUTION

Since $C(x) = x/100 + 256/x + 50 = x/100 + 256x^{-1} + 50$, we have $D_xC(x) = \frac{1}{100} - 256x^{-2}$ and $D_x^2C(x) = 512x^{-3} = 512/x^3$. As shown in the solution to Example 6.2-5, $D_xC(160) = 0$ and 160 is the only critical point of C between 10 and 300, the interval of permissible values for x. Now, since $D_x^2C(160) = 512/(160)^3 > 0$, the function C has a relative minimum value at $x = 160$, namely, $C(160) = \$53.20$. Since $D_x^2C(x) > 0$ at each x between 10 and 300, the graph of C is concave up over the interval $[10, 300]$, so that the relative minimum value of C determines the lowest point on the graph of C over $[10, 300]$ (see Fig. 6.2-14). Thus an order of 160 cans minimizes the wholesaler's average daily cost.

Since $D_x^2C(160) > 0$, the second derivative test tells us that C has a relative minimum value at the critical point $x = 160$.

In Examples 6.3-3 and 6.3-4 we were interested in the highest or lowest point on the graph of a function over a specified interval. A highest point on the graph of a function f over an interval $[a, b]$ is called

Recall from Section 6.2 that the graph of a continuous function over a closed interval has a highest point and a lowest point.

an absolute maximum point of the graph over $[a, b]$; a lowest point on the graph over $[a, b]$ is called an *absolute minimum* point of the graph over $[a, b]$. In Fig. 6.3-7, $P_1 = (x_1, f(x_1))$ is the absolute maximum point over $[a, b]$ and $P_2 = (x_2, f(x_2))$ is the absolute minimum point; we say that $f(x_1)$ is an absolute maximum value of f over $[a, b]$ and that $f(x_2)$ is an absolute minimum value of f over $[a, b]$. Notice in Fig. 6.3-7 that the point P_3 is a relative minimum but not an absolute minimum point over $[a, b]$; similarly, P_4 is a relative maximum but not an absolute maximum point over $[a, b]$.

FIGURE 6.3-7

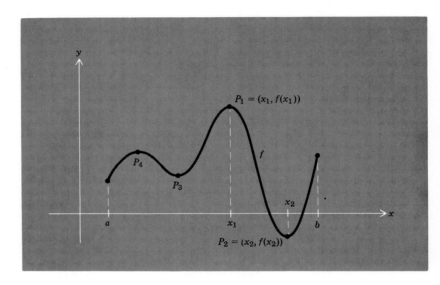

The absolute maximum value of a function f (with a smooth graph) over a closed interval $[a, b]$ might occur at either endpoint of the interval, or it might coincide with some relative maximum value of f (see Figs. 6.3-7 and 6.3-8). Thus to locate the absolute maximum value we compute $f(a)$, $f(b)$, and the relative maximum values of f over (a, b); the largest of these values is the absolute maximum value of f over $[a, b]$. Similarly, the absolute minimum value of f over $[a, b]$ is the smallest of $f(a)$, $f(b)$, and the relative minimum values of f over (a, b). If we want to discover the absolute maximum and absolute minimum values of f over $[a, b]$ we

(1)   solve $D_x f(x) = 0$ for all critical points of f

(2)   compute $f(a)$, $f(b)$, and $f(c)$ for each critical point c of f

The largest of these values is the absolute maximum, and the smallest is the absolute minimum.

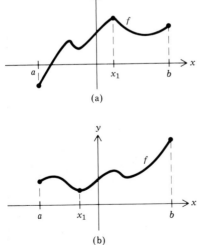

FIGURE 6.3-8   (a) The absolute maximum value is $f(x_1)$; the absolute minimum value is $f(a)$. (b) The absolute maximum value is $f(b)$; the absolute minimum value is $f(x_1)$.

EXAMPLE 6.3-5

Find the absolute maximum and absolute minimum values of the function $f(x) = x^2 \exp(x)$ over the interval $[-3, 1]$ and sketch the graph of f over this interval.

SOLUTION

First we find the critical points of f over $(-3, 1)$. Since $f(x) = x^2 \exp(x)$, we find that

$$D_x f(x) = 2x \exp(x) + x^2 \exp(x)$$

and

$$D_x^2 f(x) = 2 \exp(x) + 2x \exp(x) + 2x \exp(x) + x^2 \exp(x)$$

$$= 2 \exp(x) + 4x \exp(x) + x^2 \exp(x)$$

$$= (x^2 + 4x + 2)\exp(x)$$

Now, setting $D_x f(x) = 0$ and solving for x, we have

$$2x \exp(x) + x^2 \exp(x) = (2x + x^2)\exp(x) = 0$$

so that

$$x^2 + 2x = x(x + 2) = 0$$

Recall that since $\exp(x) \neq 0$ for all x, $(2x + x^2)\exp(x)$ can equal 0 only when $2x + x^2 = 0$.

which means $x = 0$ or $x = -2$; these are the critical points of f in $(-3, 1)$. Thus, the largest of the numbers $f(-3)$, $f(-2)$, $f(0)$, and $f(1)$ is the absolute maximum and the smallest of these numbers is the absolute minimum of f over $[-3, 1]$. Since we want to graph f, however, we will want to know about relative maximum and minimum values as well.

Since $D_x^2 f(0) = 2 \exp(0) = 2 > 0$, f has a relative minimum value at $x = 0$, namely, $f(0) = 0$; since $D_x^2 f(-2) = (4 - 8 + 2)\exp(-2) = -2 \exp(-2) < 0$, f has a relative maximum value at $x = -2$, namely, $f(-2) = 4 \exp(-2)$. The smallest of $f(-3)$, $f(1)$, and $f(0)$ is $f(0)$; this is the absolute minimum value of f over $[-3, 1]$. Similarly, the largest of $f(-3)$, $f(1)$, and $f(-2)$ is $f(1)$; this is the absolute maximum value of f over $[-3, 1]$ (see Fig. 6.3-9).

Since $D_x^2 f(0) > 0$ and $D_x^2 f(-2) < 0$, the second derivative test tells us that f has a relative maximum value at the critical point $x = -2$ and a relative minimum value at the critical point $x = 0$.

Using Table A in the back of the book, we find that $f(-3) = 9 \exp(-3) \approx 0.45$, $f(1) = \exp(1) = e \approx 2.72$, and $f(-2) = 4 \exp(-2) \approx 0.54$.

FIGURE 6.3-9

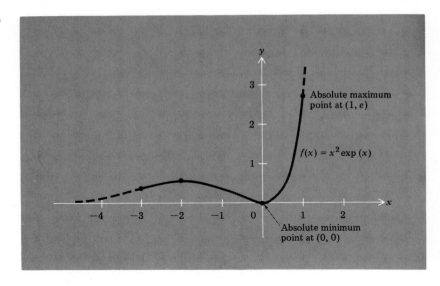

Recall from Example 4.3-5 that G is the annual total cost to a fertilizer retailer for receiving 10,000 tons of fertilizer a year in equal shipments of x tons each.

EXAMPLE 6.3-6

Find the absolute minimum value over the interval $[1, 10{,}000]$ of the cost function $G(x) = 100{,}000/x + 5x/2 + 150{,}000$ found in Example 4.3-5; sketch the graph of G over this interval.

SOLUTION

First we find the relative minimum values of G over $(1, 10{,}000)$. Since $G(x) = 100{,}000x^{-1} + 5x/2 + 150{,}000$, we see that

$$D_x G(x) = -100{,}000x^{-2} + \frac{5}{2}$$

and

$$D_x^2 G(x) = 200{,}000x^{-3}$$

Setting $D_x G(x) = 0$, we have

$$-\frac{100{,}000}{x^2} + \frac{5}{2} = 0$$

so

$$\frac{100{,}000}{x^2} = \frac{5}{2}$$

or

$$200,000 = 5x^2$$

Thus $x^2 = 40,000$, which means that $x = \pm 200$ are the critical points of G. Only the critical point $x = 200$ is in the interval $(1, 10,000)$; since $D_x^2 G(200) = 200,000(200)^{-3} > 0$, G has a relative minimum value at $x = 200$, namely, $G(200) = 100,000/200 + 5(200)/2 + 150,000 = 151,000$. Now, since

By the second derivative test, G has a relative minimum value at the critical point $x = 200$, since $D_x^2 G(200) > 0$.

$$G(1) = 100,000/1 + \tfrac{5}{2} + 150,000 = 250,002.5$$

and

$$G(10,000) = 100,000/10,000 + 5(10,000)/2 + 150,000$$

$$= 10 + 25,000 + 150,000 = 175,010$$

the smallest of $G(1)$, $G(200)$, and $G(10,000)$ is $G(200)$; this is the absolute minimum value of G over $[1, 10,000]$. Thus the cost G to the retailer is smallest when $10,000/200 = 50$ shipments of $x = 200$ tons each are ordered, and the minimum cost is $G(200) = \$151,000$ annually (see Fig. 6.3-10).

FIGURE 6.3-10

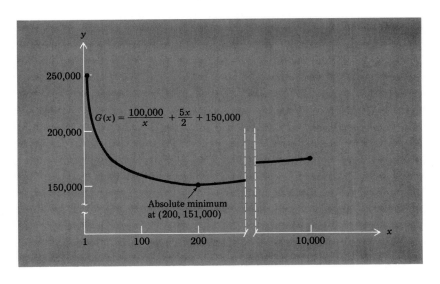

EXAMPLE 6.3-7

The County Pollution Control Board is conducting an investigation of the toxic pollutants emitted into the atmosphere by the Euphoria Machine and Foundry Manufacturing Plant. Data submitted to the board indicate that on an average work day, the quantity Q of toxic pollutants (in particles per cubic centimeter) in the smoke emitted from the factory at time t is closely approximated by the equation $Q(t) = 2t^3 - 24t^2 + 72t + 300$, where t is time in hours with $t = 0$ corresponding to 8:00 AM and $0 \le t \le 9$. Find the times of day between 9:00 AM and 5:00 PM when the quantity of pollutants in the smoke is at a maximum and when the quantity of pollutants is at a minimum. Sketch the graph of Q over the interval $[1, 9]$.

SOLUTION

We need to find the absolute maximum and minimum values of the function $Q(t) = 2t^3 - 24t^2 + 72t + 300$ over the interval $[1, 9]$. To do this, we first find the critical values of Q over $(1, 9)$ as follows. We find that $D_t Q(t) = 6t^2 - 48t + 72$ $= 6(t^2 - 8t + 12) = 6(t - 2)(t - 6)$; so setting $D_t Q(t) = 0$, we have $6(t - 2)(t - 6) = 0$, which means that the critical points of Q are $t = 2$ and $t = 6$. Thus the largest of the numbers $Q(1)$, $Q(2)$, $Q(6)$, and $Q(9)$ is the absolute maximum and the smallest of these numbers is the absolute minimum value of Q over $[1, 9]$. However, because we want to graph Q, we are also interested in relative maximum and minimum values. Since $D_t^2 Q(t) = 12t - 48 = 12(t - 4)$, we see that $D_t^2 Q(2) = 12(2 - 4) = -24 < 0$ and $D_t^2 Q(6) = 12(6 - 4)$ $= 24 > 0$. Thus Q has a relative maximum value at $t = 2$, namely, $Q(2) = 2(2)^3 - 24(2)^2 + 72(2) + 300 = 364$; similarly, Q has a relative minimum value at $t = 6$, namely, $Q(6)$ $= 2(6)^3 - 24(6)^2 + 72(6) + 300 = 300$. Now, since $Q(1)$ $= 2(1)^3 - 24(1)^2 + 72(1) + 300 = 350$ and $Q(9) = 2(9)^3 - 24(9)^2 + 72(9) + 300 = 462$, the smallest of $Q(1)$, $Q(6)$, and $Q(9)$ is $Q(6)$; this is the absolute minimum value of Q over $[1, 9]$. Also, the largest of $Q(1)$, $Q(2)$, and $Q(9)$ is $Q(9)$, so the absolute maximum value of Q over $[1, 9]$ is $Q(9)$. Thus the time of day between 9:00 AM and 5:00 PM when the smoke contains the minimum quantity of pollutants is $t = 6$ or 2:00 PM; at this time the smoke contains $Q(6) = 300$ particles per cubic centimeter. The time of day when the maximum quantity of pollutants is present is $t = 9$ or 5:00 PM, when the smoke contains $Q(9) = 462$ particles of toxic pollutants per cubic centimeter (see Fig. 6.3-11).

By the second derivative test, since $D_t^2 Q(2) < 0$, Q has a relative maximum value at $t = 2$.

Q has a relative minimum value at $t = 6$, since $D_t^2 Q(6) > 0$.

FIGURE 6.3-11

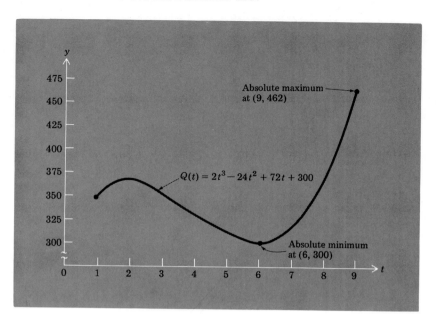

EXAMPLE 6.3-8

The Minute-Magic Company manufactures electronic micro-wave ovens. The company presently manufactures 20,000 units per month at a cost of $200 per unit. After a recent study of the company's equipment and personnel, it is estimated that an adjustment of x (thousands) of units in the company's current monthly production level will result in a production cost of $C(x) = x^4/2 + x^3 - x^2 + 202$ dollars per unit, for values of x between -3 and 2 inclusive. Find the adjustment in the monthly production level that will result in the mini-mum production cost per unit. What adjustment would pro-duce the maximum cost per unit? Sketch the graph of C.

SOLUTION

We wish to find the absolute maximum and minimum values of $C(x) = x^4/2 + x^3 - x^2 + 202$ over the interval $[-3, 2]$. To do this, we first find the critical values of C over $(-3, 2)$ as follows. We have

$$D_x C(x) = 2x^3 + 3x^2 - 2x$$

$$= x(2x^2 + 3x - 2)$$

$$= x(2x + 4)(x - \tfrac{1}{2})$$

By the second derivative test, since $D_x^2 C(0) < 0$, C has a relative maximum value at the critical point $x = 0$. Similarly, since $D_x^2 C(-2) > 0$ and $D_x^2 C(\frac{1}{2}) > 0$, C has relative minimum values at the critical points $x = -2$ and $x = \frac{1}{2}$.

so the critical points of C are the solutions to the equation $x(2x + 4)(x - \frac{1}{2}) = 0$, namely, $x = 0$, $x = -2$, and $x = \frac{1}{2}$. As in the earlier examples, the absolute maximum and minimum values could now be determined by computing immediately $C(-3)$, $C(-2)$, $C(0)$, and $C(2)$. We are going to sketch the graph of C, however, so we investigate further. Since

$$D_x^2 C(x) = 6x^2 + 6x - 2$$

we see that

$$D_x^2 C(0) = 6(0)^2 + 6(0) - 2 = -2 < 0$$

$$D_x^2 C(-2) = 6(-2)^2 + 6(-2) - 2 = 24 - 12 - 2 = 10 > 0$$

and

$$D_x^2 C\left(\tfrac{1}{2}\right) = 6\left(\tfrac{1}{2}\right)^2 + 6\left(\tfrac{1}{2}\right) - 2 = 2\tfrac{1}{2} > 0$$

Thus C has a relative maximum value at $x = 0$, a relative minimum value at $x = -2$, and a relative minimum value at $x = \frac{1}{2}$. Now, we find that

$$C(-3) = \frac{(-3)^4}{2} + (-3)^3 - (-3)^2 + 202 = 206.5$$

$$C(-2) = \frac{(-2)^4}{2} + (-2)^3 - (-2)^2 + 202 = 198$$

$$C\left(\tfrac{1}{2}\right) = \frac{\left(\tfrac{1}{2}\right)^4}{2} + \left(\frac{1}{2}\right)^3 - \left(\frac{1}{2}\right)^2 + 202 = 201.91$$

and

$$C(2) = \frac{(2)^4}{2} + (2)^3 - (2)^2 + 202 = 214$$

Thus the smallest of $C(-3)$, $C(-2)$, $C(0)$, and $C(2)$ is $C(-2) = 198$; this is the absolute minimum value of C over $[-3, 2]$. Similarly, the largest of $C(-3)$, $C(0)$, and $C(2)$ is $C(2)$; this is the absolute maximum value of C over $[-3, 2]$. This means that the minimum production cost of $C(-2) = \$198$ per unit will occur at $x = -2$, i.e., if the monthly production level is decreased by 2000 units from 20,000 to 18,000. The maximum production cost of $C(2) = \$214$ per unit will occur at $x = 2$, i.e., if the monthly production level is increased from 20,000 to 22,000 units per month (see Fig. 6.3-12).

FIGURE 6.3-12

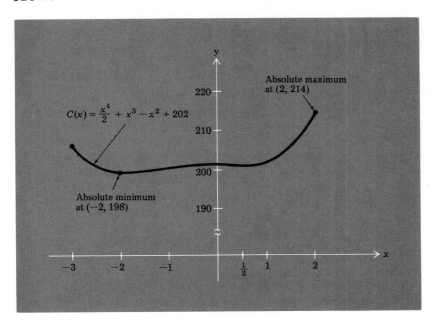

Exercises 6.3

In Exercises 1 through 12, find the intervals over which the graph of the given function is concave up and those over which the graph is concave down. Find the points of inflection (if any).

1. $f(x) = x^2 - 6$

2. $f(x) = -x^2 + 4$

3. $f(x) = x^2 - 2x + 14$

4. $f(x) = -x^2 + 6x - 10$

5. $f(x) = 2x^3 - 6x + 1$

6. $f(x) = x^3 + 3x - 2$

7. $g(t) = t^4 - 4t$

8. $v(t) = 2t^3 + 3t^2 - 12t + 1$

9. $s(z) = z^3 - 3z^2 - 9z + 10$

10. $z(t) = t + \dfrac{1}{t}$

11. $h(t) = \dfrac{\exp(t) + \exp(-t)}{2}$

12. $s(x) = x - \ln(x)$

In Exercises 13 through 24, use the second derivative test (if possible) to find the relative maximum and minimum values of the given function.

13. $f(x) = 2x^2 - 1$

14. $f(x) = -x^2 + \frac{1}{4}$

15. $f(x) = x^2 + 2x - 2$

16. $f(x) = -3x^2 + 6x - 1$

17. $g(x) = \dfrac{x^3}{3} - 3x^2 + 8x - 5$

18. $h(x) = \dfrac{x^3}{3} + \dfrac{5x^2}{2} + 4x - 2$

19. $h(t) = t + 4t^{-2}$ 20. $q(s) = s^5 - 5s$

21. $s(w) = (w^2 - 1)^2$ 22. $u(z) = \dfrac{4}{z^2 + 3}$

23. $s(t) = t^4 \exp(t)$ 24. $v(t) = [\ln(t)]^2$

Recall that if $D_x f(x)$ is positive for x in an interval (a, b), then f is increasing over (a, b); similarly, if $D_x f(x)$ is negative for x in (a, b), then f is decreasing over (a, b).

In Exercises 25 through 50, find the intervals over which the graph of the given function is increasing, those over which it is decreasing, those over which it is concave up, and those over which it is concave down. Find the points of inflection (if any) and the relative maximum and minimum values of the function; sketch the graph.

25. $f(x) = 7 - x^2$ 26. $f(x) = x^2 - 8x + 12$

27. $f(x) = (x + 4)^2$ 28. $f(x) = -(x + 4)^2$

29. $f(x) = x^2 - \dfrac{x}{2} - \dfrac{1}{2}$ 30. $f(x) = \dfrac{-3x^2}{2} + x - 1$

31. $f(x) = x^3 + 3x^2 - 24x - 4$ 32. $f(x) = -x^3 + 6x^2 + 36x - 2$

33. $f(x) = \dfrac{-x^3}{3} - \dfrac{7x^2}{2} - 12x + 1$

34. $f(x) = \dfrac{x^3}{3} - \dfrac{9x^2}{2} + 18x + 5$

35. $f(x) = x^4 - 8x^2$ 36. $f(x) = \dfrac{x^4}{4} + \dfrac{x^3}{3} - x^2 + 1$

37. $g(x) = \dfrac{x^4}{4 - 3x^3} + 9x^2 - 10$ 38. $h(t) = t^5 - \dfrac{5t^3}{3}$

39. $p(s) = s + \dfrac{3}{s^3}$ 40. $v(z) = (z^2 + 4)^{-\frac{1}{2}}$

41. $q(w) = \sqrt{w^2 + 4}$ 42. $v(z) = \exp(z^2)$

43. $z(t) = \ln(t^2 + 1)$ 44. $r(u) = 2 + \exp(u^2 + 2)$

45. $v(z) = z + \ln(z^2 + 1)$ 46. $s(t) = \ln(t^2 + 1) - t^2$

47. $p(x) = x^2 - 2 \exp(x^2)$ 48. $r(s) = \exp\left[\dfrac{-(s + 1)^2}{2} \right]$

49. $h(x) = \ln(\sqrt{x^2 + 1}\,)$ 50. $t(x) = x \exp\left(\dfrac{-x^2}{2} \right)$

In Exercises 51 through 62, find the absolute maximum and minimum values of the given function over the given interval.

51. $f(x) = x^2 - 4,\ [-1, 2]$

52. $f(x) = -3x^2 + 6x,\ [0, 3]$

53. $f(x) = 3x^2 - 6x + 1,\ [0, 3]$

54. $f(x) = -2x^2 + 8x - 6, \, [-1, 3]$

55. $f(x) = x^3 - 3x^2, \, [-\frac{1}{2}, 4]$

56. $f(x) = \dfrac{x^3}{3} - 4x + 1, \, [-3, 3]$

57. $h(x) = \dfrac{x^3}{3} - 2x^2 + 3x - 4, \, [-2, 5]$

58. $g(t) = \dfrac{-t^3}{3} - \dfrac{t^2}{2} + 2t - 1, \, [-3, 3]$

59. $z(w) = w^4 - 4w + 6, \, [0, 2]$

60. $s(t) = \sqrt{t^2 + 9}, \, [-5, \sqrt{7}\,]$

61. $p(v) = 4 \ln(v^2 + e), \, [-\sqrt{e^2 - e}, \, \sqrt{e^3 - e}\,]$

62. $r(t) = t^2 + \exp(t^2), \, [-1, 2]$

Recall that $e \approx 2.72$ and $\ln(e) = 1$.

63. Use the second derivative to verify that the parabola that is the graph of the quadratic equation $f(x) = ax^2 + bx + c$ (a $\neq 0$) opens upward if $a > 0$ and downward if $a < 0$ (see Section 2.4).

64. Let x and y be two positive numbers whose sum is 4 (thus, $x + y = 4$). Find x and y such that the product P of x^2 and y^2 is a maximum. (*Hint:* $y = 4 - x$, so $P(x) = x^2(4 - x)^2$.)

65. Porta-Vac Industries estimates that the monthly sales revenue R (in dollars) received from the production of x hundreds of units of its best portable auto vacuum cleaner is given by $R(x) = 10(-x^2 + 24x + 1656)$ for $10 \leqslant x \leqslant 15$. Find (a) the number of units the company should produce per month to maximize its monthly revenue, (b) the maximum monthly revenue, and (c) the monthly production output x ($10 \leqslant x \leqslant 15$) that yields the minimum monthly revenue.

66. The Power-Mite Company manufactures lightweight portable chain saws. The company charges \$80 per saw on orders of 20 units or less. For an order of x saws ($x > 20$), the price of each saw ordered is reduced by $(x - 20)$ dollars. Show that the revenue R received by the company for an order of x saws ($x \geqslant 20$) is given by the equation $R(x) = 100x - x^2$. What size order yields maximum revenue to the company? For an order yielding maximum revenue, what does the company charge per saw?

67. In Exercise 66, suppose that it costs the Power-Mite Company \$50 to manufacture each saw. Show that the profit P realized by the company from an order of x saws ($x \geqslant 20$) is given by the equation $P(x) = 50x - x^2$. (Recall that profit is revenue less cost.) What size order yields maximum profit? For an order yielding maximum profit, what does the company charge per saw?

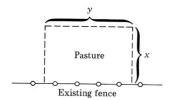

Existing fence

In Exercise 72, the volume V and surface area S are given by $V = x^2y$ and $S = 2x^2 + 4xy$.

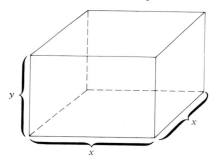

In Exercise 73, the volume V and surface area S are given by the formulas $V = 2x^2y$ and $S = 2x^2 + 6xy$.

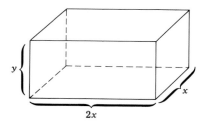

68. The Nation-Wide Bus System finds that the cost C, in dollars per mile, of operating one of its buses is given by $C(x) = 50(x^{-1} + x/2500)$, where x is the average speed (in miles per hour) of the bus. At what average speed should the company operate its buses in order to minimize the operating cost per mile? What is the minimum operating cost per mile?

69. A dairy farmer has 400 ft of fencing with which to construct a rectangular pasture adjacent to an existing 600-ft fence (see diagram at left). Find the dimensions of the pasture that will yield the maximum grazing area. (Assume that the existing fence will be used for one side of the pasture.)

70. Easy-Racer, Inc. finds that the monthly cost C (in thousands of dollars) to produce x units per month of its best quality 10-speed racing bike is given by $C(x) = x^2/1000 - 2x + 1050$ for $800 \leqslant x \leqslant 1100$. Find the number of bikes the company should produce per month to minimize its monthly production cost and the minimum monthly production cost. What monthly production output x ($800 \leqslant x \leqslant 1100$) yields the maximum monthly production cost?

71. The Crescent Candy Company estimates that its profit P from the production of x thousands of bags of its special Halloween Candy Mix is given by $P(x) = 1000(-x^2/4 + 8x - 60)$, where P is in dollars and $15 \leqslant x \leqslant 18$. How many bags of candy should the company produce in order to maximize its profit? What is the maximum profit? What production output x ($15 \leqslant x \leqslant 18$) yields the minimum profit?

72. The Capital Container Company has a contract to manufacture 10,000 cardboard boxes (with tops) that are to have a square base and contain 8 cubic ft each. Find the dimensions of the boxes that will minimize the material used in their manufacture. (*Hint*: A box of depth y ft with top and base a square x ft on a side has total surface area S given by $S = 2x^2 + 4xy$. Express S as a function of x alone and then find x such that $S(x)$ is minimum.)

73. Capital Container receives another order (see Exercise 72) for 25,000 boxes (without tops) whose bases are to be rectangles twice as long as they are wide, each box to contain 36 cubic ft. Find the dimensions of the boxes that will minimize the material used in their manufacture. (*Hint*: The volume V of such boxes is given by $V = 2x^2y$, and the surface area S is given by $S = 2x^2 + 2xy + 4xy = 2x^2 + 6xy$. See diagram.)

74. Cavalier Camping Supplies, Inc. determines the selling price S (in dollars) for its economy synthetic-insulated sleeping bag by the equation $S(x) = x^2/2 - 8.25x + 45$, where x is the number of thousands of sleeping bags manufactured per year. This equation is used for values of x between 4 and 9 inclusive. How many sleeping bags should the company produce each year in order to minimize

the selling price of each bag? What is the minimum selling price? What yearly production output x $(4 \leqslant x \leqslant 9)$ results in the maximum selling price per bag?

75. The yearly sales revenue R of the Cavalier Company from the sale of its sleeping bags (see Exercise 74) is given by $R(x) = 1000x \, S(x)$, i.e., $R(x)$ is the number of bags produced during the year ($1000x$) times the selling price $S(x)$ per bag (assuming that all bags produced are sold). Using the equation $S(x) = x^2/2 - 8.25x + 45$ from Exercise 74, write the equation that defines $R(x)$. What production output x $(4 \leqslant x \leqslant 9)$ maximizes the sales revenue R? What is the maximum yearly sales revenue? What yearly production output x $(4 \leqslant x \leqslant 9)$ results in the minimum sales revenue?

76. The yearly profit P (in dollars) of the Cavalier Company from the sale of its sleeping bags is given by $P(x) = R(x) - C(x)$, where $R(x)$ is the yearly sales revenue, $C(x)$ is the yearly production cost, and x is the number of thousands of bags produced during the year (see Exercises 74 and 75). If $C(x) = 1000(-3x^2/2 + 15x + 22.5)$ for $4 \leqslant x \leqslant 9$, find
     (a)   the equation that defines $P(x)$;
     (b)   the production output x $(4 \leqslant x \leqslant 9)$ that maximizes the yearly profit P;
     (c)   the maximum yearly profit;
     (d)   the production output x $(4 \leqslant x \leqslant 9)$ that minimizes the yearly profit P;
     (e)   The selling price S of each bag at the production level necessary to achieve maximum profit (use the equation for $S(x)$ given in Exercise 74).

77. A lake, polluted by coliform bacteria, is treated with bactericidal agents. Biologists estimate that t days after treatment, the number N of viable bacteria per milliliter will be approximated by the equation $N(t) = 100[t/10 - \ln(t/10)] - 30$ for $1 \leqslant t \leqslant 12$. How many days after treatment will the minimum number N of viable bacteria per milliliter be present? What is this minimum number? For what value of t $(1 \leqslant t \leqslant 12)$ is the number $N(t)$ a maximum?

78. The Public Health Office predicts that the percent P of the city's population that will be infected with a communicable disease t days after detection of the disease is given by the equation $P(t) = 10t \exp(-t/10)$ for $1 \leqslant t \leqslant 15$. How many days after detection will the maximum percent of the population be infected? What is the maximum percent of the population that will be infected at any given time?

79. The County Planning Commission finds that the population P (in thousands) of the county for each year since 1964 is closely approximated by the function $P(t) = 80[t - t \ln(t/8)]$, where t is time in years with $t = 1$ corresponding to 1964 and $1 \leqslant t \leqslant 12$. In

what year was the population of the county the greatest? What was the population in that year? In what year was the population the least?

In Exercise 80, the domain of the function N consists only of integer values of t between 5 and 30 inclusive. However, in order to apply the techniques of calculus we assume that the function is defined for all real numbers t such that $5 \leqslant t \leqslant 30$.

80. The Hefty Oil Company recently opened a new service station. A grand opening celebration was held during the first 10 days of operation to draw customers. The company found later that the number N of gallons of gasoline sold at the station on Day t after the station opened was closely approximated by the function $N(t) = 80[t^2 \exp(-t/5)] + 2000$ for $5 \leqslant t \leqslant 30$. On what day after opening the station was the maximum number of gallons of gasoline sold? What was this maximum number of gallons? On what day t ($5 \leqslant t \leqslant 30$) after the station opened was the minimum number of gallons of gasoline sold?

In Exercise 81, the volume V and the surface area S are given by the formulas $V = \pi r^2 h$ and $S = 2\pi r^2 + 2\pi rh$.

81. The Century Can Company receives an order from a dog food processor for tin cans which will hold 16 cubic in. of dog food. Find the dimensions of the cans that will minimize the material used in their manufacture. (*Hint:* If the radius of the can is r in. and the height is h in., then the volume V is given by $V = \pi r^2 h$ and the surface area S is given by $S = 2\pi r^2 + 2\pi rh$.)

82. An orchard contains 40 pear trees per acre; the average yield is 280 pears per tree. For each additional tree planted per acre, the yield decreases by 4 pears per tree. How many additional trees per acre should be planted in order to maximize the yield per acre? What is the maximum yield per acre?

83. Chester's Charter Service offers round trip charter flights to Europe at $450 per person for groups containing at least 50 people. Because of a recent decline in business, the airline has decided to reduce the fare per person by $3 for each person in excess of 50 making the trip. Find
   (a) an equation for the revenue R received from a charter flight for x people ($x \geqslant 50$);
   (b) the number x ($x \geqslant 50$) of people that will result in maximum revenue;
   (c) the maximum revenue the company can receive from a flight;
   (d) the fare per person when the company receives maximum revenue.

84. Atlas Aquariums, Inc. is planning to produce an 8-cubic-ft aquarium with rectangular base whose length is twice its width. The material for the base costs $0.75 per square ft, and the glass for the sides costs $1 per square ft. Find the dimensions for the aquarium that will minimize the cost of the material used in its construction. What is the minimum cost of the material for an aquarium?

85. Golden State Homes, Inc. is constructing a subdivision on a small island located 1 mi north of the shoreline (see diagram at left). The company wishes to run a sewer line from the island (point A) to an existing sewer main (point C) located 3 mi east of the point on shore nearest the island (point B). It costs $3 per ft to lay sewer pipe along the shore and $5 per ft to run the pipe under water. Find the point D on shore (between B and C) to which the pipe should be laid from point A and then to point C in order to minimize the cost. What is the minimum cost for laying the pipe?

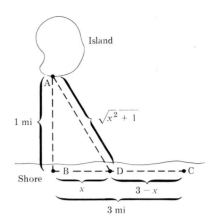

6.4. Linear Approximations: The Differential

It is evident from our discussions thus far that the notion of the derivative of a function is indeed a useful tool. We have discussed its interpretation as the slope of the tangent line to a curve, we have seen how it may be used to solve problems involving rates of change, and we have shown how it may be used to find the maximum and minimum values of a given function. In this section we investigate yet another use for the derivative, namely, how the derivative of a function f at a point x_0 can be used to approximate the value of f at a point x near x_0.

Consider, for example, the graph of the function $f(x) = \sqrt{x}$ shown in Fig. 6.4-1. Since $D_x f(x) = 1/(2\sqrt{x})$, the slope of the line l tangent to the graph of f at the point $(4, 2)$ is $D_x f(4) = 1/(2\sqrt{4}) = \frac{1}{4}$. Thus the

Recall from Section 2.3 that the equation of the line through the point $P = (x_0, y_0)$ with slope m is $y - y_0 = m(x - x_0)$.

FIGURE 6.4-1

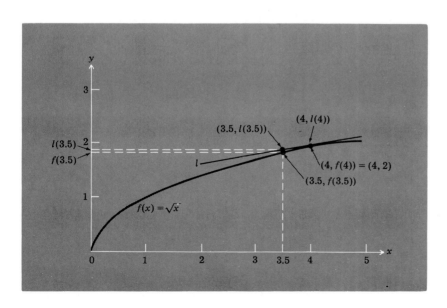

Recall that the symbol $\approx$ means "approximately equal to."

equation for the line l is $y - 2 = \frac{1}{4}(x - 4)$, i.e., $y = l(x) = x/4 + 1$. Now, suppose we wish to find an approximate value for $f(3.5) = \sqrt{3.5}$. From Fig. 6.4-1, it appears that $l(3.5)$ is approximately the same as $f(3.5)$, and $l(3.5) = (3.5)/4 + 1 = 1.875$ is certainly easier to compute than $f(3.5) = \sqrt{3.5}$. Thus we can approximate $f(3.5)$ by $l(3.5) = 1.875$ and write $f(3.5) \approx 1.875$. This approximation to $\sqrt{3.5}$ is quite good, since upon squaring 1.875 we find that $(1.875)^2 = 3.515625$. Note, however, that since the vertical separation between the tangent line l and the graph of f in Fig. 6.4-1 becomes larger as we move away from the point $(4, 2)$ on the graph, we should not expect that $l(x)$ will be a good approximation for $f(x)$ unless x is close to 4.

This procedure can be used for any function f that is differentiable at a point x_0. The derivative $D_x f(x_0)$ of f at x_0 is the slope of the line l tangent to the graph of f at $(x_0, f(x_0))$, so the equation of l is $y - f(x_0) = D_x f(x_0)(x - x_0)$, i.e., $y = l(x) = f(x_0) + D_x f(x_0)(x - x_0)$. Thus for values of x close to x_0, we have the (linear) approximation $f(x) \approx l(x) = f(x_0) + D_x f(x_0)(x - x_0)$ (see Fig. 6.4-2).

FIGURE 6.4-2

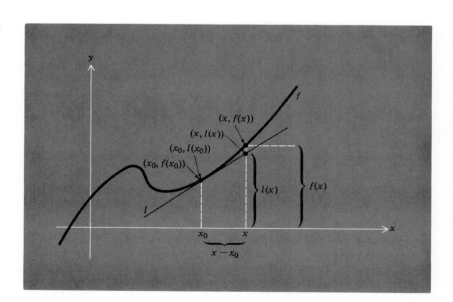

EXAMPLE 6.4-1

Find the linear approximation to the function $f(x) = (1 - x)^{-1}$ at $x_0 = 0$. Use the linear approximation to find an approximate value for $f(0.01)$.

SOLUTION

The linear approximation $l(x)$ to $f(x)$ at $x_0 = 0$ is given by

$$f(x) \approx l(x) = f(x_0) + D_x f(x_0)(x - x_0)$$

with $x_0 = 0$. Since $f(x) = (1 - x)^{-1}$, we have $f(x_0) = f(0) = (1 - 0)^{-1} = 1$. Also, since

$$D_x f(x) = -(1 - x)^{-2}(-1) = (1 - x)^{-2}$$

we find that

$$D_x f(x_0) = D_x f(0) = (1 - 0)^{-2} = 1$$

Thus

$$l(x) = 1 + 1(x - 0) = x + 1$$

and this is the linear approximation to $f(x) = (1 - x)^{-1}$ at $x_0 = 0$. Since 0.01 is close to $x_0 = 0$, we expect $l(0.01) = 0.01 + 1 = 1.01$ to be a good approximation for $f(0.01)$, and since $f(0.01) = (1 - 0.01)^{-1} = (0.99)^{-1} = 1/0.99 = 1.\overline{01}$, the approximation is in fact excellent.

Recall from Section 1.2 that $1.\overline{01} = 1.010101\ldots$.

EXAMPLE 6.4-2

Find the linear approximation to the function $f(x) = \ln(x)$ at $x_0 = 1$. Use the linear approximation to find an approximate value for $\ln(1.04)$.

SOLUTION

The linear approximation $l(x)$ is given by $f(x) \approx l(x) = f(x_0) + D_x f(x_0)(x - x_0)$ with $x_0 = 1$ and $f(x) = \ln(x)$. Since

$$D_x f(x) = D_x \ln(x) = \frac{1}{x}$$

we have $D_x f(x_0) = D_x f(1) = 1/1$. Also, $f(x_0) = f(1) = \ln(1) = 0$, so $l(x) = 0 + 1(x - 1) = x - 1$ is the linear approximation to $f(x) = \ln(x)$ at $x_0 = 1$. Now, since 1.04 is close to 1, $l(1.04)$ should be a good approximation to $f(1.04)$, and we find that $l(1.04) = 1.04 - 1 = 0.04$. Using Table A in the back of the book, we see that $f(1.04) = \ln(1.04) = 0.0392$, so the approximation is indeed good.

EXAMPLE 6.4-3

Use a linear approximation to estimate the value of $\sqrt{15}$.

SOLUTION

We wish to find an approxmate value for the function $f(x) = \sqrt{x}$ at $x = 15$. The linear approximation $l(x)$ to $f(x)$ at the point x_0 is given by $f(x) \approx l(x) = f(x_0) + D_x f(x_0)$ $(x - x_0)$, and the approximation is good provided x is close to x_0. Thus we wish to choose x_0 close to $x = 15$, and in addition we want x_0 such that $f(x_0) = \sqrt{x_0}$ and $D_x f(x_0) = 1/(2\sqrt{x_0}\cdot)$ can be readily evaluated. Evidently then, the choice for x_0 should be $x_0 = 16$, and for this choice we have

$$D_x f(x_0) = D_x f(16) = \frac{1}{2\sqrt{16}} = \frac{1}{8}$$

and

$$f(x_0) = f(16) = \sqrt{16} = 4$$

Thus

$$l(x) = f(x_0) + D_x f(x_0)(x - x_0) = 4 + \tfrac{1}{8}(x - 16) = \frac{x}{8} + 2$$

and

$$f(15) = \sqrt{15} \approx l(15) = \tfrac{15}{8} + 2 = \tfrac{31}{8} = 3.875$$

Since $(3.875)^2 = 15.015625$, our approximation of $\sqrt{15} = 3.875$ is quite good.

Let us now examine more closely some of the geometric aspects of the linear approximation $l(x) = f(x_0) + D_x f(x_0)(x - x_0)$ to $f(x)$ at the poin x_0. As shown in Fig. 6.4-3, when x_1 is close to x_0, the change $f(x_1) -$

FIGURE 6.4-3

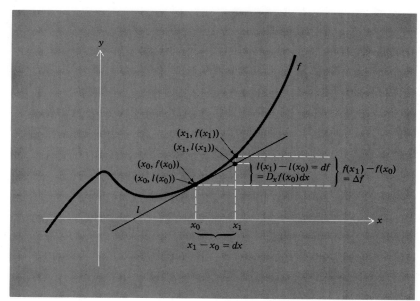

In Fig. 6.4-3, recall from Section 5.3 that $f(x_1) - f(x_0)$ is frequently denoted Δf. Thus, we see that $df \approx \Delta f$, i.e., $df(x_0, dx)$ is approximately the change Δf in f as x changes from x_0 to x_1.

$f(x_0)$ in f from $x = x_0$ to $x = x_1$ is approximately the same as the change $l(x_1) - l(x_0)$ in l. But since the slope of l is given both by $D_x f(x_0)$ and by

$$\frac{l(x_1) - l(x_0)}{x_1 - x_0}$$

we have

$$D_x f(x_0) = \frac{l(x_1) - l(x_0)}{x_1 - x_0}$$

which means that the change $l(x_1) - l(x_0)$ is equal to $D_x f(x_0)(x_1 - x_0)$. Thus

$$f(x_1) - f(x_0) \approx l(x_1) - l(x_0) = D_x f(x_0)(x_1 - x_0)$$

which means that the change in f from $x = x_0$ to $x = x_1$ is approximately equal to $D_x f(x_0)(x_1 - x_0)$. The quantity $D_x f(x_0)(x_1 - x_0)$ is called the *differential* of f at $x = x_0$ for the change $x_1 - x_0$. It is customary in this context to denote the change $x_1 - x_0$ by dx rather than Δx and the differential by $df(x_0, dx)$ or simply df; thus

$$df = df(x_0, dx) = D_x f(x_0)dx$$

With this notation, we see that df is defined at each x for which $D_x f(x)$ exists and that

$$df = D_x f(x)(x_1 - x_0) = D_x f(x)dx$$

Definition of the differential of a function f

Recall from Section 5.3 that $x_1 - x_0$ is frequently denoted by Δx.

The value of the differential $df = df(x, dx) = D_x f(x) dx$ at x depends upon two quantities, namely, x and dx.

this suggests the alternative notation

$$\frac{df}{dx}$$

for the derivative $D_x f$ (see Section 5.2). Also, the linear approximation

$$l(x) = f(x_0) + D_x f(x_0)(x - x_0)$$

to $f(x)$ at x_0 may now be written as

$$f(x) \approx f(x_0) + df(x_0, dx)$$

In the differential approximation formula $f(x) \approx f(x_0) + df(x_0, dx)$, the quantity dx is the change $x - x_0$. Thus, since $dx = x - x_0$, we have $x = x_0 + dx$. For this reason the differential approximation formula often appears as $f(x_0 + dx) \approx f(x_0) + df(x_0, dx)$.

Thus the approximation is frequently called the *differential approximation* to $f(x)$ at $x = x_0$.

EXAMPLE 6.4-4

Find the differential $df = df(x, dx)$ for the function $f(x) = 3x^2 + 2x - 1$, and evaluate df at $x = x_0 = 2$ for $dx = \frac{1}{4}$. Then use the formula $f(x_0 + dx) \approx f(x_0) + df(x_0, dx)$ to approximate the value of $f(2 + \frac{1}{4}) = f(2.25)$.

SOLUTION

Since $df = df(x, dx) = D_x f(x)dx$, we have $df = (6x + 2)dx$. Thus when $x = x_0 = 2$ and $dx = \frac{1}{4}$, $df(x_0, dx) = [6(2) + 2](\frac{1}{4}) = \frac{7}{2}$. Now, using the formula $f(x_0 + dx) \approx f(x_0) + df(x_0, dx)$, we find that $f(2 + \frac{1}{4}) = f(2.25) \approx f(2) + \frac{7}{2} = [3(2^2) + 2(2) - 1] + \frac{7}{2} = 15 + \frac{7}{2} = \frac{37}{2} = 18.5$. Computing $f(2.25)$ directly from the equation $f(x) = 3x^2 + 2x - 1$ yields $f(2.25) = 18.6875$, so our approximation is good.

EXAMPLE 6.4-5

A city in South America has been struck by an earthquake and a subsequent typhoid epidemic. The government has ordered evacuation of the city, and authorities estimate that the city's population P (in millions) t days after beginning evacuation will be given by $P(t) = \exp(0.05t - 0.01t^2)$ for $0 \leqslant t \leqslant 15$. Without using tables, estimate the city's population 11 days after the evacuation order.

SOLUTION

We wish to approximate the value of $P(11) = \exp[(0.05)(11) - (0.01)(121)] = \exp(-0.66)$. To do this, we use the formula $f(x) \approx f(x_0) + df(x_0, dx)$ to approximate the value of $f(x) = \exp(x)$ at $x = -0.66$. First we choose a point x_0 close to -0.66 for which $f(x_0) = \exp(x_0)$ and $df(x_0, dx) = D_x f(x_0) dx = \exp(x_0) dx$ can be evaluated. Thus we choose $x_0 = -1$, and for this choice we have

$$f(x_0) = \exp(x_0) = \exp(-1) = e^{-1} = \frac{1}{e} \approx \frac{1}{2.718} \approx 0.368 \qquad \text{Recall that } e \approx 2.718.$$

$$dx = x - x_0 = -0.66 - (-1) = -0.66 + 1 = 0.34$$

and

$$df(x_0, dx) = \exp(x_0) \, dx = \exp(-1)(0.34)$$

$$= (0.368)(0.34) \approx 0.125$$

Thus we have $f(-0.66) = \exp(-0.66) \approx 0.368 + 0.125 = 0.493$, which means that the population of the city 11 days after the evacuation order will be approximately 0.493 million people. Table A in the back of the book yields $f(-0.66) = \exp(-0.66) = 0.5169$, so our approximation is quite good.

EXAMPLE 6.4-6

Overland Freight, Inc. estimates that the cost c in dollars per mile of operating one of its trucks is given by $c(x) = 60(x^{-1} + x/2025)$, where x is the average speed of the truck in miles per hour. Use differentials to approximate the increase in the cost per mile when the average speed changes from $x_0 = 60$ mph to $x = 62$ mph. What is the approximate operating cost per mile at an average speed of 62 mph?

SOLUTION

The approximate change in c as the speed changes from $x_0 = 60$ mph to $x = 62$ mph is given by $dc(x_0, dx)$ with $x_0 = 60$ and $dx = x - x_0 = 62 - 60 = 2$. Since $c(x) = 60(x^{-1} + x/2025)$, we find that $D_x c(x) = 60(-x^{-2} + 1/2025)$, which means that $dc(x_0, dx) = D_x c(x_0) \, dx = 60(-x_0^{-2} + 1/2025) \, dx = 60(-1/x_0^2 + 1/2025) \, dx = 60(1/3600 + 1/2025)(2) = 120(3600 - 2025)/(3600)(2025) = 0.026.$

Thus the cost c increases by approximately \$0.026 per mile when the average speed changes from 60 to 62 mph. Now, using the formula $c(x) \approx c(x_0) + dc(x_0, dx)$ with $x = 62$, $x_0 = 60$, and $dc(x_0, dx) = 0.026$, we find that $c(62) \approx c(60) + 0.026 = 60(1/60 + 60/2025) + 0.026 = 60(0.046) + 0.026 = 2.79$. This means that the approximate operating cost per mile at an average speed of 62 mph is \$2.79. If we compute $c(62)$ directly using the equation $c(x) = 60(x^{-1} + x/2025)$, we find that $c(62) = 2.80$, so our approximation is indeed good.

Exercises 6.4

In Exercises 1 through 20, find the linear approximation $l(x) = D_x f(x_0)(x - x_0) + f(x_0)$ to the given function f at the given point x_0, and use the linear approximation to find an approximate value for $f(x)$ at the given value of x.

1. $f(x) = x^2$, $\quad x_0 = 3$, $x = 3.2$

2. $f(x) = 4x^2 - 5$, $\quad x_0 = -4$, $x = -4.2$

3. $f(x) = -3x^3$, $\quad x_0 = -2$, $x = -1.9$

4. $f(x) = 2x^3 + 4$, $\quad x_0 = 3$, $x = 2.8$

5. $f(x) = 2x^5 - 3x^2 + 1$, $\quad x_0 = 1$, $x = 1.1$

6. $f(x) = -3x^4 + 2x - 3$, $\quad x_0 = 2$, $x = 1.9$

7. $f(x) = 4x^{-2} + \dfrac{x}{2} - 1$, $\quad x_0 = 2$, $x = 2.04$

8. $f(x) = -3x^{-3} + 2x^{-1}$, $\quad x_0 = -1$, $x = -1.01$

9. $f(x) = 2\sqrt{x + 5}$, $\quad x_0 = 4$, $x = 3.8$

10. $f(x) = -\sqrt{x^2 - 5}$, $\quad x_0 = 3$, $x = 3.18$

11. $f(x) = 2x^{-3} - 3\sqrt{1 - x}$, $\quad x_0 = -3$, $x = -3.14$

12. $f(x) = \dfrac{-x^5}{\sqrt{x + 3}} + x^{-2}$, $\quad x_0 = 1$, $x = 0.98$

13. $f(x) = \dfrac{\sqrt{6 - x}}{\sqrt{1 - x}}$, $\quad x_0 = -3$, $x = -3.02$

14. $f(x) = \dfrac{-6}{\sqrt{7 - x}}$, $\quad x_0 = -2$, $x = -1.8$

15. $f(x) = (x + 4)^6$, $x_0 = -2$, $x = -2.01$

16. $f(x) = (x - 2)^5(x - 5)$, $x_0 = 3$, $x = 2.92$

17. $f(x) = \exp(x) + 3 \ln(x + 1)$, $x_0 = 0$, $x = -0.1$

18. $f(x) = (x^2 + \sqrt{x}) \exp(x - 1)$, $x_0 = 1$, $x = 1.02$

19. $f(x) = \sqrt{x} \ln(x) + 2$, $x_0 = 1$, $x = 0.98$

20. $f(x) = \dfrac{2 \ln(x - 3)}{\sqrt{x}}$, $x_0 = 4$, $x = 3.95$

In Exercises 21 through 36, use a linear approximation to estimate the value of the given quantity.

21. $\sqrt{26}$ 22. $\sqrt{37}$

23. $\sqrt{8}$ 24. $\sqrt{63}$

25. $\sqrt{50.1}$ 26. $\sqrt{80.2}$

27. $(1.96)^3$ 28. $(-2.04)^4$

29. $(-0.98)^7$ 30. $(2.14)^5$

31. $(1.87)^{-4}$ 32. $(-1.16)^{-3}$

33. $\exp(-0.07)$ 34. $\ln(1.14)$

35. $\exp(0.17)$ 36. $\ln(0.89)$

In Exercises 37 through 50, find the differential $df = df(x, dx)$ for the given function and evaluate df at $x = x_0$ for the given values of x_0 and dx. Then use the formula $f(x_0 + dx) \approx f(x_0) + df(x_0, dx)$ to approximate the value of $f(x_0 + dx)$.

37. $f(x) = 2x + 3$, $x_0 = 3$, $dx = \frac{1}{2}$

38. $f(x) = -5x + 2$, $x_0 = -2$, $dx = \frac{1}{4}$

39. $f(x) = 5x^2 - 2x + 3$, $x_0 = 2$, $dx = -\frac{1}{6}$

40. $f(x) = -3x^2 - 4x + 5$, $x_0 = -3$, $dx = -\frac{1}{7}$

41. $f(x) = 2x^5 - 3x^2 + x - 1$, $x_0 = -1$, $dx = \frac{1}{3}$

42. $f(x) = 4x^4 + 2x^3 - x + 2$, $x_0 = -2$, $dx = \frac{1}{5}$

43. $f(x) = 4x^{-4} + 3\sqrt{x}$, $x_0 = 4$, $dx = \frac{1}{2}$

44. $f(x) = -3x^{-2} + \sqrt{x - 1}$, $x_0 = 5$, $dx = -\frac{1}{8}$

45. $f(x) = \frac{1}{4}x^{-3} + \sqrt{x^2 + 3}$, $x_0 = 1$, $dx = -\frac{1}{5}$

46. $f(x) = 5(x^2 + 1) \ln(x)$, $x_0 = 1$, $dx = \frac{1}{6}$

47. $f(x) = (x^3 + 2x)\exp(x^2)$, $x_0 = -1$, $dx = -\frac{1}{7}$

Recall that $e \approx 2.718$ and $\ln(e) = 1$.

48. $f(x) = \dfrac{2x^3}{\ln(x)}$, $x_0 = e$, $dx = -\frac{1}{2}$

49. $f(x) = \sqrt{x + 2}\, \exp(x - 2) - 5\ln(x^2 - 3)$, $x_0 = 2$, $dx = \frac{1}{8}$

50. $f(x) = \dfrac{\sqrt{x}\, \exp(2x^2)}{\ln(x + 1) + 1}$, $x_0 = 0$, $dx = \frac{1}{4}$

51. Given two functions f and g, verify that $d[f + g] = df + dg$ and that $d(f - g) = df - dg$.

52. Verify that the formula $d[fg] = g\,df + f\,dg$ is valid for finding the differential of the product of two functions f and g.

53. Verify that the formula $d[f/g] = [g\,df - f\,dg]/g^2$ is valid for finding the differential of the quotient of two functions f and g.

54. The Capital Container Company receives an order for 10,000 cubical cardboard boxes measuring 8.15 in. on a side. Use the differential approximation formula $V(x) \approx V(x_0) + dV(x_0, dx)$ with $x_0 = 8$ to approximate the volume of each of the boxes.

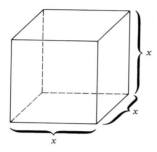

55. The Faultless Aluminum Foundry is manufacturing cubical aluminum boxes with an interior measurement of exactly 11 in. along each edge. The sides, top, and bottom of each box are each 0.4-in. thick. Use differentials to approximate the volume of metal used in the construction of each box.

In Exercise 56, the quantity $dC(x_0, dx)$ is the approximate change in C as the production level changes from $x_0 = 12$ thousands of units to $x = 12.24$ thousands of units.

56. Creative Confections, Inc. estimates that its production cost C (in dollars) to produce x thousands of boxes of its miniature chocolate Easter Bunnies is given by $C(x) = -4x^2 + 500x + 500$, for values of x between 10 and 20, inclusive. When the level of production changes from $x_0 = 12$ to $x = 12.24$, what is the value of $dC(x_0, dx)$? Using the formula $C(x) \approx C(x_0) + dC(x_0, dx)$, what is the approximate production cost for manufacturing 12,240 boxes of Bunnies?

57. If Creative Confections (see Exercise 56) sells each box of Bunnies for $0.75, then the revenue R received from the manufacture of x thousands of boxes of the candy is $R(x) = (0.75)(1000x) = 750x$ (assuming that the company can sell all boxes of candy manufactured). Find the equation that gives the company's profit $P(x)$ from the manufacture and sale of x thousands of boxes of candy for $10 \leqslant x \leqslant 20$ (recall that profit is revenue less cost). When the production level changes from $x_0 = 18$ to $x = 18.45$, what is the value of $dP(x_0, dx)$? Using the formula $P(x) \approx P(x_0) + dP(x_0, dx)$, what is the approximate profit P from the manufacture and sale of 18,450 boxes of candy?

58. The monthly sales revenue R received from the production of x hundreds of shirts at the Superior Shirt Company is given by $R(x) = -8x^2 + 240x + 15,000$ for $12 \leqslant x \leqslant 15$. When the production level changes from $x_0 = 13$ to $x = 13.15$, what is the value

of $dR(x_0, dx)$? Use the formula $R(x) \approx R(x_0) + dR(x_0, dx)$ to approximate the revenue R for a month during which 1315 shirts are produced.

59. The Carter Cutlery Company finds that the monthly demand D for its four-piece steak knife set is given by $D(x) = 100[4 - \ln(x^2/400)]$, where x is the selling price (in dollars) per set and $20 \leqslant x \leqslant 40$. Find the value of $dD(x_0, dx)$ when $x_0 = 30$ and $dx = 1.65$. Use the formula $D(x) \approx D(x_0) + dD(x_0, dx)$ to approximate the monthly demand D when the selling price per set is \$31.65. (Use Table A in the back of the book to find the value of $\ln(2.25)$.)

60. A large department store estimates that by the end of a week-long advertising campaign, its average daily sales revenue will increase by \$5000. Store officials also predict that t days after the end of the campaign, the extra daily sales revenue R from the advertising will be given by $R(t) = 5000 \exp(-t/4)$ for values of t between 0 and 10 inclusive. Use differentials to approximate the decrease in the extra sales revenue as t changes from 8 to 9. Approximately how much extra sales revenue will the store receive on the ninth day after the advertising campaign ends? (Do not use tables.)

Recall that $e \approx 2.72$.

61. Steel ball bearings 4 cm in diameter for railroad locomotives are found to decrease 0.05 cm in diameter during a year's use. Use differentials to approximate the decrease in the volume of a bearing during a year. (The volume V of a sphere of radius r is given by $V = 4\pi r^3/3$.)

62. A small leak in a spherical weather balloon has caused the radius of the balloon to decrease from 6 ft to 5.75 ft. Use differentials to approximate the decrease in the surface area of the balloon. (The surface area S of a sphere of radius r is given by $S = 4\pi r^2$.)

$V = \dfrac{4\pi r^3}{3}$, where $\pi \approx 3.14$.

63. A paper mill discharges waste sulfuric acid into a river. On an average work day, the amount Q (in gallons) of acid discharged into the river during t hr of plant operation is given by the equation $Q(t) = 10(t^2 + 4)\ln(t)$ for $0 \leqslant t \leqslant 9$ with $t = 0$ corresponding to 8:00 AM. Use differentials to approximate the number of gallons of acid discharged into the river between 3:00 and 3:30 PM on an average day. (Use Table A in the back of the book to find the value of $\ln(7)$.)

64. The population P of a certain culture of bacteria is given by $P(t) = P_0 \exp(0.5t) + 500t$, where P_0 is the population at time $t = 0$ and t is measured in hours. A biologist needs to know what the population of the culture will be $4\frac{1}{4}$ hr from now. If the culture presently contains 20,000 bacteria, use differentials to approximate the population at $t = 4.25$. (Do not use tables.)

65. In a study on learning behavior, a group of psychologists finds that the percent P of nonsense syllables from a list that are retained t min after the end of a learning period is given by $P(t) = 700/$

$[\ln(t/10) + 10]$ for $t \geqslant 1$. Use differentials to approximate the percent of the syllables that are forgotten as the time changes from $t_0 = 10$ to $t = 10.25$. Approximately what percent of the syllables are retained 10.25 min after the end of the learning period?

6.5.   Antiderivatives

So far in this chapter we have been interested in obtaining information about a function f from its derivative $D_x f$. However, in many instances it may be desirable to take a given function f and produce another function F such that $D_x F = f$. For example, if $f(x) = 2x$, then $F(x) = x^2$ is a function such that $D_x F = f$, and the function $G(x) = x^2 - 3$ is another function for which $D_x G = f$. In general, if the derivative of a function F is f, we call F an *antiderivative* of f. Thus the functions $F(x) = x^2$ and $G(x) = x^2 - 3$ are each antiderivatives of $f(x) = 2x$.

An important fact in calculus is that if two differentiable functions F and G have the same derivative, then $G(x) = F(x) + C$ for some real number C. This means that if F is an antiderivative of f, then every other antiderivative G of f has the form $G(x) = F(x) + C$. For example, we observed above that $F(x) = x^2$ and $G(x) = x^2 - 3$ are each antiderivatives of $f(x) = 2x$; here $G(x) = F(x) + C$ with $C = -3$.

An antiderivative of a function f is denoted by

$$\int f(x)\, dx$$

Thus if F is a function such that $D_x F = f$, then since every antiderivative G of f has the form $G(x) = F(x) + C$, we have

$$\int f(x)\, dx = F(x) + C$$

For example,

$$\int 2x\, dx = x^2 + C$$

and

$$\int (3x + 2)\, dx = \frac{3x^2}{2} + 2x + C$$

> **Definition of an antiderivative of a function f**

> The elongated S in the symbol $\int f(x)dx$ is called an *integral* sign, and an antiderivative $\int f(x)dx$ is sometimes called an *indefinite integral*.

EXAMPLE 6.5-1

Find $\int \exp(x)\, dx$.

SOLUTION

Since $D_x \exp(x) = \exp(x)$, we have $\int \exp(x)\, dx = \exp(x) + C$.

EXAMPLE 6.5-2

Find $\int (1/t)\, dt$.

SOLUTION

If $t > 0$, then

$$D_t \ln(t) = \frac{1}{t}$$

If $t < 0$, then

$$D_t \ln(-t) = \left(\frac{1}{-t} \right)(-1) = \frac{1}{t}$$

Thus since $|t| = t$ if $t > 0$ and $|t| = -t$ if $t < 0$, we have $D_t \ln(|t|) = 1/t$ for $t \neq 0$. Thus $\int (1/t)\, dt = \ln(|t|) + C$.

If a function f has values $f(t)$ instead of $f(x)$, then an antiderivative of f is $\int f(t)dt$ instead of $\int f(x)dx$. Similarly, if the values of g are $g(z)$, then an antiderivative of g is denoted $\int g(z)dz$, etc.

The rules of differentiation discussed in Sections 5.2 and 5.4 yield corresponding rules for finding antiderivatives. For example, since $D_x[x^n] = nx^{n-1}$, we have

$$D_x \left(\frac{x^{n+1}}{n+1} \right) = \frac{(n+1)x^n}{n+1} = x^n$$

provided that $n \neq -1$; thus

$$\int x^n\, dx = \frac{x^{n+1}}{n+1} + C$$

To test that $\int f(x)dx = F(x) + C$ we need only show that $D_x[F(x) + C] = f(x)$.

for $n \neq -1$. This rule and several others for finding antiderivatives are listed in Table 6.5-1. Each entry in the table can be verified by finding the derivitive of the function in the right-hand column and noting that it is equal to the corresponding function in the left column. Thus to check rules (8) and (9), we have

$$D_x \left[a \int g(x)\, dx \right] = a D_x \left[\int g(x)\, dx \right] = ag(x)$$

Since $\int g(x)dx = G(x) + C$ where $D_x G(x) = g(x)$, we have $D_x[\int g(x)dx] = g(x)$.

and

$$D_x \left[\int g(x)\, dx + \int h(x)\, dx \right] = D_x \left[\int g(x)\, dx \right] + D_x \left[\int h(x)\, dx \right]$$

$$= g(x) + h(x)$$

TABLE 6.5-1

Rule	$f(x)$	$\int f(x)\,dx$
(1)	a	$ax + C$
(2)	$x^n, n \neq -1$	$\dfrac{x^{n+1}}{n+1} + C$
(3)	$a_n x^n + a_{n-1}x^{n-1} + \cdots + a_0$	$\dfrac{a_n x^{n+1}}{(n+1)} + \dfrac{a_{n-1}x^n}{n} + \cdots + a_0 x + C$
(4)	$\dfrac{1}{x},\ x \neq 0$	$\ln(|x|) + C$
(5)	$\sqrt{x}$	$\frac{2}{3}(\sqrt{x}^{\,3}) + C$
(6)	$\dfrac{1}{2\sqrt{x}}$	$\sqrt{x} + C$
(7)	$\exp(ax)$	$\dfrac{\exp(ax)}{a} + C$
(8)	$ag(x)$	$a\int g(x)\,dx$
(9)	$g(x) + h(x)$	$\int g(x)\,dx + \int h(x)\,dx$

Rule (8), which states that $\int ag(x)dx = a\int g(x)dx$, applies only for a constant a. For example $\int x\sqrt{x-1}\,dx \neq x\int \sqrt{x-1}\,dx$.

Since $D_x[\ln(x) + x^2 + C] = 1/x + 2x$, we see that

$$\int\left(\frac{1}{x} + 2x\right)dx = \ln(x) + x^2 + C$$

If we apply rule (9) to

$$\int\left(\frac{1}{x} + 2x\right)dx$$

we obtain

$$\int\left(\frac{1}{x} + 2x\right)dx = \int \frac{1}{x}\,dx + \int 2x\,dx$$

$$= \ln(|x|) + C + x^2 + C$$

$$= \ln(|x|) + x^2 + 2C$$

The answers obtained by these two methods, namely $\ln(|x|) + x^2 + C$ and $\ln(|x|) + x^2 + 2C$, look different at first, but the difference is only *apparent*. The point is that if *any number* is added to $\ln(|x|) + x^2$, the resulting function is an antiderivative for $1/x + 2x$. For example, the function $\ln(|x|) + x^2 + 25$ is of the form $\ln(|x|) + x^2 + C$ (with $C = 25$), but it is also of the form $\ln(|x|) + x^2 + 2C$ (with $C = 12\frac{1}{2}$). When applying rule (9), confusion will be most easily avoided by writing each of the individual antiderivatives without using C and then adding C to the resulting sum. The following examples illustrate this technique.

EXAMPLE 6.5-3

Find $\int(4x^5 - 6x^2 + 2x - 10)\, dx$.

SOLUTION

Using rule (3) of Table 6.5-1, we find that

$$\int(4x^5 - 6x^2 + 2x - 10)\, dx = \frac{4x^6}{6} - \frac{6x^3}{3} + \frac{2x^2}{2} - 10x + C$$

$$= \frac{2x^6}{3} - 2x^3 + x^2 - 10x + C.$$

EXAMPLE 6.5-4

Find $\int[4\exp(5z) + 6z^2 - 3/z]\, dz$.

SOLUTION

Using rule (9) of Table 6.5-1, we have

$$\int\left[4\exp(5z) + 6z^2 - \frac{3}{z}\right] dz$$

$$= \int[4\exp(5z)]\, dz + \int 6z^2\, dz + \int \frac{-3}{z}\, dz$$

Now, by rule (8),

$$\int[4\exp(5z)]\, dz = 4\int[\exp(5z)]\, dz,$$

$$\int 6z^2 \, dz = 6 \int z^2 \, dz$$

and

$$\int \frac{-3}{z} \, dz = -3 \int \frac{1}{z} \, dz$$

Thus, using rules (2), (4), and (7), we find that

$$\int \left[4 \exp(5z) + 6z^2 - \frac{3}{z} \right] dz$$

$$= 4 \int \left[\exp(5z) \right] dz + 6 \int z^2 \, dz - 3 \int \frac{1}{z} \, dz$$

$$= \frac{4 \exp(5z)}{5} + \frac{6z^3}{3} - 3 \ln(|z|) + C$$

$$= \tfrac{4}{5} \exp(5z) + 2z^3 - 3 \ln(|z|) + C$$

EXAMPLE 6.5-5

Find $\int \left[(x^2 - 16)/(x + 4) \right] dx$.

SOLUTION

Since $(x^2 - 16)/(x + 4) = (x - 4)(x + 4)/(x + 4) = x - 4$ for $x \neq -4$, we have

$$\int \frac{x^2 - 16}{x + 4} \, dx = \int (x - 4) \, dx$$

$$= \frac{x^2}{2} - 4x + C.$$

EXAMPLE 6.5-6

Purity Pumps, Inc. manufactures a heavy-duty bilge pump that uses $\frac{1}{2}$ gal of gasoline and takes $\frac{1}{4}$ hr to reach its peak pumping capacity. After $\frac{1}{4}$ hr of operation, the pump uses gasoline at the rate of 1 gal per hour. Find an equation that gives the number of gallons of gasoline G required to run the pump for t hr, $t \geqslant \frac{1}{4}$.

SOLUTION

Since the rate of gasoline consumption is the derivative of G with respect to t, we must have $D_t G(t) = 1$ gal per hour for $t \geqslant \frac{1}{4}$. Thus

$$G(t) = \int 1 \, dt = t + C \qquad (t \geqslant \tfrac{1}{4})$$

Now, since $G(\frac{1}{4}) = \frac{1}{2}$ gal, we must have $G(\frac{1}{4}) = \frac{1}{4} + C = \frac{1}{2}$, which means that $C = \frac{1}{4}$. Thus the number of gallons of gasoline G required to run the pump for t hr, $t \geqslant \frac{1}{4}$, is given by the equation $G(t) = t + \frac{1}{4}$.

EXAMPLE 6.5-7

Aztec Optics, Inc. manufactures photographic laboratory equipment. The company has determined that its marginal cost M_c (in dollars per unit produced) at a yearly production level of x units for its economy model photograph enlarger is given by the equation $M_c(x) = -x/500 + 40$ for $6000 \leqslant x \leqslant 8000$. Find an equation that gives the yearly cost $P_c(x)$ for producing x enlargers ($6000 \leqslant x \leqslant 8000$), given that the cost of producing 6000 enlargers is $204,400.

SOLUTION

Since the marginal cost $M_c(x)$ is the derivative of the production cost $P_c(x)$, we have

$$P_c(x) = \int M_c(x) dx$$

$$= \int \left(\frac{-x}{500} + 40 \right) dx$$

$$= \frac{-1}{500} \left(\frac{x^2}{2} \right) + 40x + C$$

$$= \frac{-x^2}{1000} + 40x + C$$

Now to determine C, we use the fact that $P_c(6000) =$ \$204,400. This means we must have C such that

$$P_c(6000) = \frac{-(6000)^2}{1000} + 40(6000) + C = 204,400$$

which means that $-36,000 + 240,000 + C = 204,400$, or $C = 400$. Thus the yearly cost $P_c(x)$ for producing x units is given by $P_c(x) = -x^2/1000 + 40x + 400$.

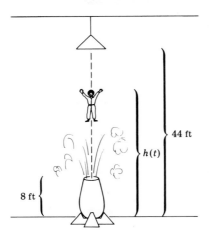

44 ft

$h(t)$

8 ft

EXAMPLE 6.5-8

A circus performer is shot vertically upward from a cannon to a trapeze suspended 44 ft above the ground. If the speed v (in feet per second) of the performer t sec after the cannon is fired is given by $v(t) = -32t + 48$ for $0 \leqslant t \leqslant 1.5$ and if the muzzle of the cannon is 8 ft high, find an equation that gives the height $h(t)$ of the performer at time t after firing the cannon ($0 \leqslant t \leqslant 1.5$). How high is the performer 1 sec after the cannon is fired?

SOLUTION

Since the speed $v(t)$ of the performer is the rate of change of distance traveled, (in this case, height) with respect to time, we have $v(t) = D_t h(t)$. Thus

$$h(t) = \int v(t)\, dt$$

$$= \int (-32t + 48)\, dt$$

$$= -\frac{32t^2}{2} + 48t + C$$

$$= -16t^2 + 48t + C$$

To determine the value of C, we use the fact that at time $t = 0$ the height $h(0)$ of the performer is 8 ft (the height of the cannon). This means that we want $h(0) = 8$, i.e., we want C

such that $h(0) = -16(0^2) + 48(0) + C = 8$. Thus $C = 8$, and the equation that gives the performer's height h at time t is $h(t) = -16t^2 + 48t + 8$. Since $h(1) = -16(1^2) + 48(1) + 8 = -16 + 48 + 8 = 40$, the performer is 40 ft high 1 sec after the cannon is fired.

EXAMPLE 6.5-9

Medical researchers estimate that the number N of cancer cells present in a certain tissue sample is increasing at the rate of $r(t) = 10 \exp(0.05t)$ cells per day, where t is time in days $(0 \le t \le 10)$ and 400 cancer cells are present in the sample at time $t = 0$. Find an equation that gives the number $N(t)$ of cancer cells present in the tissue at time t $(0 \le t \le 10)$. How many cells are present at time $t = 10$?

SOLUTION

Since the rate $r(t)$ of increase of cancer cells per day is the rate of change of the number $N(t)$ of cells present with respect to t, we have $D_t N(t) = r(t)$. Thus

$$N(t) = \int r(t) \, dt$$

$$= \int \left[10 \exp(0.05t) \right] dt$$

$$= 10 \int \exp(0.05t) dt$$

$$= 10 \frac{\exp(0.05t)}{0.05} + C$$

$$= 200 \exp(0.05t) + C$$

Since 400 cells are present at time $t = 0$, we want C such that $N(0) = 200 \exp[(0.05)(0)] + C = 400$, which means that $C = 200$. Thus we have $N(t) = 200 \exp(0.05t) + 200$. Since $N(10) = 200 \exp[(0.05)(10)] + 200 = 200 \exp(0.5) + 200 = 200(1.6487) + 200 \approx 530$, there are approximately 530 cells present in the sample at time $t = 10$.

Recall that $\exp(0) = 1$.

From Table A in the back of the book, $\exp(0.5) = 1.6487$.

Exercises 6.5

In Exercises 1 through 30, find the indicated antiderivative.

1. $\displaystyle\int 5\,dx$ 2. $\displaystyle\int 2x\,dx$

3. $\displaystyle\int (8x + 3)\,dx$ 4. $\displaystyle\int 3x^2\,dx$

5. $\displaystyle\int 6x^{-3}\,dx$ 6. $\displaystyle\int -5x^{-6}\,dx$

7. $\displaystyle\int (15x^4 - 4x)\,dx$ 8. $\displaystyle\int (6x^5 + 2x)\,dx$

9. $\displaystyle\int (7x^3 + 2x - 3)\,dx$ 10. $\displaystyle\int (-6x^2 - 4x - 1)\,dx$

11. $\displaystyle\int (10x^8 - 7x^5 + 2x^3 - 4)\,dx$

12. $\displaystyle\int (13x^7 - 8x^4 - 3x^2 + x)\,dx$

13. $\displaystyle\int \left(\frac{x^2}{4} - 2x + x^{-2} \right) dx$

14. $\displaystyle\int \left(\frac{x^3}{5} - \frac{x^2}{2} - \frac{1}{5} \right) dx$

15. $\displaystyle\int \left(\frac{3x^7}{5} - \frac{8x^3}{3} + \frac{2x^{-2}}{3} \right) dx$

16. $\displaystyle\int \left(\frac{-5x^9}{6} + \frac{4x^7}{5} - \frac{3x^2}{7} + \frac{1}{10} \right) dx$

17. $\displaystyle\int \left(2x^{-3} + 4x^{-2} + \frac{5}{2\sqrt{x}} \right) dx$

18. $\displaystyle\int \left(\frac{2x^5}{3} - \frac{7}{2\sqrt{x}} + 3x^{-3} - 2 \right) dx$

19. $\displaystyle\int [4x(x + 3)^2]\,dx$

20. $\displaystyle\int \frac{x + 2}{x^4}\,dx$

21. $\displaystyle\int \frac{x^2 - 9}{x + 3}\,dx$

22. $\displaystyle\int [t^2(t - 2)^2 + 4t^{-7} - 3]\,dt$

23. $\displaystyle\int (3z^4 - \frac{8}{z^4} + \frac{2}{z^3} - 3)\,dz$

24. $\displaystyle\int (11x^{-1} + 4x^{-3} - 8)\,dx$

25. $\displaystyle\int [7\,\exp(x) + 2x^{-5} - \tfrac{1}{6}]\,dx$

26. $\displaystyle\int [-3\,\exp(2v) + 3v^{-1} + \tfrac{1}{3}]\,dv$

27. $\displaystyle\int [2\,\exp(-6w) - \dfrac{7}{w} + \tfrac{2}{3}]\,dw$

28. $\displaystyle\int \left[4u^{-4} - \dfrac{3}{2u} + \exp\!\left(\dfrac{4u}{5} \right) \right] du$

29. $\displaystyle\int \left[\dfrac{15}{16s} - \dfrac{3\,\exp(2s/7)}{10} + \dfrac{6}{9} \right] ds$

30. $\displaystyle\int \left[\dfrac{2r^{11}}{5} - \dfrac{7}{9r} + \dfrac{5\,\exp(4r/11)}{7} - \dfrac{18r^{-6}}{19} \right] dr$

31. Murray Manufacturing, Inc. produces automobile seat covers. The company finds that the marginal cost M_c (in dollars per unit produced) at a monthly production level of x units for its best quality seat covers is given by the equation $M_c(x) = -x/600 + 30$ for $600 \leqslant x \leqslant 1200$. Find an equation that gives the monthly production cost $P_c(x)$ for producing x units ($600 \leqslant x \leqslant 1200$) given that the cost of producing 600 units is $18,200.

32. The Quality-Craft Furniture Company manufactures office desks. The company estimates that the marginal revenue M_R (in dollars per unit) at a monthly production level of x units for its executive model office desk is given by the equation $M_R(x) = -3x^2/25 + 16x + 3$ for $100 \leqslant x \leqslant 120$. Find an equation that gives the monthly revenue $R(x)$ for producing x units of the desk ($100 \leqslant x \leqslant 120$) if the revenue R from the production of 100 desks is $40,350.

Recall that if $R(x)$ is the revenue received at a production level of x units, then the marginal revenue M_R is defined by $M_R(x) = D_x R(x)$.

33. Everwrite, Inc. finds that the monthly marginal demand M_d (in units of the commodity per dollar) for its deluxe retractible ball-point pen is given by the equation $M_d(x) = -500\,\exp(x - 1)$, where x is the selling price (in dollars) per unit for the pens and $0.5 \leqslant x \leqslant$ and 2. Find an equation that gives the monthly demand $d(x)$ for the pens when they are priced at $x each given that the monthly demand is 3500 when $x = 1$.

Recall that if $d(x)$ is the demand for a product priced at $x per unit, then the marginal demand M_d is defined by $M_d(x) = D_x d(x)$.

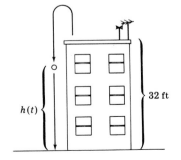

34. A ball is thrown vertically upward from the top of a 32-ft high building. The speed v of the ball (in feet per second) t sec after it is released is given by the equation $v(t) = -32t + 16$ for $0 \leqslant t \leqslant 2$. Find an equation that gives the height $h(t)$ of the ball t sec after it is thrown ($0 \leqslant t \leqslant 2$). When is the ball at its highest point? How high does the ball travel?

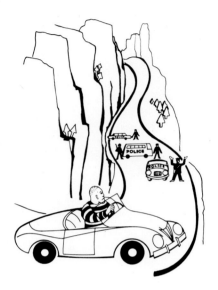

35. An automobile radiator has developed a leak. The owner estimates that the rate at which the volume V of water in the radiator is decreasing t min after the radiator is filled is given by $D_t V(t) = - t/45$ qts per minute. If the capacity of the radiator is 10 qts, find an equation that gives the number of quarts V in the radiator t min after it is filled. If it will take 25 min to drive to the nearest garage, can the owner fill the radiator with water and drive to the garage before the radiator empties? How long does it take the radiator to empty after it is filled?

36. An escaped convict is speeding down the highway at 90 mph in a stolen car. Rounding a bend in the road, the convict sees a barricade 200 ft ahead and applies the brakes; the speed v of the car t sec after the brakes are applied is $v(t) = 132 - 33t$ ft per second. Find an equation that gives the distance s the car will travel t sec after the brakes are applied. How long will it take the car to stop? Will the convict hit the barricade?

37. Suppose you are paddling a canoe up a stream that is flowing at 10 mph. As you become tired, your (still water) speed v (in miles per hour) slows according to the equation $v(t) = 20/\sqrt{t + 1}$, where t is time in hours with $t = 0$ corresponding to the time when you began paddling. Find the equation that gives the distance s (in miles) you will have traveled after t hr of paddling. How long will it be before you are slowed to a stop? How far upstream will you travel before being slowed to a stop?

38. A biologist estimates that the rate of increase of the population P of a certain culture of bacteria is given by $D_t P(t) = 400 \exp(0.4t)$ bacteria per day, where t is time in days ($0 \leqslant t \leqslant 12$). If the culture contains 2000 bacteria at time $t = 0$, find an equation that gives the population $P(t)$ of the culture at time t. What is the population of the sample at time $t = 4$ days?

39. Forestry officials find that the percent P of diseased trees x mi from the center of a large industrial city decreases at the rate of $D_x P(x) = -3/\sqrt{x}\ \%$ per mile, for $10 \leqslant x \leqslant 100$. If 40% of the trees are diseased 25 mi from the city, find an equation that gives the percent of diseased trees x mi from the city. What percent of the trees are diseased 64 mi from the city?

1. Find an approximate value for $\sqrt{9.5}$

2. Find the derivatives of the following functions.

   a. $f(x) = x^2 \exp(x) - \ln(x^2)$

   b. $g(x) = \dfrac{2x^3 - 1}{\exp(5x + 2)}$

   c. $h(x) = [\ln(2x - \sqrt{x}\,)](x^3 - 7x)$

   d. $s(x) = \sqrt{\exp(-x^3)}$

3. Find the equation of the line tangent to the graph of $f(x) = \ln(x)$ at the point $(1, 0)$.

4. Find the intervals over which the graph of the given function is increasing, those over which it is decreasing, those over which it is concave up, and those over which it is concave down. Find the points of inflection (if any) and the relative maximum and minimum values of the function; sketch the graph.

   a. $f(x) = 2x^3 + 9x^2 + 12x + 6$

   b. $g(x) = 3x^4 - 8x^3 + 6x^2 + 2$

5. During the tourist season, the Buena Vista Motel can rent all 60 of its rooms each night at $16 per night. However, for each dollar the room rate is raised, an average of three rooms per night will remain vacant. At what rate should the rooms be rented in order for the motel to maximize its average daily rental income? How many rooms, on the average, will be vacant each night? What is the maximum average daily rental income?

6. Medical researchers find that the percent concentration P of an experimental drug in the bloodstream t min after oral administration is increasing at the rate of $D_t P(t) = -t/200 + 0.5$ percent per minute for $10 \leqslant t \leqslant 60$. If the concentration is 20% after 60 min, find an equation that gives the concentration $P(t)$ at time t min after administration of the drug. What is the concentration 20 min after administration?

7. The Skillcraft Company produces electronic model airplane guidance systems. The company determines that the total weekly revenue R (in dollars) that it receives from the production of x units of its product per week is given by $R(x) = (\sqrt{x} + x)^2 - x(x + 20)$, for $200 \leqslant x \leqslant 350$. Find the marginal revenue (rate of change of R with respect to x) at production outputs of 225 units per week and of 289 units per week.

The Integral and its Applications

PART IV

Area and the Integral

7.1. Introduction

The two central topics in calculus are the *derivative* of a function and the *integral* of a function. We are ready now to develop the second of these topics, the integral, and discuss some of its applications in such disciplines as economics, medicine, business, biology, psychology, and sociology. We shall see that the integral rivals the derivative in importance in these applications.

An antiderivative of a function, as discussed in Section 6.5, is often called an *indefinite integral*. This term is not to be confused with the notion of the integral (more precisely called the *definite integral*) of a function; but as we shall see, the indefinite integral, the derivative, and the integral are closely connected concepts. The connection between these concepts is of such importance that the theorem that describes it is called *The Fundamental Theorem of Calculus*.

The integral, like the derivative, can be interpreted geometrically. Whereas the derivative of a function f yields the *slope* of the tangent line at a point on the graph of f, the integral of f gives the *area* under the graph of f and above the x axis (provided $f(x) \geqslant 0$ for each x). Thus we begin our development of the integral by discussing the concept of area.

Throughout this book, the term "integral" will always mean the *definite* integral, *not* an indefinite integral (antiderivative).

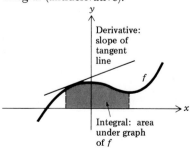

317

7.2. The Concept of Area

Recall that we spent some time in Section 1.2 describing how to calibrate a line with a chosen unit length. We are accustomed to using calibrated lengths, such as yardsticks and tape measures, to measure lengths, but what about the measurement of areas? There are no such familiar measuring devices for areas. Indeed, measuring an area is somewhat more complex than measuring a length, especially if we want to find the area of an irregularly shaped surface.

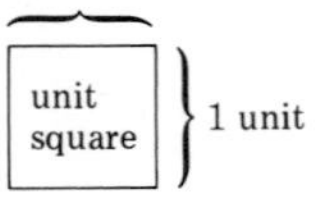

To measure area, we begin (as we did in measuring lengths) by choosing a *unit of area* with which to work. A convenient unit of area to use is a square with sides of some chosen unit length (see Fig. 7.2-1); as in measuring lengths, any convenient length may be chosen for the sides of this square. This square is called *one square unit of area*. Finding the area of a surface now means determining how many unit squares would be needed to cover that surface exactly.

Suppose we wish to measure the area of a rectangle of length 4 units and width 3 units. We simply observe that the rectangle can be covered with 12 unit squares (see Fig. 7.2-2), so the area of the rectangle is 12 square units. Note that in covering the rectangle we put down three rows of four unit squares for a total of $3 \cdot 4 = 12$ unit squares. Similarly, if a rectangle is 5 units wide and 7 units long, we can cover it with unit squares by putting down five unit squares seven times for a total of $5 \cdot 7 = 35$ unit squares, so the area of the rectangle is 35 square units. In general, a rectangle whose sides are a units and b units long has an area of $a \cdot b$ square units (see Fig. 7.2-3).

Now, if we take a rectangle whose sides are a units and b units long and divide the rectangle in half with a diagonal line (see Fig. 7.2-4), we produce a pair of right triangles each having half the area of the rectangle, i.e., $\frac{1}{2}(a \cdot b)$. Thus we conclude that a right triangle of base b and altitude a has area $ab/2$ square units. In fact, given *any* triangle with base b and altitude a, we can divide the triangle into two right triangles (see Fig. 7.2-5) and determine that the area of the triangle is $ab/2$.

Suppose now we have a region in the plane that has line segments (not curves) for sides, as in Fig. 7.2-6. Then we can divide the region into triangles and add up the areas of the triangles to determine the area of the region. So far, so good! However, if we want to measure the area of a region in the plane with curved sides we are in difficulty; even if

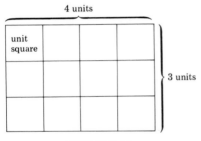

FIGURE 7.2-1. The area of the unit square is called *one square unit* of area.

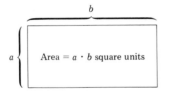

FIGURE 7.2-2

FIGURE 7.2-3

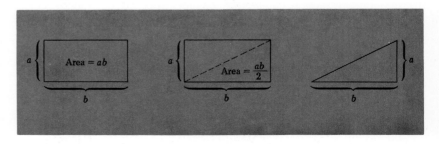

FIGURE 7.2-4

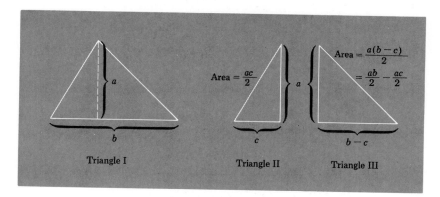

FIGURE 7.2-5 The area of Triangle I = area of Triangle II + area of Triangle III = $ac/2 + a(b-c)/2$ = $ac/2 + ab/2 - ac/2 = ab/2$.

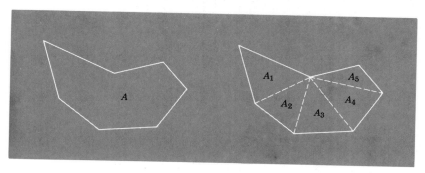

FIGURE 7.2-6 The area A = the sum of the areas $A_1 + A_2 + A_3 + A_4 + A_5$.

the region is a rectangle or a triangle with one side exchanged for a curved side (see Fig. 7.2-7), we do not have any way of measuring its area.

The regions shown in Fig. 7.2-7 are typical of regions that lie beneath the graph of a function f, above the x axis, and between the lines $x = a$ and $x = b$ (see Fig. 7.2-8). Solving the problem of computing such areas is a major step toward being able to determine the area of any region in the plane. Moreover, the techniques used lead to the definition of the integral of a function f, as we shall see in the next section.

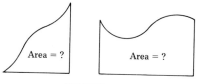

FIGURE 7.2-7

FIGURE 7.2-8

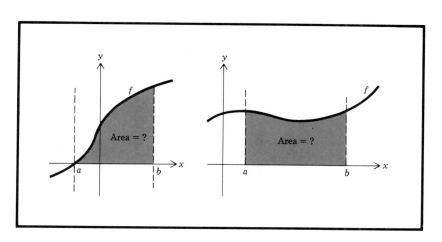

Exercises 7.2

In Exercises 1 through 12, compute the area of the shaded region.

1.

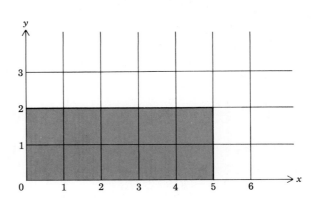

2.

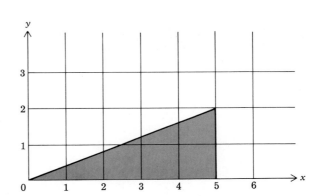

3.

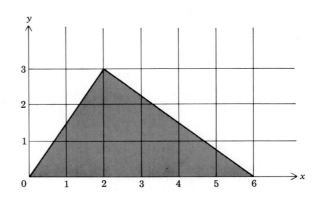

4.

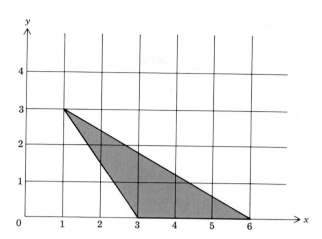

5.

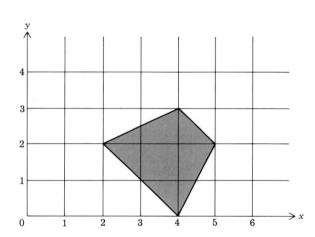

6.

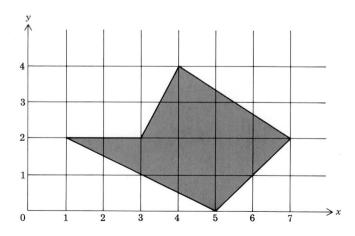

7.

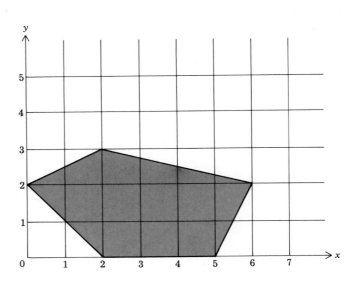

8.

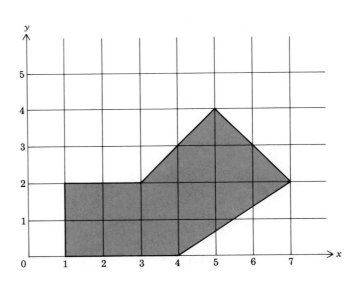

9.

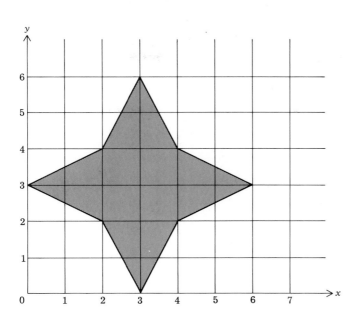

10.

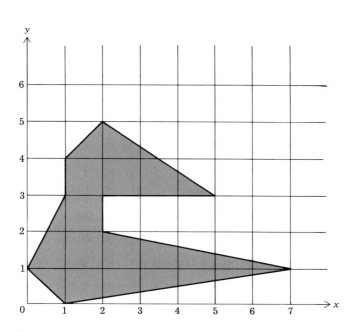

11.

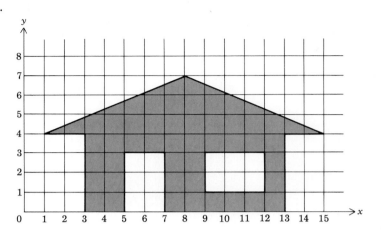

12.

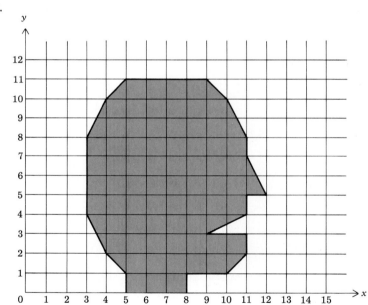

For any given number a, the line $x = a$ is the vertical line through the point $(a, 0)$.

13.   Sketch the region that lies below the graph of the function $f(x) = 5$, above the x axis, and between the lines $x = 2$ and $x = 6$. Find the area of this region.

14.   Sketch the region that lies below the graph of the function $f(x) = x - 1$, above the x axis, and between the lines $x = 1$ and $x = 4$. Compute the area of this region.

15.   Given that the area of a circle of radius r is πr^2, compute the area
      of a window that is 4 ft wide and 4 ft high and whose top is a
      semicircle (see accompanying diagram).

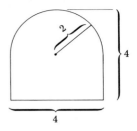

7.3.   The Area under $f(x) = x^2$

In this section we compute a measurement for the area A of the region
under the graph of $f(x) = x^2$, above the x axis and between the lines
$x = 0$ and $x = 2$ (see Fig. 7.3-1).

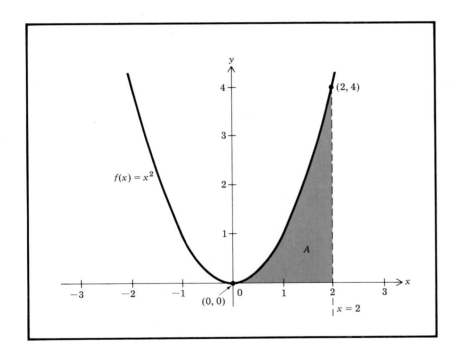

FIGURE 7.3-1

We begin by observing that the area A is less than the area of a
rectangle of height $f(2) = 4$ and width 2 (see Fig. 7.3-2). Thus $A < 4 \cdot 2$
$= 8$ square units. Next we obtain a closer approximation to the area A
by dividing the interval $[0, 2]$ into two smaller intervals, $[0, 1]$ and $[1, 2]$,
each of length 1, and considering the corresponding division of A into
smaller areas A_1 and A_2, as in Fig. 7.3-3(a). Then the area A_1 is less than
the area of the rectangle of height $f(1) = 1^2 = 1$ and width 1 (Fig.
7.3-3(b)); so

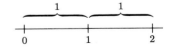

$$A_1 < 1 \cdot 1 = 1$$

FIGURE 7.3-2

Approximation of A with one rectangle: $A < 8$.

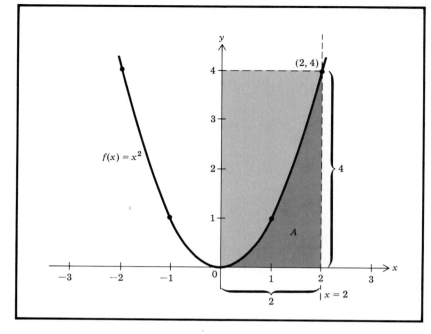

FIGURE 7.3-3

Approximation of A with two rectangles: $A < 5$.

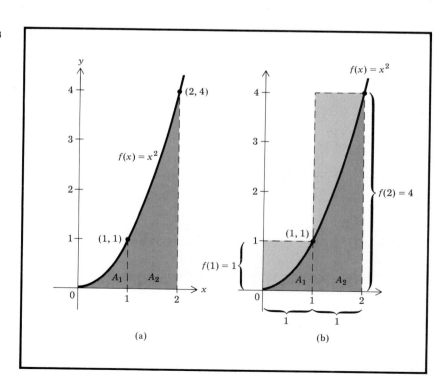

Similarly, the area A_2 is less than the area of the rectangle of height $f(2) = 4$ and width 1, so

$$A_2 < 4 \cdot 1 = 4$$

Thus, since

$$A = A_1 + A_2$$

(see Fig. 7.3-3(a)), we see that

$$A = A_1 + A_2 < 1 + 4 = 5$$

so $A < 5$ square units. This is certainly a better approximation of A than the one obtained by using one rectangle.

We obtain yet a closer approximation to the area A by dividing the interval $[0, 2]$ into three still smaller intervals, each of length $\frac{2}{3}$, and considering the corresponding division of A into three smaller areas A_1, A_2, and A_3 as in Fig. 7.3-4(a). Then the area A_1 is less than the area of the rectangle of height $f(\frac{2}{3}) = \frac{4}{9}$ and width $\frac{2}{3}$ (see Fig. 7.3-4(b)); so

$$A_1 < \frac{4}{9}\left(\frac{2}{3}\right) = \frac{8}{27}$$

FIGURE 7.3-4

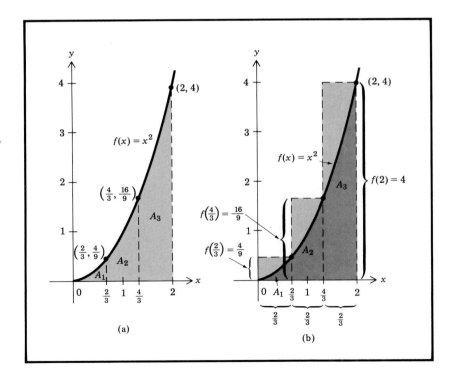

Approximation of A with three rectangles: $A < 4\frac{4}{27}$.

Similarly, the area A_2 is less than the area of the rectangle of height $f\left(\frac{4}{3}\right) = \frac{16}{9}$ and width $\frac{2}{3}$, so

$$A_2 < \tfrac{16}{9}\left(\tfrac{2}{3}\right) = \tfrac{32}{27}$$

Finally, a similar computation shows that

$$A_3 < \tfrac{8}{3}$$

Thus, since

$$A = A_1 + A_2 + A_3$$

(see Fig. 7.3-4(a)), we find that

$$A = A_1 + A_2 + A_3 < \tfrac{8}{27} + \tfrac{32}{27} + \tfrac{8}{3} = \tfrac{112}{27} = 4\tfrac{4}{27}$$

so $A < 4\tfrac{4}{27}$. This is a better approximation of A than the one obtained by using two rectangles.

Apparently, this procedure of dividing the interval $[0, 2]$ into smaller and smaller intervals of equal length and adding the areas of the corresponding rectangles produces better and better approximations of the area A. If we divide the interval $[0, 2]$ into n smaller intervals of equal length (where n is any positive integer), then each interval is of length $(2 - 0)/n = 2/n$ units (see Fig. 7.3-5). Now, considering the corresponding division of A into n smaller areas $A_1, A_2, \ldots, A_n$ (see Fig. 7.3-6), we have that

$$A = A_1 + A_2 + \cdots + A_n$$

FIGURE 7.3-5

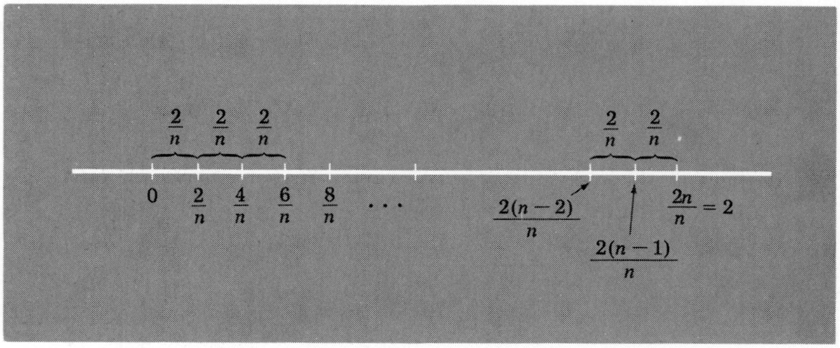

Thus we need only approximate the value of the sum $A_1 + A_2 + \cdots + A_n$ to get an approximation of the area A. But A_1 is less than the area of the rectangle of height $f(2/n) = 4/n^2$ and width $2/n$, so

$$A_1 < \frac{8}{n^3}$$

FIGURE 7.3-6

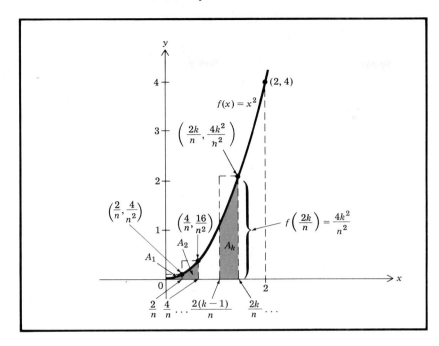

(see Fig. 7.3-6). Similarly, A_2 is less than the area of the rectangle of height $f(4/n) = 16/n^2$ and width $2/n$, so

$$A_2 < \frac{32}{n^3}$$

In general, A_k (for $1 \leqslant k \leqslant n$) is less than the area of the rectangle of height $f(2k/n) = 4k^2/n^2$ and width $2/n$, so

$$A_k < \frac{8k^2}{n^3}$$

(see Fig. 7.3-6). Thus we find that

$$A = A_1 + A_2 + \cdots + A_n$$

$$< \frac{8}{n^3} + \frac{8 \cdot 2^2}{n^3} + \frac{8 \cdot 3^2}{n^3} + \cdots + \frac{8 \cdot n^2}{n^3}$$

$$= \frac{8}{n^3}\left(1 + 2^2 + 3^2 + \cdots + n^2\right)$$

Now, it is known from elementary algebra that

$$1 + 2^2 + 3^2 + \cdots + n^2 = \frac{n(n+1)(2n+1)}{6}$$

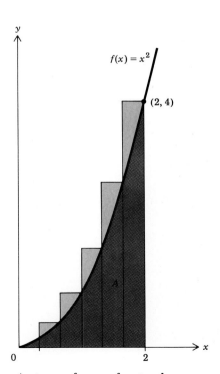

$A <$ sum of areas of rectangles.

so that

Approximation of A with 10 rectangles: $A < 3\frac{2}{25}$.

$$A < \frac{8}{n^3}\left(\frac{n(n+1)(2n+1)}{6}\right) = \frac{4}{3}\left(1 + \frac{1}{n}\right)\left(2 + \frac{1}{n}\right)$$

Consequently, if $n = 10$ we have

$$A < \frac{4}{3}\left(1 + \frac{1}{10}\right)\left(2 + \frac{1}{10}\right) = 3\tfrac{2}{25}$$

and if $n = 20$, we find that

Approximation of A with 20 rectangles: $A < 2\frac{87}{100}$.

$$A < \frac{4}{3}\left(1 + \frac{1}{20}\right)\left(2 + \frac{1}{20}\right) = 2\tfrac{87}{100}$$

Since, regardless of how many rectangles are employed (regardless of the value of n), the area A is less than the sum of the areas of the rectangles, and since the sum of the areas of the rectangles seems to be getting closer and closer to the area A as n increases, we set

The area A is defined as the limit of the sum of the areas of n approximating rectangles as n approaches infinity.

$$A = \lim_{n \to \infty} (\text{sum of areas of } n \text{ rectangles})$$

$$= \lim_{n \to \infty} (4/3)\left(1 + \frac{1}{n}\right)\left(2 + \frac{1}{n}\right)$$

$$= (4/3)(1)(2)$$

$$= \tfrac{8}{3} = 2\tfrac{2}{3}$$

square units.

Thus we have developed a technique for measuring the area A of the region under the graph of $f(x) = x^2$, above the x axis, and between the lines $x = 0$ and $x = 2$ (see Fig. 7.3-1); we found that $A = 2\frac{2}{3}$ square units. But by this time you may have decided never to try to compute the area of such a region again! However, we shall see shortly that there are techniques that make the computation of the areas of such regions quite simple. In fact, it will turn out that the area A computed here is the value of the integral of $f(x) = x^2$ from $x = 0$ to $x = 2$, which is denoted

$$\int_0^2 x^2 \, dx$$

and we shall find that such integrals are quite easy to evaluate.

Exercises 7.3

1. (a) Sketch the region R below the graph of the function $f(x) = x$, above the x axis, and between the lines $x = 0$ and $x = 2$.
    (b) Divide the interval $[0, 2]$ into n intervals of equal length, and justify to yourself that each of these smaller intervals is of length $2/n$. Also, justify to yourself that the right-hand endpoint of the kth interval $(1 \leqslant k \leqslant n)$ is the point $2k/n$.
    (c) Sketch the corresponding division of the area A of the region R into n smaller areas $A_1, A_2, \ldots, A_n$ such that $A = A_1 + A_2 + \cdots + A_n$.
    (d) Convince yourself that for each k $(1 \leqslant k \leqslant n)$, the area A_k is less than the area of a rectangle of height $2k/n$ and width $2/n$. Thus $A_k < 4k/n^2$.
    (e) Use the formula $1 + 2 + 3 + \cdots + n = n(n + 1)/2$ to show that $A = A_1 + A_2 + \cdots + A_n < 2(1 + 1/n)$.
    (f) Convince yourself that $A = \lim_{n \to \infty} 2(1 + 1/n)$ and evaluate this limit to find A. Check this value of A by using the formula for the area of a triangle.

2. Repeat steps (a) through (c) of Exercise 1 for the function $f(x) = 2x + 1$. Then continue with the following steps.
    (d) Convince yourself that for each k $(1 \leqslant k \leqslant n)$, the area A_k is less than the area of the rectangle of height $(4k/n) + 1$ and width $2/n$. Thus $A_k < (8k/n^2) + 2/n$.
    (e) Use the formula $1 + 2 + 3 + \cdots + n = n(n + 1)/2$ to show that $A = A_1 + A_2 + \cdots + A_n < 4(1 + (1/n)) + 2$.
    (f) Convince yourself that $A = \lim_{n \to \infty} [4(1 + (1/n)) + 2]$ and evaluate this limit to find A. Check this value of A by using formulas for the areas of triangles and rectangles.

3. Repeat steps (a) through (c) of Exercise 1 for the function $f(x) = x^2 + 2$. Then continue with the following steps.
    (d) Convince yourself that for each k $(1 \leqslant k \leqslant n)$, the area A_k is less than the area of the rectangle of height $(4k^2/n^2) + 2$ and width $2/n$. Thus $A_k < (8k^2/n^3) + 4/n$.
    (e) Use the formula $1^2 + 2^2 + \cdots + n^2 = n(n + 1)(2n + 1)/6$ to show that $A = A_1 + A_2 + \cdots + A_n < \frac{4}{3}(1 + (1/n))(2 + (1/n)) + 4$.
    (f) Convince yourself that $A = \lim_{n \to \infty} [\frac{4}{3}(1 + (1/n))(2 + (1/n)) + 4]$ and evaluate this limit to find A.

4. (a) Sketch the region R below the graph of the function $f(x) = 2x^2$, above the x axis, and between the lines $x = 1$ and $x = 2$.
    (b) Divide the interval $[1, 2]$ into n intervals of equal length, and justify to yourself that each of these smaller intervals is of length $1/n$. Also, justify to yourself that the right-hand endpoint of the kth interval $(1 \leqslant k \leqslant n)$ is the point $1 + (k/n)$.

(c) Sketch the corresponding division of the area A of the region R into n smaller areas $A_1, A_2, \ldots, A_n$ such that $A = A_1 + A_2 + \cdots + A_n$.

(d) Convince yourself that for each k ($1 \leqslant k \leqslant n$), the area A_k is less than the area of a rectangle of height $2(1 + k/n)^2$ and width $1/n$. Thus $A_k < (2/n) + (4k/n^2) + (2k^2/n^3)$.

(e) Use the formulas $1 + 2 + 3 + \cdots + n = n(n + 1)/2$ and $1^2 + 2^2 + 3^2 + \cdots + n^2 = n(n + 1)(2n + 1)/6$ to show that

$$A = A_1 + A_2 + \cdots + A_n$$

$$< 2 + 2\left(1 + \frac{1}{n}\right) + \frac{1}{3}\left(1 + \frac{1}{n}\right)\left(2 + \frac{1}{n}\right).$$

(f) Convince yourself that $A = \lim_{n \to \infty}[2 + 2(1 + (1/n)) + \frac{1}{3}(1 + (1/n))(2 + (1/n))]$ and evaluate this limit to find A.

We shall see in the next section that for any continuous function f whose graph lies above the x axis over the interval $[a, b]$, the area A of the region R below the graph of f and above the x axis between the lines $x = a$ and $x = b$ is given by the definite integral $\int_a^b f(x)\,dx$ (see Fig. 7.3-7). In Exercises 5 through 14, *use this fact* (by computing areas of rectangles and triangles) to evaluate the definite integral $\int_a^b f(x)\,dx$ for the given function f and the given values of a and b.

5. $f(x) = 6$, $a = 1$, $b = 2$

6. $f(x) = 4$, $a = -2$, $b = 2$

7. $f(x) = 6x$, $a = 0$, $b = 4$

8. $f(x) = 2x + 2$, $a = -1$, $b = 2$

9. $f(x) = \dfrac{x}{2} + 1$, $a = -2$, $b = 4$

10. $f(x) = 4x - 1$, $a = 1$, $b = 2$

11. $f(x) = 3x + 2$, $a = 2$, $b = 5$

12. $f(x) = \dfrac{x}{3} - 2$, $a = 9$, $b = 12$

13. $f(x) = cx$, where $c > 0$, $a = 2$, $b = 4$

14. $f(x) = x + d$, where $d > 0$, $a = 0$, $b = 4$

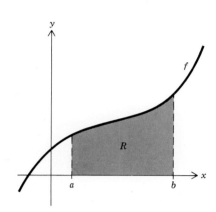

FIGURE 7.3-7   The area A of the region R is given by $A = \int_a^b f(x)\,dx$.

7.4.   Area and the Definite Integral

Having successfully measured the area under the graph of $f(x) = x^2$ between $x = 0$ and $x = 2$ (in Section 7.3), we are now ready to employ similar techniques to define the area under the graph of *any* continuous function f between $x = a$ and $x = b$, provided that over the interval $[a, b]$ the graph of f does not go below the x axis.

That the graph of f does not go below the x axis over $[a, b]$ means that $f(x) \geqslant 0$ for each x in $[a, b]$.

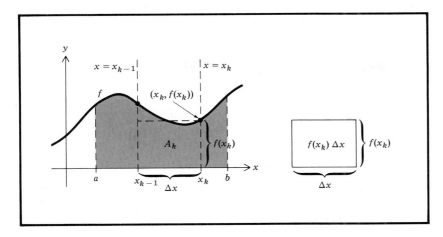

The length of $[x_{k-1}, x_k]$ is Δx $= x_k - x_{k-1}$. Corresponding to the interval $[x_{k-1}, x_k]$ is a rectangle of height $f(x_k)$ and width Δx.

FIGURE 7.4-1 A_k is approximated by the rectangular area $f(x_k)\Delta x$.

In Fig. 7.4-1 we show the graph of a continuous function f over an interval $[a, b]$. The points x_{k-1} and x_k represent any two points in the interval $[a, b]$, and A_k denotes the area under the graph of f (above the x axis) and between the lines $x = x_{k-1}$ and $x = x_k$. Now, to obtain the area A under the graph of f between the lines $x = a$ and $x = b$ we proceed as follows.

(1)   Divide the interval $[a, b]$ into n smaller intervals each of length $(b - a)/n$. Denote the n smaller intervals by

$$[x_0, x_1], [x_1, x_2], \ldots, [x_{k-1}, x_k], \ldots, [x_{n-1}, x_n]$$

where

$$x_0 = a$$

$$x_1 = a + (b - a)/n$$

$$x_2 = a + 2(b - a)/n$$

$$\vdots$$

$$x_k = a + k(b - a)/n$$

$$\vdots$$

$$x_n = a + n(b - a)/n = b$$

(see Fig. 7.4-2).

The length of $[a, b]$ is $b - a$; hence when we divide this length into n equal parts, each part has length $(b - a)/n$.

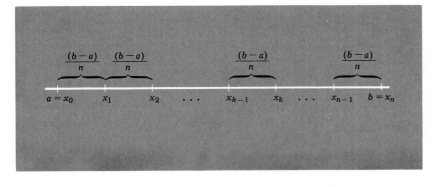

Here each interval $[x_{k-1}, x_k]$ has length $\Delta x = (b - a)/n$.

FIGURE 7.4-2   The point x_1 is obtained by adding an interval of length $(b - a)/n$ to a, x_2 is obtained by adding two intervals each of length $(b - a)/n$ to a, etc.

(2) Sum the areas of the rectangles corresponding to these intervals (see Fig. 7.4-3) to obtain an approximation S_n to the area A (i.e., $A \approx S_n = f(x_1)\Delta x + f(x_2)\Delta x + \cdots + f(x_n)\Delta x$).

Recall that the symbol $\approx$ means "is approximately equal to."

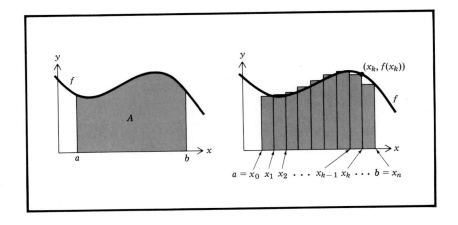

FIGURE 7.4-3 The rectangle corresponding to the interval $[x_{k-1}, x_k]$ has area $f(x_k)\Delta x$. Thus A is approximated by the sum $S_n = f(x_1)\Delta x + f(x_2)\Delta x + \cdots + f(x_n)\Delta x$.

(3) Define the area A to be the limit of the sequence of approximations S_n as n increases, i.e., $A = \lim_{n\to\infty} S_n$.

The usual notation for $\lim_{n\to\infty} S_n$ is

$$\int_a^b f(x)\, dx$$

$\int_a^b f(x)\, dx$ is read "the integral of f from a to b."

which is read "the integral of the function f from a to b." Thus if f is a continuous function whose graph, over the interval $[a, b]$, does not go below the x axis, then the integral of f from a to b is the area under the graph (above the x axis) between the lines $x = a$ and $x = b$.

The procedure just described can be performed for any continuous function f whether or not the graph of f remains above the x axis over the interval $[a, b]$. In step (2), we can still compute $S_n = f(x_1)\Delta x + f(x_2)\Delta x + \cdots + f(x_n)\Delta x$; but if the graph of f goes below the x axis over $[a, b]$, then some of the values $f(x_k)$ may be negative, and we must forego the interpretation of S_n as an area approximation. In fact, if the graph of f is entirely below the x axis over $[a, b]$, then S_n will be negative for each n. In any case, it turns out that $\lim_{n\to\infty} S_n$ still exists, so for any continuous function f we define $\lim_{n\to\infty} S_n$ to be the *definite integral of f from a to b* and denote this limit by $\int_a^b f(x)\, dx$.

The integral as the area under a graph

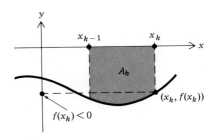

If $f(x) < 0$ for x in $[x_{k-1}, x_k]$ then $f(x_k)\Delta x$ approximates $-A_k$.

Definition of the definite integral $\int_a^b f(x)\, dx$

It is important to remember that if $f(x) \geqslant 0$ for each x in $[a, b]$, then $\int_a^b f(x)\, dx$ gives the area under the graph of f and between the lines $x = a$ and $x = b$. On the other hand, if $f(x) \leqslant 0$ for each x in $[a, b]$,

then $\int_a^b f(x)\ dx$ will give the *negative* of the area above the graph of f
(below the x axis) between the lines $x = a$ and $x = b$ (see Fig. 7.4-4).

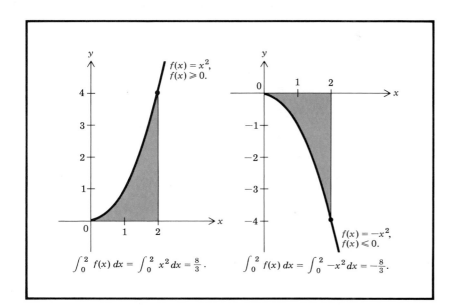

FIGURE 7.4-4

Recall from Section 7.3 that

$$\int_0^2 x^2 dx = \tfrac{8}{3}.$$

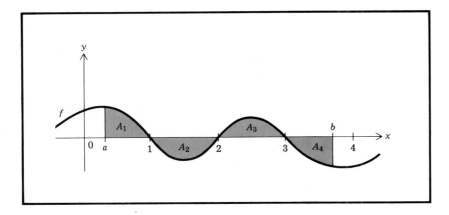

FIGURE 7.4-5

If the graph of a function f is as shown in Fig. 7.4-5, sometimes above
the x axis and sometimes below it, then $\int_a^1 f(x)\ dx = A_1$, $\int_1^2 f(x)\ dx =$
$- A_2$, $\int_2^3 f(x)\ dx = A_3$, and $\int_3^b f(x)\ dx = - A_4$, where A_1, A_2, A_3, and

A_4 are the (positive) shaded areas. The definite integral of f from a to b gives the area above the x axis and below the graph minus the area below the x axis and above the graph. Thus in Fig. 7.4-5,

$$\int_a^b f(x)\,dx = A_1 + A_3 - A_2 - A_4$$

The following mnemonic formula (for a function f continuous over $[a, b]$) is often helpful.

$$\int_a^b f(x)\,dx = (\text{area above the } x \text{ axis}) - (\text{area below the } x \text{ axis})$$

EXAMPLE 7.4-1

From the information given in Fig. 7.4-6, evaluate $\int_{-1}^{3} f(x)\,dx$.

SOLUTION

$\int_{-1}^{3} f(x)\,dx = (\text{area above the } x \text{ axis}) - (\text{area below the } x$ axis) $= (3 \text{ square units}) - ((1 + \frac{3}{2}) \text{ square units}) = \frac{1}{2}$ square unit. The positive result indicates that there is more area above the x axis than below.

FIGURE 7.4-6

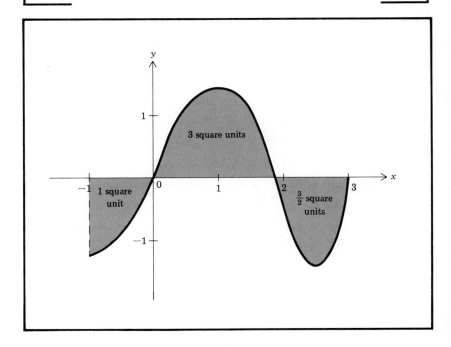

EXAMPLE 7.4-2

For the function $f(x) = 2x$, determine $\int_{-1}^{0} f(x)\, dx$, $\int_{0}^{1} f(x)\, dx$

and $\int_{-1}^{1} f(x)\, dx$.

SOLUTION

The graph of $f(x) = 2x$ is shown in Fig. 7.4-7. Thus, since the graph lies below the x axis over the interval $[-1, 0]$ and above the x axis over the interval $[0, 1]$, we have $\int_{-1}^{0} f(x)\, dx =$

$\int_{-1}^{0} 2x\, dx = -A_1$ and $\int_{0}^{1} f(x)\, dx = \int_{0}^{1} 2x\, dx = A_2$, where A_1 and A_2 are the areas of the regions shown in Fig. 7.4-7. But A_1 and A_2 are each the area of a triangle of base 1 and altitude 2, so $A_1 = A_2 = 1$. Thus

Using known areas to evaluate integrals

FIGURE 7.4-7

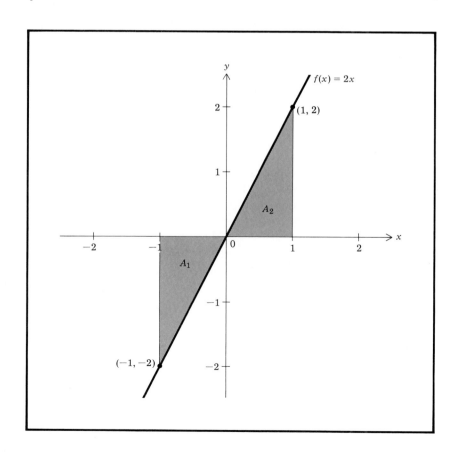

$$\int_{-1}^{0} f(x)\,dx = \int_{-1}^{0} 2x\,dx = -A_1 = -1$$

and

$$\int_{0}^{1} f(x)\,dx = \int_{0}^{1} 2x\,dx = A_2 = 1$$

Finally, since $\int_{-1}^{1} f(x)\,dx$ is $A_2 - A_1$, we have $\int_{-1}^{1} 2x\,dx$
$= 1 - 1 = 0.$

Recall that $\int_{a}^{b} f(x)\,dx$
$= \lim_{n\to\infty} S_n.$

As we have seen, $\int_{a}^{b} f(x)\,dx$ is defined as the value of a limit that is often difficult to evaluate. Fortunately, however, there is a remarkable and convenient connection between the indefinite integral $\int f(x)\,dx$ and the definite integral $\int_{a}^{b} f(x)\,dx$ that will enable us to evaluate definite integrals easily for a large class of functions f. This connection is given in what is called *The Fundamental Theorem of Calculus*, the topic of the next section.

Exercises 7.4

In Exercises 1 through 18, sketch the graph of the given function f and shade the region R between the graph of f and the x axis and between the given lines $x = a$ and $x = b$. Then, by computing the areas of appropriate portions of R, find the values of the indicated integrals.

1. $f(x) = 5x,\ x = 0,\ x = 2.$ Find $\int_{0}^{2} f(x)\,dx$

2. $f(x) = -x,\ x = 0,\ x = 4.$ Find $\int_{0}^{4} f(x)\,dx$

3. $f(x) = -2,\ x = -2,\ x = 1.$ Find $\int_{-2}^{1} f(x)\,dx$

4. $f(x) = 3x,\ x = -2,\ x = 3.$ Find $\int_{-2}^{0} f(x)\,dx,\ \int_{0}^{3} f(x)\,dx,$ and
$\int_{-2}^{3} f(x)\,dx$

5. $f(x) = x + 1$, $x = -2$, $x = 1$. Find $\int_{-2}^{-1} f(x)\,dx$, $\int_{-1}^{1} f(x)\,dx$, and

   $\int_{-2}^{1} f(x)\,dx$

6. $f(x) = -x - 1$, $x = -3$, $x = 0$. Find $\int_{-3}^{-1} f(x)\,dx$, $\int_{-1}^{0} f(x)\,dx$,

   and $\int_{-3}^{0} f(x)\,dx$

7. $f(x) = \dfrac{x}{2} + 2$, $x = 0$, $x = 4$. Find $\int_{0}^{1} f(x)\,dx$, $\int_{1}^{4} f(x)\,dx$, and

   $\int_{0}^{4} f(x)\,dx$

8. $f(x) = -\dfrac{3x}{4} + 3$, $x = -2$, $x = 6$. Find $\int_{-2}^{4} f(x)\,dx$, $\int_{4}^{6} f(x)\,dx$,

   and $\int_{-2}^{6} f(x)\,dx$

9. $f(x) = 2x - 4$, $x = 0$, $x = 4$. Find $\int_{0}^{1} f(x)\,dx$, $\int_{1}^{4} f(x)\,dx$, and

   $\int_{0}^{4} f(x)\,dx$

10. $f(x) = -4x + 8$, $x = 1$, $x = 4$. Find $\int_{1}^{3} f(x)\,dx$, $\int_{3}^{4} f(x)\,dx$, and

    $\int_{1}^{4} f(x)\,dx$

11. $f(x) = |x|$, $x = 0$, $x = 4$. Find $\int_{0}^{4} f(x)\,dx$ and $\int_{3}^{4} f(x)\,dx$

12. $f(x) = |x| - 2$, $x = -3$, $x = 2$. Find $\int_{-3}^{-2} f(x)\,dx$, $\int_{-2}^{0} f(x)\,dx$,

    $\int_{-2}^{2} f(x)\,dx$, and $\int_{-3}^{2} f(x)\,dx$

13. $f(x) = |x| + 2$, $x = -4$, $x = 1$. Find $\int_{-4}^{0} f(x)\,dx$, $\int_{0}^{1} f(x)\,dx$, and

    $\int_{-4}^{1} f(x)\,dx$

14. $f(x) = 2|x| + 1$, $x = -2$, $x = 1$. Find $\int_{-2}^{0} f(x)\,dx$, $\int_{0}^{1} f(x)\,dx$, and

    $\int_{-2}^{1} f(x)\,dx$

15. $f(x) = |x - 2|$, $x = 0$, $x = 3$. Find $\int_{0}^{2} f(x)\,dx$, $\int_{2}^{3} f(x)\,dx$,

    and $\int_{0}^{3} f(x)\,dx$

16. $f(x) = |x - 2| - 2$, $x = -2$, $x = 4$. Find $\int_{-2}^{0} f(x)\, dx$, $\int_{0}^{4} f(x)\, dx$,

and $\int_{-2}^{4} f(x)\, dx$

17. $f(x) = |x| + x + 1$, $x = -2$, $x = 2$. Find $\int_{-2}^{0} f(x)\, dx$, $\int_{0}^{2} f(x)\, dx$,

and $\int_{-2}^{2} f(x)\, dx$

18. $f(x) = |x| - x - 2$, $x = -4$, $x = 2$. Find $\int_{-4}^{-2} f(x)\, dx$,

$\int_{-2}^{2} f(x)\, dx$, and $\int_{-4}^{2} f(x)\, dx$

7.5. The Fundamental Theorem of Calculus

As was pointed out in the previous section, we can frequently circumvent the tedious and inefficient method of computing the limit of a sequence of approximating sums in order to evaluate a definite integral. The theorem which enables us to do this is called, because of its importance, the

Fundamental Theorem of Calculus   Suppose that f is a continuous function on a closed interval $[a, b]$ and that F is an antiderivative of f. Then

$$\int_{a}^{b} f(x)\, dx = F(b) - F(a)$$

It is customary to denote $F(b) - F(a)$ by $F(x)\big|_{a}^{b}$; thus the Fundamental Theorem may be expressed by the equation

$$\int_{a}^{b} f(x)\, dx = F(x)\big|_{a}^{b}$$

where $D_x F(x) = f(x)$.

EXAMPLE 7.5-1

Using Table 6.5-1 to find the appropriate antiderivatives, we see from the Fundamental Theorem of Calculus that

(a) $\int_{1}^{2} (2x + 1)\, dx = (x^2 + x)\big|_{1}^{2} = (2^2 + 2) - (1^2 + 1) = 4$,

(b) $\int_{0}^{1} x^3\, dx = \dfrac{x^4}{4}\bigg|_{0}^{1} = \dfrac{1^4}{4} - \dfrac{0^4}{4} = \dfrac{1}{4}$,

(c) $\int_{-1}^{1} (3x^2 + x - 1)\, dx = \left(x^3 + \dfrac{x^2}{2} - x \right) \Big|_{-1}^{1}$

$= \left(1^3 + \dfrac{1^2}{2} - 1 \right) - \left[(-1)^3 + \dfrac{(-1)^2}{2} - (-1) \right] = 0,$

(d) $\int_{1}^{2} \left(\dfrac{1}{x} \right) dx = \ln(x) \Big|_{1}^{2} = \ln(2) - \ln(1) = \ln(2) - 0$

$= \ln(2),$

(e) $\int_{9}^{15} \dfrac{1}{2\sqrt{x}}\, dx = \sqrt{x}\ \Big|_{9}^{15} = \sqrt{15} - \sqrt{9} = \sqrt{15} - 3$

$\approx 3.873 - 3 = 0.873,$

Since $D_x(\sqrt{x}) = 1/(2\sqrt{x})$, we see that $F(x) = \sqrt{x}$ is an anti-derivative of $f(x) = 1/(2\sqrt{x})$.

(f) $\int_{0}^{1} e^{2x}\, dx = \dfrac{e^{2x}}{2}\ \Big|_{0}^{1} = \dfrac{e^{2(1)}}{2} - \dfrac{e^{2(0)}}{2} = \dfrac{e^2}{2} - \dfrac{1}{2}$

$= \tfrac{1}{2}(e^2 - 1).$

In addition to the Fundamental Theorem of Calculus, there are several properties of the definite integral that are frequently helpful in evaluating integrals. These properties are given in Table 7.5-1.

TABLE 7.5-1

(1) $\displaystyle\int_{a}^{a} f(x)\, dx = 0$

(2) $\displaystyle\int_{a}^{b} kf(x)\, dx = k\int_{a}^{b} f(x)\, dx$ for any constant k

(3) $\displaystyle\int_{a}^{b} [f(x) \pm g(x)]\, dx = \int_{a}^{b} f(x)\, dx \pm \int_{a}^{b} g(x)\, dx$

(4) $\displaystyle\int_{a}^{b} f(x)\, dx = \int_{a}^{c} f(x)\, dx + \int_{c}^{b} f(x)\, dx$

(5) $\displaystyle\int_{a}^{b} f(x)\, dx = -\int_{b}^{a} f(x)\, dx$

Property (4) of Table 7.5-1 can be nicely illustrated geometrically, as shown in Fig. 7.5-1, for a continuous function f whose graph lies above the interval $[a, b]$ with c between a and b.

FIGURE 7.5-1   The area A of the entire shaded region is $\int_a^b f(x)\,dx = A_1 + A_2$

$$= \int_a^c f(x)\,dx + \int_c^b f(x)\,dx.$$

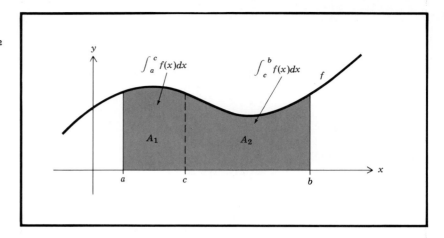

EXAMPLE 7.5-2

Evaluate (a) $\displaystyle\int_1^2 (3/x)\,dx$,   (b) $\displaystyle\int_1^2 (1/x + 2x + 1)\,dx$, and

(c) $\displaystyle\int_0^2 x^3\,dx$.

SOLUTION

(a) By property (2) of Table 7.5-1 we have

$$\int_1^2 \frac{3}{x}\,dx = 3\int_1^2 \frac{1}{x}\,dx$$

and from Example 7.5-1(d),

$$\int_1^2 \frac{1}{x}\,dx = \ln(2)$$

Thus

$$\int_1^2 \frac{3}{x}\,dx = 3\int_1^2 \frac{1}{x}\,dx = 3\ln(2)$$

(b) By property (3) of Table 7.5-1,

$$\int_1^2 ((1/x) + 2x + 1)\,dx = \int_1^2 \frac{1}{x}\,dx + \int_1^2 (2x + 1)\,dx$$

and from Example 7.5-1(a) and (d),

•
$$\int_1^2 (2x + 1)\, dx = 4 \quad \text{and} \quad \int_1^2 \frac{1}{x}\, dx = \ln(2)$$

Thus,

$$\int_1^2 \left(\frac{1}{x} + 2x + 1 \right) dx = \int_1^2 \frac{1}{x}\, dx + \int_1^2 (2x + 1)\, dx$$

$$= \ln(2) + 4$$

(c)
$$\int_0^2 x^3\, dx = \frac{x^4}{4} \Big|_0^2 = \frac{2^4}{4} - \frac{0^4}{4} = 4$$

Alternatively, by property (4) of Table 7.5-1 we have

$$\int_0^2 x^3\, dx = \int_0^1 x^3\, dx + \int_1^2 x^3\, dx$$

and from Example 7.5-1(b),

$$\int_0^1 x^3\, dx = \frac{1}{4}$$

Thus

$$\int_0^2 x^3\, dx = \int_0^1 x^3\, dx + \int_1^2 x^3\, dx$$

$$= \frac{1}{4} + \frac{x^4}{4} \Big|_1^2$$

$$= \frac{1}{4} + \left(\frac{2^4}{4} - \frac{1^4}{4} \right) = \frac{1}{4} + 4 - \frac{1}{4} = 4$$

Since $F(x) + C = \int f(x)\, dx$ means that $D_x F(x) = f(x)$, and since $D_x F(x)$ is the rate of change of F with respect to x, we can use the Fundamental Theorem of Calculus to gain information about F from given information about $f = D_x F$.

Integration of rate of change

Since $F(x) + C = \int f(x)\, dx$ means that $D_x F(x) = f(x)$, we see that $F(x) + C = \int D_x F(x)\, dx$,

i.e., $F(x) + C$ is an antiderivative of $D_x F(x)$.

City of Smartvil
Water Dept.

← Leak

EXAMPLE 7.5-3

The Smartville Water Department discovers that the city's water storage tank has sprung a leak and is losing water at the rate of $2t - 1000$ gal per hour, where t is the number of hours since the tank began to leak. How much water will be lost between $t = 10$ and $t = 20$?

SOLUTION

Let $G(t)$ be the number of gallons of water in the tank t hr after the tank begins to leak. Then $D_t G(t)$ is the rate of change of G with respect to t, which is given to be $2t - 1000$ gal per hour, i.e.,

$$D_t G(t) = 2t - 1000$$

Now, $G(10)$ = number of gallons in the tank after 10 hr and $G(20)$ = number of gallons in the tank after 20 hr, so $G(10) - G(20)$ is the amount of water lost between $t = 10$ and $t = 20$, the quantity we want to find. But by the Fundamental Theorem,

$$G(10) - G(20) = \int_{20}^{10} (2t - 1000)\, dt$$

$$= (t^2 - 1000t)\Big|_{20}^{10}$$

$$= \left[10^2 - 1000(10) \right] - \left[20^2 - 1000(20) \right]$$

$$= -9900 - (-19{,}600) = 9700 \text{ gal}$$

Note that although the information given in Example 7.5-3 allows us to find how much water will be lost between $t = 10$ and $t = 20$, it does not allow us to find how much water remains in the tank at $t = 10$. For that we need more information.

EXAMPLE 7.5-4

In Example 7.5-3, suppose there were 250,000 gal of water in the tank when the leak began; how many gallons will there be in the tank after 10 hr?

SOLUTION

Again, let $G(t)$ be the number of gallons of water in the tank t hr after the tank begins to leak; we want to find $G(10)$. We know that $D_t G(t) = 2t - 1000$, so

$$G(t) + C = \int (2t - 1000)\, dt = t^2 - 1000t$$

Thus $G(t) = t^2 - 1000t - C$, and we need only determine C and then evaluate $G(10)$. But $G(0)$ is the number of gallons of water in the tank when it begins to leak; thus

$$G(0) = 250{,}000 = 0^2 - 1000(0) - C$$

so $C = -250{,}000$. Thus, $G(t) = t^2 - 1000t - C = t^2 - 1000t + 250{,}000$ and

$$G(10) = 10^2 - 1000(10) + 250{,}000 = 240{,}100$$

so there are 240,100 gal in the tank after 10 hr.

The two examples above illustrate some general principles involving integrals of rates of change. Given a function f that gives the rate of change of $F(x)$ with respect to x (i.e., $D_x F(x) = f(x)$), we can find the change in $F(x)$ from $x = a$ to $x = b$ with the equation $F(b) - F(a) = \int_a^b f(x)\, dx$. In addition, we can find an equation that gives $F(x_0)$ for any value of $x = x_0$ if we are given the value $F(a)$ at some point $x = a$, because

The *change* in $F(x)$ from $x = a$ to $x = b$ is $F(b) - F(a)$.

$$F(x_0) - F(a) = \int_a^{x_0} f(x)\, dx$$

so

$$F(x_0) = \int_a^{x_0} f(x)\, dx + F(a)$$

EXAMPLE 7.5-5

The Half-Cal Candy Company estimates that on the tth day after it opens its new plant, it will be producing sugarless candy at the rate of $(3t/10) + 2000$ lb per day. How many pounds of candy will the company produce between the twentieth and thirtieth days after the new plant opens?

SOLUTION

$D_t P(t)$ is the rate of change of P with respect to t, so $D_t P(t)$ is the number of pounds produced per day, i.e., the production rate.

Let $P(t)$ be the number of pounds of candy the company will produce during the first t days after opening the new plant. Then $D_t P(t)$ is the production rate (in pounds per day) on the tth day after the plant opens. Thus

$$D_t P(t) = \frac{3t}{10} + 2000$$

Now, since $P(30)$ is the number of pounds of candy produced during the first 30 days and $P(20)$ is the number of pounds produced during the first 20 days, we wish to find $P(30) - P(20)$, i.e., the change in $P(t)$ from $t = 20$ to $t = 30$. But

$$P(30) - P(20) = \int_{20}^{30} D_t P(t)\, dt$$

$$= \int_{20}^{30} \left(\frac{3t}{10} + 2000 \right) dt$$

so

$$P(30) - P(20) = \frac{3}{10} \int_{20}^{30} t\, dt + \int_{20}^{30} 2000\, dt$$

$$= \frac{3}{10} \left(\frac{t^2}{2} \Big|_{20}^{30} \right) + 2000 t \Big|_{20}^{30}$$

$$= \frac{3}{10} \left(\frac{30^2}{2} - \frac{20^2}{2} \right) + (2000)(30) - (2000)(20)$$

$$= \frac{3}{10} (450 - 200) + 20{,}000 = 20{,}075 \text{ lb}$$

EXAMPLE 7.5-6

In Example 7.5-5, find an equation which gives the number of pounds $P(t)$ of candy the company will produce during the first t days, of plant operation, given that on the first day 2010.15 lb are produced.

SOLUTION

During t_0 days of operation the plant will produce $P(t_0)$ lb of candy, and

$$P(t_0) - P(1) = \int_1^{t_0} D_t P(t)\, dt = \int_1^{t_0} \left(\frac{3t}{10} + 2000 \right) dt$$

Thus, since $P(1) = 2010.15$, we have

$$P(t_0) = \int_1^{t_0} \left(\frac{3t}{10} + 2000 \right) dt + P(1)$$

$$= \frac{3}{10} \int_1^{t_0} t\, dt + \int_1^{t_0} 2000\, dt + P(1)$$

$$= \frac{3}{10} \left(\frac{t^2}{2} \Big|_1^{t_0} \right) + 2000\, t \Big|_1^{t_0} + 2010.15$$

$$= \frac{3}{10} \left(\frac{t_0^2}{2} - \frac{1}{2} \right) + 2000 t_0 - 2000 + 2010.15$$

$$= 3 \frac{t_0^2}{20} - \frac{3}{20} + 2000 t_0 + 10.15$$

$$= 3 \frac{t_0^2}{20} + 2000 t_0 + 10$$

Since this equation is valid for any positive value t_0, we see that during the first t days of plant operation the company will produce $P(t) = 3t^2/20 + 2000t + 10$ lb of candy. We check this by noting that $D_t P(t) = 3t/10 + 2000$ and $P(1) = 2010.15$.

Another common use of the integral is to determine the "average value" or "mean value" of a function f over an interval $[a, b]$. Recall that the average of a finite collection of numbers $x_1, x_2, \ldots, x_n$ is

$$\frac{x_1 + x_2 + \ldots + x_n}{n}$$

The integral $\int_a^b f(x)\,dx$ is sometimes thought of as the "sum" of all the values $f(x)$ for x in $[a, b]$; this leads us to define the *average* or *mean value m* of f over $[a, b]$ as

Definition of the mean value m of a function f over an interval $[a, b]$

$$m = \frac{\int_a^b f(x)\,dx}{b - a}$$

Geometrically, for a continuous function f whose graph is above the x axis over $[a, b]$, m is the height required for a rectangle of width $b - a$ to have an area equal to $\int_a^b f(x)\,dx$ (see Fig. 7.5-2). Note that when $f(x)$

If $F(x) + C = \int f(x)\,dx$, then $D_x[F(x) + C] = f(x)$, so $f(x)$ is the rate of change of $F(x) + C$ with respect to x.

is interpreted as the rate of change of $F(x) + C = \int f(x)\,dx$, then the average rate of change of $F(x) + C$ from $x = a$ to $x = b$ is

Recall that average rate of change was discussed in Section 5.3.

$$\frac{F(b) - F(a)}{b - a} = \frac{\int_a^b f(x)\,dx}{b - a}$$

FIGURE 7.5-2 The solid shaded area is $\int_a^b f(x)\,dx$ and is equal to $m(b - a)$, the area of the rectangle.

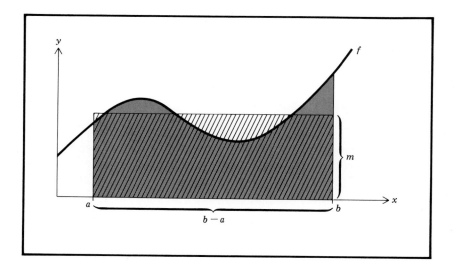

An important theoretical fact necessary to establish the Fundamental Theorem of Calculus is that the mean value m of a continuous function f over $[a, b]$ is the value of f at some point c in (a, b), that is, $m = f(c)$ (see Fig. 7.5-2). The precise statement of the result is

The Mean Value Theorem for Integrals If f is continuous on $[a, b]$, then there exists a number c, with $a < c < b$, such that $\int_a^b f(x)\, dx = f(c)[b - a]$.

EXAMPLE 7.5-7

The Timely Taxi Company finds that after one of its cabs has been driven a distance x (measured in hundreds of thousands of miles), the repair cost C (in dollars per mile) necessary to continue operating the cab is given by $C(x) = 0.04x^2 + 0.01$. If a cab has already been driven 100,000 mi, find the average repair cost per mile for operating it an additional 200,000 mi.

SOLUTION

We wish to find the average or mean value m of $C(x) = 0.04x^2 + 0.01$ over the interval $[1, 3]$, which is given by

$$m = \frac{\int_1^3 C(x)\, dx}{3 - 1}$$

Thus, since

$$\int_1^3 C(x)\, dx = \int_1^3 (0.04x^2 + 0.01)\, dx$$

$$= \left(\frac{0.04x^3}{3} + 0.01x \right)\Big|_1^3$$

$$= \left[\frac{0.04(3^3)}{3} + (0.01)3 \right] - \left[\frac{0.04(1^3)}{3} + (0.01)1 \right]$$

$$= (0.36 + 0.03) - \left(\frac{0.04}{3} + 0.01 \right)$$

$$= 0.39 - \frac{0.07}{3} \approx 0.37$$

we see that the average repair cost per mile is

$$\frac{\int_1^3 C(x)\,dx}{3-1} \approx \frac{0.37}{2} = 0.185$$

dollars, or 18.5 cents per mile.

EXAMPLE 7.5-8

In Example 7.5-3, find the average rate of water loss between $t = 10$ and $t = 20$ days after the tank began to leak.

SOLUTION

Let $G(t)$ be the number of gallons of water in the tank t hr after the tank began to leak. Then $D_t G(t) = 2t - 1000$ is the rate of change of G with respect to t (the rate of water loss), and $(G(20) - G(10))/(20 - 10)$ is the average rate of change of G (i.e., average rate of water loss) from $t = 10$ to $t = 20$. But

$$G(20) - G(10) = \int_{10}^{20} (2t - 1000)\,dt$$

$$= t^2 - 1000t \Big|_{10}^{20}$$

$$= (-19{,}600) - (-9900) = -9700$$

Thus the average rate of water loss from $t = 10$ to $t = 20$ is

$$\frac{G(20) - G(10)}{20 - 10} = \frac{\int_{10}^{20} (2t - 1000)\,dt}{10}$$

$$= \frac{-9700}{10} = -970$$

gal per hour (the negative sign indicates that the amount G of water in the tank is decreasing as t increases).

EXAMPLE 7.5-9

Use the Mean Value Theorem for Integrals to prove the Fundamental Theorem of Calculus.

SOLUTION

We are given that f is continuous on $[a, b]$. We define G as the function with domain $[a, b]$ given by the equation

$$G(x) = \int_a^x f(t)\, dt$$

In following this proof it may be helpful to think of f as positive-valued so that $G(x)$ gives the area under the graph of f over the interval $[a, x]$ (see Fig. 7.5-3(a)). Now we compute the derivative $D_x G(x_0)$ at a point x_0 in $[a, b]$. To do this, we must evaluate

$$\lim_{x \to x_0} \frac{G(x) - G(x_0)}{x - x_0}$$

The difference $G(x) - G(x_0)$ (for $x > x_0$) is the area A under f over the interval $[x_0, x]$ (see Fig. 7.5-3(b)). By the Mean Value Theorem for Integrals this area A is also equal to $f(c)(x - x_0)$ for some c in (x_0, x); consequently

$$G(x) - G(x_0) = f(c)(x - x_0)$$

FIGURE 7.5-3

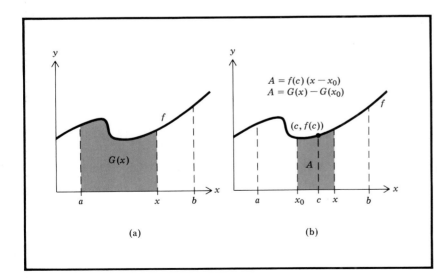

(a)

(b)

that is,

$$\frac{G(x) - G(x_0)}{x - x_0} = f(c)$$

A similar argument shows that this equation is also true when $x < x_0$; we leave it to the student to verify this. Now, taking limits of both sides, we have

$$D_x G(x_0) = \lim_{x \to x_0} \frac{G(x) - G(x_0)}{x - x_0} = \lim_{x \to x_0} f(c)$$

Since c is between x and x_0, c also approaches x_0 as x approaches x_0, so

$$\lim_{x \to x_0} f(c) = \lim_{c \to x_0} f(c) = f(x_0)$$

since f is continuous. Thus, $D_x G(x_0) = f(x_0)$, i.e., $D_x G(x) = f(x)$ for x in $[a, b]$, which means that G is an antiderivative of f. Thus if F is *any* antiderivative of f, then $F(x) = G(x) + C$ for some constant C (see Section 6.5). But

$$G(b) - G(a) = \int_a^b f(t)\, dt - \int_a^a f(t)\, dt = \int_a^b f(t)\, dt - 0$$

so

$$F(b) - F(a) = \left[G(b) + C \right] - \left[G(a) + C \right]$$

$$= G(b) - G(a)$$

$$= \int_a^b f(t)\, dt$$

This is the desired result.

Exercises 7.5

In Exercises 1 through 30, use the Fundamental Theorem of Calculus to evaluate the given definite integral.

1. $\displaystyle\int_1^2 3\, dx$

2. $\displaystyle\int_{-1}^2 -5\, dx$

3. $\displaystyle\int_1^4 4x\, dx$

4. $\displaystyle\int_{-2}^1 -3x\, dx$

5. $\int_0^2 (2x + 2)\, dx$

6. $\int_{-1}^3 (-4x - 1)\, dx$

7. $\int_{-1}^1 3x^2\, dx$

8. $\int_{-2}^1 - 8x^3\, dx$

9. $\int_0^2 (3x^2 + 2x - 1)\, dx$

10. $\int_2^3 (-4x^2 - 2x - 2)\, dx$

11. $\int_1^2 4x^{-5}\, dx$

12. $\int_1^2 - 5x^{-6}\, dx$

13. $\int_{-1}^1 (4x^3 - 3x^2 + 2x + 4)\, dx$

14. $\int_0^1 (-8x^3 + 6x^2 - 4x + 1)\, dx$

15. $\int_{-2}^1 (10x^4 - 8x)\, dx$

16. $\int_1^2 (15x^5 + 6x^2)\, dx$

17. $\int_4^9 \frac{5}{2\sqrt{x}}\, dx$

18. $\int_1^2 \left(\frac{x^3}{2} - x^{-4} \right) dx$

19. $\int_4^{16} \left(\frac{4}{x} + \frac{1}{2\sqrt{x}} \right) dx$

20. $\int_1^2 \left(\exp(x) - \frac{2}{x} \right) dx$

21. $\int_0^2 \left(\frac{x^2}{2} - e^{3x} \right) dx$

22. $\int_0^1 \left(\frac{x^4}{5} + e^{-x} \right) dx$

23. $\int_0^1 \left(\frac{2x^3}{5} - \frac{3x^2}{2} + \frac{1}{2} \right) dx$

24. $\int_1^2 (2t^4 - 7t^{-2} + e^t)\, dt$

25. $\int_1^3 \left(\frac{4}{z} - \exp\left(\frac{2z}{3} \right) \right) dz$

26. $\int_1^2 \left(2 \exp(-6w) + \frac{2}{w} - 1 \right) dw$

27. $\int_{-1}^1 \left(\frac{7s^6}{2} - \frac{s^3}{2} + 1 \right) ds$

28. $\int_{-2}^{-1} \left(\frac{6r^5}{5} - \frac{r^{-3}}{4} + \frac{1}{r} \right) dr$

29. $\int_2^1 \left(8u^7 - \exp\left(\frac{4u}{3} \right) + \frac{3}{4u} \right) du$

30. $\int_{-1}^{-2} \left(-5v^{-6} + 2 \exp\left(\frac{3v}{2} \right) - \frac{1}{4v} + 1 \right) dv$

In Exercises 31 through 40, find the area of the region bounded by the graph of the given function, the x axis, and the lines $x = a$ and $x = b$ for the given values of a and b.

31. $f(x) = x^2$, $a = -2, b = 1$

32. $f(x) = 2x^4 + 1$, $a = 0, b = 2$

33. $f(x) = 2x^3 + 3x + 1$, $a = 1, b = 3$

34. $f(x) = \exp(3x), \quad a = 0, b = 2$

35. $f(x) = \dfrac{2}{x} + 1, \quad a = 1, b = 2$

36. $f(x) = \dfrac{1}{2\sqrt{x}}, \quad a = 2, b = 4$

37. $f(x) = 2x - 1, \quad a = -1, b = 2$

38. $f(x) = -x^4 - 1, \quad a = 1, b = 2$

39. $f(x) = x^3 + 1, \quad a = -2, b = 2$

40. $f(x) = x^4 - 1, \quad a = -2, b = 2$

In Exercises 41 through 50, use the Fundamental Theorem of Calculus and the given information to find an equation that defines $F(x)$.

41. $D_x F(x) = 2x + 1, \quad F(1) = 3$

42. $D_x F(x) = 4x - 2, \quad F(-1) = 6$

43. $D_x F(x) = 3x^2 - 2x + 1, \quad F(2) = 8$

44. $D_x F(x) = 6x^2 - 4x + 2, \quad F(-1) = -2$

45. $D_x F(x) = -2x^4 + x^3, \quad F(0) = 2$

46. $D_x F(x) = 3x^2 + \dfrac{x}{2} - 1, \quad F(-2) = 1$

47. $D_x F(x) = \exp(2x) + x^2, \quad F(0) = 1$

48. $D_x F(x) = \dfrac{2}{x} - x, \quad F(1) = 0$

49. $D_x F(x) = \dfrac{1}{2\sqrt{x}} + 2\exp(2x), \quad F(1) = 2 + e^2$

50. $D_x F(x) = \dfrac{1}{x} - \dfrac{1}{2\sqrt{x}} + e^x, \quad F(1) = e$

In Exercises 51 through 56, find the mean value m of the given function f over the given interval.

51. $f(x) = 2x - 3, \quad [1, 4]$

52. $f(x) = -4x + 2, \quad [-2, 0]$

53. $f(x) = 3x^2 + 2, \quad [0, 2]$

54. $f(x) = -4x^3 - 2x, \quad [-1, 2]$

55. $f(x) = \exp(2x) - x^2, \quad [0, 2]$

56. $f(x) = \dfrac{4}{x} + \dfrac{1}{2\sqrt{x}}, \quad [1, 4]$

57. The Euphoria City Planning Commission estimates that x mo from now, the city's population will be increasing at the rate of $10 + x$ people per month for $0 \leqslant x \leqslant 12$. What will be the increase in the population P of the city between $x = 4$ and $x = 8$ mo from now?

58. In Exercise 57, if the population of Euphoria is now 6000 people, find a formula for the population $P(x)$ of the city x mo from now ($0 \leqslant x \leqslant 12$). What will the city's population be 10 mo from now?

59. A rancher's beef cattle are gaining weight, on the average, at the rate of $20 + (16/x^2)$ lb per month at age x mo for $1 \leqslant x \leqslant 12$. How much weight, on the average, do the cattle gain between $x = 2$ and $x = 8$ mo after birth?

60. In Exercise 59, if the average weight of the rancher's cattle is 100 lb 2 mo after birth, find a formula for the weight $W(x)$ of the cattle x mo after birth ($1 \leqslant x \leqslant 12$). What is the average weight of the cattle 4 mo after birth?

61. The Murcer Company manufactures industrial lighting fixtures. The company finds that the marginal cost M_c (in dollars per unit produced) at a monthly production level of x units is given by the equation $M_c(x) = (-x/50) + 40$ for $500 \leqslant x \leqslant 1000$. Find the increase in the total monthly production cost P_c if the production level is increased from 600 to 800 units per month.

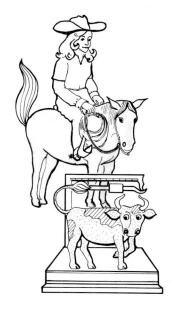

Recall that if $P_c(x)$ is the production cost at a production level of x units, then the marginal cost M_c is defined by $M_c(x) = D_x P_c(x)$.

62. In Exercise 61, find an equation that gives the total monthly production cost $P_c(x)$ at a monthly production level of x units ($500 \leqslant x \leqslant 1000$), given that the cost of producing 800 units is $30,000. What is the total monthly production cost P_c at a production level of 1000 units?

63. A manufacturer finds that the marginal revenue M_R (in dollars per unit) at a monthly production level of x units of its product is given by the equation $M_R(x) = (-3x^2/10) + 30x + 20$ for $80 \leqslant x \leqslant 100$. Find the decrease in the monthly revenue R if the monthly production level is cut from 100 units to 90 units.

Recall that if $R(x)$ is the revenue received at a production level of x units, then the marginal revenue M_R is defined by $M_R(x) = D_x R(x)$.

64. In Exercise 63, find an equation that gives the monthly revenue $R(x)$ at a monthly production level of x units ($80 \leqslant x \leqslant 100$) given that the revenue received from the monthly production of 80 units is $50,000. What is the monthly revenue received at a production level of 100 units?

65. The monthly marginal demand M_d for a product is given by $M_d(x) = -200 \exp(x/2)$, where x is the selling price per unit (in dollars) and $2 \leqslant x \leqslant 4$. Find the decrease in the monthly demand d for the product if the selling price is increased from $2 per unit to $4 per unit.

Recall that if $d(x)$ is the demand for a product priced at x dollars per unit, then the marginal demand M_d is given by $M_d(x) = D_x d(x)$.

66. In Exercise 65, find an equation that gives the monthly demand $d(x)$ for the product when it is sold at $\$x$ per unit ($2 \leqslant x \leqslant 4$) given that the demand is 4000 units when $x = 2$. What is the monthly demand when the product is priced at $3 per unit? (Use Table A in the back of the book to find the value of $e^{1.5}$.)

67. Suppose you are paddling up a river in a canoe. Your speed v (in miles per hour relative to the banks of the river) is given by the equation $v(t) = 5 - t/2 - t^2/16$, where t is the time in hours since you began paddling and $0 \leqslant t \leqslant 4$. How far do you travel during the first hour? During the second hour? During the first 4 hours?

68. The Martin Mining Company estimates that the rate (in dollars per year) at which its loading conveyors depreciate in value is given by $400x - 8000$, where x is the age of the conveyor in years. How much does a conveyor depreciate in value between the second and fifth years of use? If a conveyor costs $60,000 new, find a formula for the value $v(x)$ of the conveyor after x years of use. How much is a conveyor worth after 4 yr of use?

In Exercise 69, use Table A in the back of the book to find the values of ln(5) and ln(12).

69. The Midland Dairy Association estimates that the retail price of 1 gal of milk x mo from now ($x \geqslant 1$) will be increasing at the rate of $1 + (2/x)$ cents per month. How much will 1 gal of milk increase in price between $x = 1$ and $x = 5$ mo from now? If milk sells at $\$1.50$ per gallon 1 mo from now, find a formula for the retail price $P(x)$ of 1 gal of milk x mo from now. What will be the selling price of milk 12 mo from now?

70. The percent concentration of a drug in the bloodstream is increasing at the rate of $(-t/40 + 2)\%$ per minute, t min after administration of the drug for $10 \leqslant t \leqslant 30$. What is the increase in the percent concentration between $t = 10$ and $t = 20$ min after administration? If the percent concentration is 30% after 30 min, find an equation that gives the percent concentration $P(t)$ at time t min after administration of the drug. What is the percent concentration 20 min after administration?

In Exercise 71, use Table A in the back of the book to find the values of exp(3) and exp(2).

71. The population of a certain culture of bacteria is increasing at the rate of $500 \exp(0.5t)$ bacteria per day, where t is time in days with $t = 0$ corresponding to the present time. How much will the population increase between $t = 4$ and $t = 6$ days from now? If at $t = 0$ the population contains 5000 bacteria, find an equation that gives the population $P(t)$ at time t days from now. What will the population be 4 days from now?

In Exercise 72, use Table A in the back of the book to find the values of ln(10), ln(20), and ln(40).

72. Forestry officials find that x mi from the edge of an industrial city, the average height of mature ponderosa pine trees changes at the rate of $0.5 + 10/x$ ft per mile, for $10 \leqslant x \leqslant 40$. How much does the average height of the trees decrease between $x = 10$ and $x = 20$ mi from the city? If the average height of the trees 20 mi from the city is 60 ft, find a formula for the height $h(x)$ of the trees x mi from the city. What is the average height 40 mi from the city?

Techniques of Integration

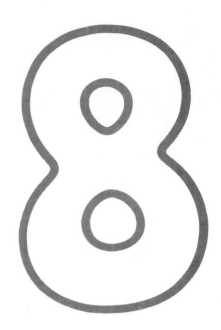

8.1. The Area between Two Curves

In this section we show how the methods developed thus far can be used to produce a formula for computing the area of a region between two graphs.

EXAMPLE 8.1-1

Compute the area A between the graphs of the functions $f(x) = 2 - x^2$ and $g(x) = x^2$ (see Fig. 8.1-1).

SOLUTION

To find where the graphs intersect, we set $f(x) = g(x)$ and solve for x. From

$$2 - x^2 = x^2$$

we have

$$2x^2 - 2 = 0$$

so

$$2(x^2 - 1) = 0 \quad \text{or} \quad 2(x - 1)(x + 1) = 0$$

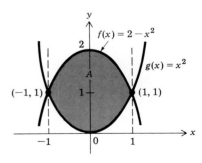

FIGURE 8.1-1

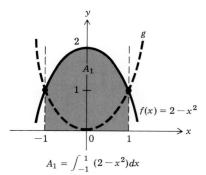

$$A_1 = \int_{-1}^{1} (2 - x^2)\, dx$$

FIGURE 8.1-2

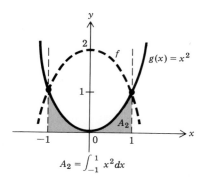

$$A_2 = \int_{-1}^{1} x^2\, dx$$

FIGURE 8.1-3

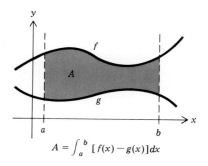

$$A = \int_{a}^{b} [f(x) - g(x)]\, dx$$

Area between the graphs of
two functions

Thus $x = 1$ or $x = -1$, so the graphs intersect at the points $(-1, 1)$ and $(1, 1)$. The area A_1 (Fig. 8.1-2) under f between $x = -1$ and $x = 1$ is

$$\int_{-1}^{1} (2 - x^2)\, dx = \left(2x - \frac{x^3}{3}\right)\Bigg|_{-1}^{1}$$

$$= \left[2(1) - \frac{1^3}{3}\right] - \left[2(-1) - \frac{(-1)^3}{3}\right]$$

$$= \tfrac{5}{3} + \tfrac{5}{3} = \tfrac{10}{3}$$

The area A_2 (Fig. 8.1-3) under g between $x = -1$ and $x = 1$ is

$$\int_{-1}^{1} x^2\, dx = \frac{x^3}{3}\Bigg|_{-1}^{1} = \tfrac{1}{3} - \left(-\tfrac{1}{3}\right) = \tfrac{2}{3}$$

Thus the desired area A (Fig. 8.1-1) is

$$A_1 - A_2 = \tfrac{10}{3} - \tfrac{2}{3} = \tfrac{8}{3}$$

Note that

$$A = A_1 - A_2$$

$$= \int_{-1}^{1} (2 - x^2)\, dx - \int_{-1}^{1} x^2\, dx$$

$$= \int_{-1}^{1} \left[(2 - x^2) - x^2\right] dx$$

$$= \int_{-1}^{1} \left[f(x) - g(x)\right] dx$$

because of property (3) of integrals in Table 7.5-1.

The reasoning employed in Example 8.1-1 yields the following general formula. If f and g are continuous functions over the interval $[a, b]$ and if the graph of f is above the graph of g over this interval ($f(x) \geqslant g(x)$ for x in $[a, b]$), then the area A between the graphs of f and g is given by the formula

$$A = \int_{a}^{b} \left[f(x) - g(x)\right] dx$$

Note that since $f(x) \geqslant g(x)$ for all x in $[a, b]$, $f(x) - g(x) \geqslant 0$ for all x in $[a, b]$. Thus the graph of $f - g$ does not go below the x axis on $[a, b]$, so the integral $\int_a^b [f(x) - g(x)]\, dx$ is always positive or zero.

EXAMPLE 8.1-2

Because of rising food costs, Daisy Dairy Farms finds that the cost C of producing 1 gal of milk is now as much as the revenue received from the sale of the milk. The company also estimates that the cost $C(t)$ of producing 1 gal of milk is increasing at the rate of $g(t) = D_t C(t) = \$t^2$ per year, where t is time in years with $t = 0$ corresponding to the present time; moreover, the revenue $R(t)$ (in dollars per gallon) is increasing at the rate of $f(t) = D_t R(t) = \$2t$ per year. Find the profit P per gallon that the company can expect at the end of 1 yr, given that 1 gal of milk now costs \$1 to produce and yields \$1 in revenue. Graph the functions f and g and illustrate the profit P as an area on the graph.

SOLUTION

Since $D_t C(t) = t^2$ and $D_t R(t) = 2t$, we see that

$$C(1) - C(0) = \int_0^1 t^2\, dt \quad \text{and} \quad R(1) - R(0) = \int_0^1 2t\, dt$$

Thus

$$C(1) = \int_0^1 t^2\, dt + C(0) = \frac{t^3}{3}\Big|_0^1 + C(0) = \frac{1}{3} + 1 = \$1\frac{1}{3}$$

per gallon, and

$C(0) = 1$, since $C(0)$ is the cost of producing 1 gal of milk now. Similarly, $R(0) = 1$.

$$R(1) = \int_0^1 2t\, dt + R(0) = t^2\Big|_0^1 + 1 = \$2$$

per gallon. Thus the profit P per gallon after 1 yr is

$$P(1) = R(1) - C(1) = 2 - 1\frac{1}{3} = \$\frac{2}{3}$$

per gallon. Since the profit

$$P(1) = \left(\int_0^1 2t\, dt + 1 \right) - \left(\int_0^1 t^2\, dt + 1 \right)$$

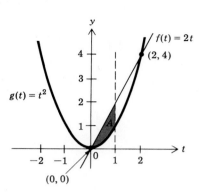

$g(t) = t^2$

$f(t) = 2t$

$(2, 4)$

A

$(0, 0)$

FIGURE 8.1-4 $A = \int_0^1 [f(t) - g(t)]\, dt.$

$$= \int_0^1 2t\, dt - \int_0^1 t^2\, dt$$

$$= \int_0^1 (2t - t^2)\, dt$$

$$= \int_0^1 \left[\, f(t) - g(t) \,\right] dt$$

it can be represented as the area A between the graphs of f and g over the interval $[0, 1]$ as shown in Fig. 8.1-4.

EXAMPLE 8.1-3

Find the profit P per gallon of milk that Daisy Dairy Farms can expect to be making at the end of 2 yr (see Example 8.1-2).

SOLUTION

Since $P(2)$ is the profit per gallon after 2 yr and $P(2) = R(2) - C(2)$, we need only find $R(2)$ and $C(2)$. But

$$R(2) - R(0) = \int_0^2 2t\, dt \quad \text{and} \quad C(2) - C(0) = \int_0^2 t^2\, dt$$

so

$$R(2) = \int_0^2 2t\, dt + R(0) \quad \text{and} \quad C(2) = \int_0^2 t^2\, dt + C(0)$$

We know that $R(0) = C(0) = 1$. Thus we find that

$$P(2) = \left(\int_0^2 2t\, dt + 1 \right) - \left(\int_0^2 t^2\, dt + 1 \right)$$

$$= \int_0^2 2t\, dt - \int_0^2 t^2\, dt$$

$$= \int_0^2 (2t - t^2)\, dt$$

$$= \left(t^2 - \frac{t^3}{3} \right)\Big|_0^2$$

$$= \left(2^2 - \frac{2^3}{3} \right) - \left(0^2 - \frac{0^3}{3} \right)$$

$$= 4 - \frac{8}{3} = \frac{4}{3}$$

dollars per gallon. The profit $P(2) = \int_0^2 (2t - t^2)\, dt = \frac{4}{3}$ is the shaded area shown in Fig. 8.1-5.

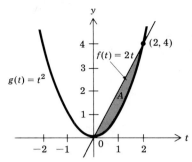

FIGURE 8.1-5 $A = \int_0^2 (2t - t^2)\, dt.$

EXAMPLE 8.1-4

Find a formula for the profit $P(t)$ per gallon of milk that Daisy Dairy Farms can expect to make t yr from now (see Example 8.1-2.). How many years will it take for the rising production cost to reduce the profit P per gallon to \$0 again?

SOLUTION

The profit $P(t_0)$ per gallon t_0 yr from now is $P(t_0) = R(t_0) - C(t_0)$, and

$$R(t_0) - R(0) = \int_0^{t_0} 2t\, dt$$

and

$$C(t_0) - C(0) = \int_0^{t_0} t^2\, dt$$

Thus

$$P(t_0) = \left(\int_0^{t_0} 2t\, dt + 1 \right) - \left(\int_0^{t_0} t^2\, dt + 1 \right)$$

$$= \int_0^{t_0} 2t\, dt - \int_0^{t_0} t^2\, dt$$

$$= \int_0^{t_0} (2t - t^2)\, dt$$

$$= \left(t^2 - \frac{t^3}{3} \right)\Big|_0^{t_0}$$

$$= t_0^2 - \frac{t_0^3}{3}$$

Thus t yr from now the profit will be $P(t) = t^2 - t^3/3$ dollars per gallon. To find when the profit P will be 0, we set $P(t) = 0$ and solve for t. Thus we have

$$t^2 - \frac{t^3}{3} = 0$$

so

$$t^2\left(1 - \frac{t}{3}\right) = 0$$

which means that $t = 0$ or $t = 3$. This means that 3 yr from now the profit per gallon will again be reduced to 0. Graphically, this represents the fact that the area A_1 between the graphs of f and g over $[0, 2]$ equals the area A_2 between the graphs of f and g over $[2, 3]$ (see Fig. 8.1-6). Note that

$$\int_0^2 [f(t) - g(t)] \, dt = \frac{4}{3} = A_1$$

and

$$\int_2^3 [f(t) - g(t)] \, dt = -\frac{4}{3} = -A_2$$

Thus

$$P(3) = \int_0^3 [f(t) - g(t)] \, dt$$

$$= \int_0^2 [f(t) - g(t)] \, dt + \int_2^3 [f(t) - g(t)] \, dt$$

$$= A_1 + (-A_2)$$

$$= \frac{4}{3} + \left(-\frac{4}{3}\right) = 0$$

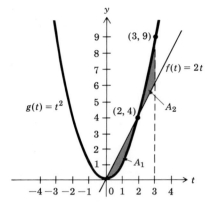

FIGURE 8.1-6 $A_1 = A_2$ so

$$\int_0^3 [f(t) - g(t)] \, dt = A_1 + (-A_2) = 0.$$

EXAMPLE 8.1-5

Find the area between the graphs of $f(x) = x$ and $g(x) = x^3$.

SOLUTION

To find where the graphs intersect, we set $f(x) = g(x)$ and solve for x. Thus we have

$$x^3 = x \quad \text{or} \quad x^3 - x = 0$$

so

$$x(x^2 - 1) = 0$$

which means that $x = 0$ or $x = \pm 1$. Now, over the interval $[-1, 0]$ the graph of g is above the graph of f (see Fig. 8.1-7), so the area A_1 between the graphs is given by

$$\int_{-1}^{0} [\, g(x) - f(x)\,]\, dx = \int_{-1}^{0} (x^3 - x)\, dx$$

$$= \left(\frac{x^4}{4} - \frac{x^2}{2} \right)\Big|_{-1}^{0}$$

$$= \left(\frac{0^4}{4} - \frac{0^2}{2} \right) - \left(\frac{(-1)^4}{4} - \frac{(-1)^2}{2} \right)$$

$$= 0 - \left(-\tfrac{1}{4} \right) = \tfrac{1}{4}$$

Over $[0, 1]$ the graph of f is above that of g, so the area A_2 between the graphs is

Finding the area between the graphs of f and g when the graph of f is not always above that of g

FIGURE 8.1-7

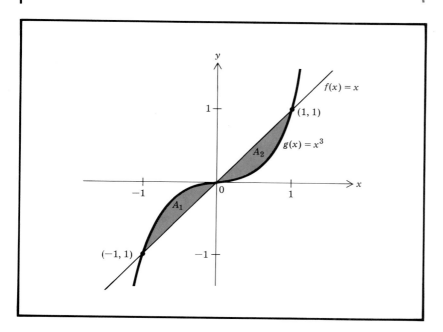

$$\int_0^1 [f(x) - g(x)]\, dx = \int_0^1 (x - x^3)\, dx$$

$$= \left(\frac{x^2}{2} - \frac{x^4}{4} \right)\Big|_0^1$$

$$= \left(\frac{1^2}{2} - \frac{1^4}{4} \right) - \left(\frac{0^2}{2} - \frac{0^4}{4} \right) = \frac{1}{4}$$

Thus the total area between the graphs of f and g is $A_1 + A_2$
$= \frac{1}{4} + \frac{1}{4} = \frac{1}{2}$.

Exercises 8.1

In Exercises 1 through 15, compute the area A of the region between
the graphs of the given functions.

1. $f(x) = 4, \quad g(x) = x^2$

2. $f(x) = x, \quad g(x) = x^2$

3. $f(x) = x + 2, \quad g(x) = x^2$

4. $f(x) = x^2, \quad g(x) = x^3$

5. $f(x) = 2x^2, \quad g(x) = 3x$

6. $f(x) = 4 - x^2, \quad g(x) = x^2 - 4$

7. $f(x) = x^2 + x, \quad g(x) = 2x$

8. $f(x) = x^2 - 2, \quad g(x) = -x^2 + 2x - 3$

9. $f(x) = x^2 - 2x - 3, \quad g(x) = 2x - 3$

10. $f(x) = x^2 - x - 6, \quad g(x) = x + 2$

11. $f(x) = x^3, \quad g(x) = 4x$

12. $f(x) = x, \quad g(x) = x^5$

13. $f(x) = x^3 + 1, \quad g(x) = x + 1$

14. $f(x) = x^5 - 1, \quad g(x) = x - 1$

15. $f(x) = x^3 + 2, \quad g(x) = 4x + 2$

16. As a result of using a new nutritional supplement for hogs, a farmer expects the revenue R from each hog to increase for the next 10 yr at the rate of $f(t) = D_t R(t) = -0.2t^2 + 4t + 5$ dollars per year, where t is time in years and $t = 0$ corresponds to the present. Moreover, the cost C of raising each hog is increasing at the rate of $g(t) = D_t C(t) = \$2t$ per year. If each hog now costs \$400 to raise and yields \$400 in revenue, find the profit P per hog that the farmer can expect at the end of 5 yr. Graph the functions f and g and illustrate the profit P as an area on the graph.

17. In Exercise 16, find a formula for the profit $P(t)$ per hog that the farmer can expect to make t yr from now $(0 \leq t \leq 10)$. What will the profit per hog be 8 yr from now?

8.2. Integration by Substitution

Now we would like to extend our integration techniques so that we may evaluate certain definite integrals that are still inaccessible to us. A simple example will provide some insight and motivation for our discussion.

EXAMPLE 8.2-1

Evaluate $\displaystyle\int_0^1 (x + 1)^3 \, dx$.

SOLUTION

We expand $(x + 1)^3$ to get

$$(x + 1)^3 = x^3 + 3x^2 + 3x + 1$$

Thus

$$\int_0^1 (x + 1)^3 \, dx = \int_0^1 (x^3 + 3x^2 + 3x + 1) \, dx$$

$$= \left(\frac{x^4}{4} + x^3 + \frac{3x^2}{2} + x \right)\Big|_0^1$$

$$= \left(\tfrac{1}{4} + 1 + \tfrac{3}{2} + 1 \right) - 0 = 3\tfrac{3}{4}$$

Evaluation of definite integrals depends on being able to find appropriate antiderivatives.

Now suppose we want to evaluate $\int_0^1 (x + 1)^{100} \, dx$. We would certainly shrink from the task of expanding $(x + 1)^{100}$ and duplicating the process of Example 8.2-1. If, however, we set $t = x + 1$, then as x varies from 0 to 1, $t = x + 1$ varies from 1 to 2. Thus we might guess that

$$\int_0^1 (x + 1)^{100} \, dx = \int_1^2 t^{100} \, dt$$

$$= \frac{t^{101}}{101} \Big|_1^2$$

$$= \frac{2^{101}}{101} - \frac{1}{101} = \frac{2^{101} - 1}{101}$$

This, in fact, turns out to be correct!

The substitution method gets its name from the substitution of t for $g(x)$.

This procedure, called *integration by substitution*, rests on the following basic formula for antiderivatives.

> (1) $\int f(g(x)) D_x g(x) \, dx = \int f(t) \, dt + C$, where $t = g(x)$

This formula is just a disguised form of the chain rule; its interpretation is illustrated in the examples that follow.

EXAMPLE 8.2-2

Evaluate $\int 2(2x + 1)^9 \, dx$ using formula (1).

SOLUTION

Letting $g(x) = 2x + 1$, $t = g(x)$, and $f(t) = t^9$, we have $D_x g(x) = 2$ and $f(g(x)) = (2x + 1)^9$. Thus

$$\int (2x + 1)^9 (2) \, dx = \int f(g(x)) D_x g(x) \, dx$$

and from formula (1),

$$\int f(g(x)) D_x g(x) \, dx = \int f(t) \, dt + C$$

$$= \int t^9 \, dt + C = \frac{t^{10}}{10} + C$$

Now, replacing t by $2x + 1$, we have

$$\int (2x + 1)^9 (2) \, dx = \frac{(2x + 1)^{10}}{10} + C$$

The substitution formula can be easily remembered and used through the differential notation of Section 6.4. Recall that the differential of the function g is given by $dg = D_x g(x) \, dx$; thus, writing $t = g(x)$ suggests the notation $dt = D_x g(x) \, dx$. Substituting $t = g(x)$ and $dt = D_x g(x) \, dx$ into $\int f(g(x)) D_x g(x) \, dx$ yields $\int f(t) \, dt$. This converts the indefinite integral involving x into a simpler one involving t and avoids the explicit identification of the functions f and g (as was done in Example 8.2-2). We repeat Example 8.2-2 using this technique.

EXAMPLE 8.2-3

Evaluate $\int 2(2x + 1)^9 \, dx$.

SOLUTION

Let $t = 2x + 1$; then $dt = D_x(2x + 1) \, dx = 2 \, dx$. Therefore,

$$\int 2(2x + 1)^9 \, dx = \int (2x + 1)^9 (2) \, dx$$

$$= \int t^9 \, dt = \frac{t^{10}}{10} + C$$

$$= \frac{(2x + 1)^{10}}{10} + C$$

EXAMPLE 8.2-4

Evaluate $\int (2x + 1)^9\, dx$.

SOLUTION

Again let $t = 2x + 1$; then $dt = D_x(2x + 1)\, dx = 2\, dx$. Therefore, $\frac{1}{2}\, dt = dx$ and

$$\int (2x + 1)^9\, dx = \int t^9 \left(\tfrac{1}{2} \right) dt = \tfrac{1}{2} \int t^9\, dt$$

$$= \tfrac{1}{20} t^{10} + C = \tfrac{1}{20} (2x + 1)^{10} + C$$

EXAMPLE 8.2-5

Evaluate $\int (ax + b)^9\, dx$ for $a \neq 0$.

SOLUTION

Let $t = ax + b$; then $dt = D_x(ax + b)\, dx = a\, dx$. Therefore, $(1/a)\, dt = dx$ and

$$\int (ax + b)^9\, dx = \int t^9 \left(\frac{1}{a} \right) dt$$

$$= \frac{1}{a} \int t^9\, dt$$

$$= \frac{1}{a} \left[\frac{t^{10}}{10} + C \right]$$

$$= \frac{t^{10}}{10a} + \frac{C}{a} = \frac{(ax + b)^{10}}{10a} + C_1$$

(where C_1 replaces C/a).

EXAMPLE 8.2-6

Evaluate $\int e^{3x+2}\, dx$.

SOLUTION

Let $t = 3x + 2$; then $dt = 3\, dx$, so $\frac{1}{3} dt = dx$. Thus

$$\int e^{3x+2}\, dx = \int e^{t} \frac{1}{3}\, dt$$

$$= \frac{1}{3} \int e^{t}\, dt$$

$$= \frac{1}{3} \left[e^{t} + C \right]$$

$$= \frac{1}{3} e^{t} + \frac{1}{3} C = \frac{1}{3} e^{3x+2} + C_1$$

We can check this, using the chain rule. $D_x(\frac{1}{3} e^{3x+2} + C_1)$ $= \frac{1}{3} e^{3x+2} D_x(3x + 2) = e^{3x+2}$.

EXAMPLE 8.2-7

Evaluate $\int [e^{x}/(1 + e^{x})]\, dx$.

SOLUTION

Let $t = 1 + e^{x}$; then $dt = D_x(1 + e^{x})\, dx = e^{x}\, dx$. Therefore,

$$\int \left[e^{x}/(1 + e^{x}) \right] dx = \int \left[1/(1 + e^{x}) \right] e^{x}\, dx$$

$$= \int (1/t)\, dt = \ln|t| + C$$

$$= \ln(1 + e^{x}) + C$$

To check, note that $D_x[\ln(1 + e^{x}) + C] = [1/(1 + e^{x})] \cdot D_x(1 + e^{x}) = e^{x}/(1 + e^{x})$.

EXAMPLE 8.2-8

Evaluate $\int (x^2 + 1)x \, dx$.

SOLUTION

Let $t = x^2 + 1$, then $dt = D_x(x^2 + 1) \, dx = 2x \, dx$. Thus, $\frac{1}{2} dt = x \, dx$, and

$$\int (x^2 + 1)x \, dx = \int t \frac{1}{2} \, dt$$

$$= \frac{1}{2} \int t \, dt$$

$$= \frac{1}{2} \left(\frac{t^2}{2} + C \right)$$

$$= \frac{t^2}{4} + \frac{C}{2} = \frac{(x^2 + 1)^2}{4} + C_1$$

To check, note that $D_x[(x^2 + 1)^2/4] = \frac{1}{4}(2)(x^2 + 1)D_x(x^2 + 1) = \frac{1}{4}(2)(x^2 + 1)(2x) = (x^2 + 1)x$.

EXAMPLE 8.2-9

Evaluate $\int e^{x^2} \, dx$.

SOLUTION

Let $t = x^2$; then $dt = 2x \, dx$. If we write $(1/2x) \, dt = dx$, then

$$\int e^{x^2} \, dx = \int e^t (1/2x) \, dt$$

The integral on the right contains both an x and a t, which results in confusion, not simplification. The point to be made here is that substitution will not always work. There is no simple way to evaluate this indefinite integral.

The trick in each of the above examples is to spot the product of a function $g(x)$ and its derivative $D_x g(x)$ in the integral to be evaluated; then we let $t = g(x)$ and write the integral as $\int f(t)\, dt$ for an appropriate function f, making use of formula (1). Finally, once an antiderivative is found for $f(t)$, we replace t by $g(x)$ to obtain an antiderivative in terms of x. When using the method of substitution to evaluate definite integrals this last step is frequently omitted; instead the function $f(t)$ is integrated over an adjusted interval according to the formula

$$(2) \quad \int_a^b f(g(x))\, D_x g(x)\, dx = \int_{g(a)}^{g(b)} f(t)\, dt, \text{ where } t = g(x)$$

The substitution formula for definite integrals

EXAMPLE 8.2-10

Evaluate $\displaystyle\int_2^7 2(2x + 1)^9\, dx$.

SOLUTION (1)

In Example 8.2-2 we found that $\int 2(2x + 1)^9\, dx = (2x + 1)^{10}/10 + C$. Therefore,

$$\int_2^7 2(2x + 1)^9\, dx = \frac{(2x + 1)^{10}}{10}\Big|_2^7$$

$$= \frac{(2(7) + 1)^{10}}{10} - \frac{(2(2) + 1)^{10}}{10}$$

$$= \frac{15^{10}}{10} - \frac{5^{10}}{10}$$

SOLUTION (2)

In Example 8.2-2 we found that $\int 2(2x + 1)^9\, dx = \int t^9\, dt + C$, where $t = 2x + 1 = g(x)$. Now, using formula (2), we find that $x = a = 2$ gives $t = 2x + 1 = 2a + 1 = 5$ and that $x = b = 7$ gives $t = 2x + 1 = 2b + 1 = 15$. Consequently,

$$\int_2^7 2(2x + 1)^9\, dx = \int_5^{15} t^9\, dt = \frac{t^{10}}{10}\Big|_5^{15} = \frac{15^{10}}{10} - \frac{5^{10}}{10}$$

EXAMPLE 8.2-11

Evaluate $\int_1^5 [e^x/(1 + e^x)]\, dx$.

SOLUTION

From Example 8.2-7 we know that $\int [e^x/(1 + e^x)]\, dx$
$= \int (1/t)\, dt + C$, where $t = 1 + e^x = g(x)$. Now $x = a = 1$
gives $t = 1 + e^x = 1 + e^a = 1 + e$, and $x = b = 5$ gives
$t = 1 + e^x = 1 + e^b = 1 + e^5$. Thus from formula (2),

$$\int_1^5 \frac{e^x}{1 + e^x}\, dx = \int_{1+e}^{1+e^5} \frac{1}{t}\, dt$$

$$= \ln|t| \Big|_{1+e}^{1+e^5}$$

$$= \ln(1 + e^5) - \ln(1 + e)$$

EXAMPLE 8.2-12

Evaluate $\int_0^1 (x^2 + 1)x\, dx$.

SOLUTION

From Example 8.2-8 we know that

$$\int (x^2 + 1)x\, dx = \tfrac{1}{2} \int t\, dt + C_1$$

where $t = x^2 + 1 = g(x)$. Thus $x = a = 0$ gives $t = x^2 + 1$
$= 0^2 + 1 = 1$, and $x = b = 1$ gives $t = x^2 + 1 = 1^2 + 1$
$= 2$. By formula (2),

$$\int_0^1 (x^2 + 1)x\, dx = \tfrac{1}{2} \int_1^2 t\, dt$$

$$= \tfrac{1}{2} \left(\frac{t^2}{2} \Big|_1^2 \right) = \tfrac{1}{2}(2 - \tfrac{1}{2}) = \tfrac{3}{4}$$

Exercises 8.2

In Exercises 1 through 26, use the method of substitution to find the indicated antiderivatives.

1. $\int (x + 2)^5 \, dx$

2. $\int (x - 4)^7 \, dx$

3. $\int (x - 4)^{-3} \, dx$

4. $\int (2x - 1)^4 \, dx$

5. $\int (-6x + 2)^5 \, dx$

6. $\int (4 - 5x)^{-3} \, dx$

7. $\int 2x(x^2 + 1)^{11} \, dx$

8. $\int 4x^3(x^4 - 2)^9 \, dx$

9. $\int xe^{x^2} \, dx$

10. $\int 4x^3 e^{x^4 + 2} \, dx$

11. $\int (2x - 2)(x^2 - 2x + 1)^7 \, dx$

12. $\int (-6x + 4)(-3x^2 + 4x - 5)^9 \, dx$

13. $\int \dfrac{2x}{x^2 - 1} \, dx$

14. $\int \dfrac{2e^{2x}}{4 + e^{2x}} \, dx$

15. $\int \dfrac{3x^2}{2\sqrt{x^3 - 1}} \, dx$

16. $\int e^{1 - 3x} \, dx$

17. $\int x(4x^2 - 1) \, dx$

18. $\int x^2(6 - x^3)^5 \, dx$

19. $\int (2x^2 + 1)(6x^3 + 9x - 1)^6 \, dx$

20. $\int (-2x + 1)(-3x^2 + 3x - 1)^8 \, dx$

21. $\int (x^2 + 2)e^{x^3 + 6x} \, dx$

22. $\displaystyle\int \frac{4x^3 - 3x}{2x^4 - 3x^2 + 2} \, dx$

23. $\displaystyle\int \frac{\ln(3x)}{x} \, dx$

24. $\displaystyle\int \frac{e^{\sqrt{x}}}{\sqrt{x}} \, dx$

25. $\displaystyle\int \frac{x^2 + 1}{\sqrt{x^3 + 3x - 1}} \, dx$

26. $\displaystyle\int \frac{-2x^3 + 3x^2 - 1}{\sqrt{-x^4 + 2x^3 - 2x}} \, dx$

In Exercises 27 through 38, evaluate the given definite integral, using the method of substitution and integrating over an appropriate adjusted interval (see Formula (2) and Examples 8.2-10 through 8.2-12).

27. $\displaystyle\int_0^1 3(3x - 1)^4 \, dx$

28. $\displaystyle\int_1^2 2x(x^2 + 1)^3 \, dx$

29. $\displaystyle\int_{-1}^1 3x^2 e^{x^3} \, dx$

30. $\displaystyle\int_0^1 4x^3(x^4 - 4)^2 \, dx$

31. $\displaystyle\int_1^2 (4x - 1)(2x^2 - x)^3 \, dx$

32. $\displaystyle\int_2^3 \frac{3x^2}{x^3 - 2} \, dx$

33. $\displaystyle\int_0^1 \frac{3e^{3x}}{e^{3x} + 2} \, dx$

34. $\displaystyle\int_0^1 e^{4 - 2x} \, dx$

35. $\displaystyle\int_1^2 x^2(4 - x^3)^3 \, dx$

36. $\displaystyle\int_0^1 (2x^3 + x)e^{x^4 + x^2} \, dx$

37. $\displaystyle\int_{\frac{1}{2}}^{\frac{e}{2}} \frac{\ln(2x)}{x} \, dx$

38. $\displaystyle\int_1^4 \frac{e^{2\sqrt{x}}}{\sqrt{x}} \, dx$

39. The Planning Commission of the city of Pleasantdale estimates that t yr from now, the city's population will be increasing at the rate of $3000t^2/(t^3 + 2)$ people per year, for $1 \leqslant t \leqslant 5$. How much will the population of the city increase between $t = 2$ and $t = 5$ yr from now?

40. The Tyler Textile Manufacturing Company estimates that its looms depreciate in value at the rate of $-1000e^{-t+1}$ dollars per year, for $0 \leqslant t \leqslant 4$. How much does a loom depreciate between $t = 2$ and $t = 4$ yr of use? If a loom is worth \$4000 new, find a formula for the value of the loom after t yr of use $(0 \leqslant t \leqslant 4)$.

8.3.   Integration by Parts

In this section we investigate another technique for evaluating definite integrals. As usual, we find a formula for finding antiderivatives and apply this formula in evaluating the definite integral. If we let $F(x) = f(x)g(x)$ in the formula $F(x) = \int D_x F(x) \, dx + C$, we have

$$f(x)g(x) = \int D_x[\, f(x)g(x)\,] \, dx + C$$

On the other hand,

$$D_x[\, f(x)g(x)\,] = f(x)D_x g(x) + g(x)D_x f(x)$$

by rule (5) of Table 5.4-2; hence,

$$\int D_x[\, f(x)g(x)\,] \, dx = \int [\, f(x)D_x g(x) + g(x)D_x f(x)\,] \, dx$$

$$= \int f(x)D_x g(x) \, dx + \int g(x)D_x f(x) \, dx$$

by rule (9) of Table 6.5-1. Consequently,

$$f(x)g(x) = \int D_x[\, f(x)g(x)\,] \, dx + C$$

$$= \int [\, f(x)D_x g(x) + g(x)D_x f(x)\,] \, dx + C$$

$$= \int f(x)D_x g(x) \, dx + \int g(x)D_x f(x) \, dx + C$$

From these equalities (with $C = 0$) we obtain the following formula for the method of *integration by parts*.

The formula for integration by parts

$$\int f(x)D_x g(x)\, dx = f(x)\, g(x) - \int g(x)D_x f(x)\, dx$$

EXAMPLE 8.3-1

The integration by parts formula using the differential notation of Section 6.4 becomes

$$\int f\, dg = fg - \int g\, df$$

Find $\int xe^x\, dx$.

SOLUTION

Let $f(x) = x$ and $D_x g(x) = e^x$. Then

$$g(x) = \int D_x g(x)\, dx = \int e^x\, dx = e^x + C$$

and $D_x f(x) = 1$. Thus, choosing $C = 0$, $g(x) = e^x$ and

$$\int xe^x\, dx = \int f(x)D_x g(x)\, dx$$

$$= f(x)\, g(x) - \int g(x)D_x f(x)\, dx$$

$$= xe^x - \int e^x(1)\, dx = xe^x - \int e^x\, dx$$

$$= xe^x - (e^x + C) = xe^x - e^x + C_1$$

It is instructive to compute $D_x[xe^x - e^x + C_1]$ directly and verify that the result is xe^x.

EXAMPLE 8.3-2

Find $\int x(x + 1)^6\, dx$.

SOLUTION

Let $f(x) = x$ and $D_x g(x) = (x + 1)^6$; then

$$g(x) = \int (x + 1)^6 \, dx$$

$$= \int t^6 \, dt = \frac{t^7}{7} + C$$

$$= \frac{(x + 1)^7}{7} + C$$

where $t = x + 1$. Again, $D_x f(x) = 1$ and (choosing $C = 0$)

$$\int x(x + 1)^6 \, dx = \int f(x) D_x g(x) \, dx$$

$$= f(x) g(x) - \int g(x) D_x f(x) \, dx$$

$$= \frac{x(x + 1)^7}{7} - \int \frac{(x + 1)^7}{7} \, dx$$

$$= \frac{x(x + 1)^7}{7} - \frac{1}{7} \int (x + 1)^7 \, dx$$

As above,

$$\int (x + 1)^7 \, dx = \int t^7 \, dt = \frac{t^8}{8} + C = \frac{(x + 1)^8}{8} + C$$

where $t = x + 1$. Consequently,

$$\int x(x + 1)^6 = \frac{x(x + 1)^7}{7} - \frac{1}{7} \left(\frac{(x + 1)^8}{8} + C \right)$$

$$= \frac{(x + 1)^7}{7} \left(x - \frac{x + 1}{8} \right) + C_1$$

$$= \frac{(x + 1)^7}{7} \left(\frac{7x}{8} - \frac{1}{8} \right) + C_1$$

$$= (x + 1)^7 \left(\frac{x}{8} - \frac{1}{56} \right) + C_1$$

Integration of $f(x) = \ln(x)$

EXAMPLE 8.3-3

Find $\int \ln(x)\, dx$.

SOLUTION

This requires some clever manipulation: letting $f(x) = \ln(x)$ and $g(x) = x$, we have $D_x g(x) = 1$, so

$$\ln(x) = \ln(x)(1) = f(x)D_x g(x)$$

Since $D_x f(x) = 1/x$, we have

$$\int \ln(x)\, dx = \int f(x)D_x g(x)\, dx$$

$$= f(x)\,g(x) - \int g(x)D_x f(x)\, dx$$

$$= \left[\ln(x)\right](x) - \int x\left(\frac{1}{x}\right) dx$$

$$= x\ln(x) - \int 1\, dx$$

$$= x\ln(x) - x + C$$

EXAMPLE 8.3-4

Find $\int x^2 e^x\, dx$.

SOLUTION

Let $f(x) = x^2$ and $D_x g(x) = e^x$. Then $D_x f(x) = 2x$ and $g(x) = e^x + C$. Thus (choosing $C = 0$)

$$\int x^2 e^x\, dx = \int f(x)D_x g(x)\, dx$$

$$= f(x)\,g(x) - \int g(x)D_x f(x)\, dx$$

$$= x^2 e^x - \int 2x e^x\, dx = x^2 e^x - 2\int x e^x\, dx$$

Now, using the method of integration by parts again, we find that $\int xe^x\,dx = xe^x - e^x + C$ as shown in Example 8.3-1. Thus

$$\int x^2 e^x\,dx = x^2 e^x - 2\int xe^x\,dx$$

$$= x^2 e^x - 2(xe^x - e^x + C)$$

$$= x^2 e^x - 2xe^x + 2e^x + C_1$$

EXAMPLE 8.3-5

Evaluate $\int_1^2 xe^x\,dx$.

SOLUTION

From Example 8.3-1 we have

$$\int_1^2 xe^x\,dx = (xe^x - e^x)\Big|_1^2$$

$$= (2e^2 - e^2) - (e^1 - e^1)$$

$$= e^2 - 0 = e^2 \approx 7.3891$$

EXAMPLE 8.3-6

Evaluate $\int_{-1}^0 x(x+1)^6\,dx$.

SOLUTION

From Example 8.3-2 we find that

$$\int_{-1}^0 x(x+1)^6\,dx = (x+1)^7\left[\frac{x}{8} - \frac{1}{56}\right]\Big|_{-1}^0$$

$$= 1^7 \left[\frac{0}{8} - \frac{1}{56} \right] - 0\left(-\frac{1}{8} - \frac{1}{56} \right)$$

$$= -\frac{1}{56}$$

EXAMPLE 8.3-7

Evaluate $\displaystyle\int_1^e \ln(x)\, dx$.

SOLUTION

Using the results of Example 8.3-3, we have

$$\int_1^e \ln(x)\, dx = (x \ln(x) - x)\Big|_1^e$$

$$= (e \ln(e) - e) - (1 \ln(1) - 1)$$

$$= (e - e) - (0 - 1) = 1$$

It should be pointed out here that the process of evaluating definite integrals is essentially one of trial and error. Given an integral to evaluate, we simply try the techniques available to us in an attempt to find one that will work. There are many functions that cannot be integrated using either substitution or integration by parts. Some of these functions may be integrated using more sophisticated techniques or tables of integration formulas found in more advanced calculus books, or it may be necessary to revert to the definition of the definite integral and employ approximation techniques.

Exercises 8.3

In Exercises 1 through 15, use the method of integration by parts to find the indicated antiderivative.

1. $\int xe^{4x}\, dx$

2. $\int xe^{-x}\, dx$

3. $\int x(x-2)^7\, dx$

4. $\int x\ln(x)\, dx$

5. $\int (x+4)e^{2x}\, dx$

6. $\int (3x-1)\ln(x)\, dx$

7. $\int x^2 e^{-x}\, dx$

8. $\int x^2 \ln(x)\, dx$

9. $\int x^2(x-4)^8\, dx$

10. $\int (x^2+x)e^x\, dx$

11. $\int (3x^2-2x)\ln(x)\, dx$

12. $\int (5x^2+3x)e^{-4x}\, dx$

13. $\int x\sqrt{x+1}\, dx$

14. $\int \dfrac{x}{\sqrt{x+1}}\, dx$

15. $\int x(\ln(x))^2\, dx$

In Exercises 16 through 25, evaluate the given definite integral.

16. $\int_0^2 xe^{2x}\, dx$

17. $\int_{-1}^1 x(x+1)^5\, dx$

18. $\int_{-2}^0 xe^{-3x}\, dx$

19. $\int_{-1}^0 (x+1)e^{-x}\, dx$

20. $\int_e^{e^2} (x-6)\ln(x)\, dx$

21. $\int_1^e 4x^2 \ln(x)\, dx$

22. $\int_0^1 x^2 e^{-2x}\, dx$

23. $\int_1^e (x^2 - 1)\ln(x)\, dx$

24. $\int_0^4 x\sqrt{x + 4}\, dx$

25. $\int_0^7 \dfrac{x}{\sqrt{x + 9}}\, dx$

26. The Triton Manufacturing Company determines that its marginal cost $M_c(x)$ at a production level of x units of its product is given by $M_c(x) = 16x\,\ln(x)$. If the cost of producing one unit is \$56, find a formula for the cost $C(x)$ to produce x units.

27. A race driver experiences engine trouble as he is approaching his pit stop. His engine quits running when he is 200 ft from the pits, and t sec after his engine stops his speed v is given by $v(t) = e^{t/3}(9 - t)$ ft per second. Find a formula for the distance $S(t)$ that the car will travel during the first t sec after the engine quits. Will the driver be able to coast into the pits?

*8.4.   Improper Integrals (*Optional*)

This section may present something of a surprise to those who feel quite knowledgable now about the area under the graph of a function. Let us consider the graph of $f(x) = 1/x^2$ (Fig. 8.4-1) and ask whether it is reasonable to try to measure the area A under the graph of f, above the x axis, and to the right of the line $x = 1$. The initial response to this question is very often an emphatic no, reflecting a feeling that there must be an infinite amount of area in the shaded region of Fig. 8.4-1, since that region extends indefinitely to the right. However, if we pick a number b to the right of 1, it poses no problem to compute the area A_1 under the graph of $f(x) = 1/x^2$ from 1 to b (Fig. 8.4-2(a)). We know that

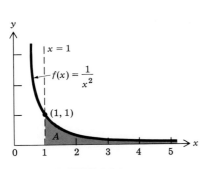

FIGURE 8.4-1

$$A_1 = \int_1^b \frac{1}{x^2}\, dx = \int_1^b x^{-2}\, dx$$

$$= (-x^{-1})\Big|_1^b = -\frac{1}{b} - (-1) = 1 - \frac{1}{b}$$

Thus for $b = 3$ we have

$$\int_1^3 \frac{1}{x^2}\, dx = 1 - \frac{1}{3} = \frac{2}{3}$$

For $b = 10$ we have

$$\int_1^{10} \frac{1}{x^2}\, dx = 1 - \frac{1}{10} = \frac{9}{10}$$

FIGURE 8.4-2

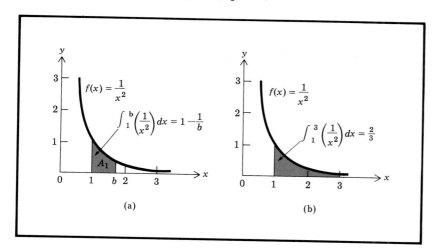

(a) (b)

Evidently, for any $b > 1$,

$$\int_1^b \frac{1}{x^2}\,dx = 1 - \frac{1}{b}$$

This number, $1 - 1/b$, is always less than 1, so no matter how far to the right of 1 we go, the area under the graph of f to this point is not only finite, but less than 1. Thus it seems entirely reasonable to let b get larger and larger and define the area under the graph of f and to the right of the line $x = 1$ to be

For $b > 1$, $1/b$ is positive and less than 1, so if we decrease 1 by this amount, we find that $0 < 1 - 1/b < 1$.

$$\lim_{b \to \infty} \int_1^b \frac{1}{x^2}$$

if the limit exists and is finite. Letting $g(b) = 1 - 1/b$, we differentiate to obtain $D_b g(b) = (1/b^2)$, which is positive; hence g is an increasing function for $b > 1$. The graph of g is shown in Fig. 8.4-3, and from the graph we see that

$$\lim_{b \to \infty} g(b) = \lim_{b \to \infty} \left(1 - \frac{1}{b}\right) = 1$$

Thus,

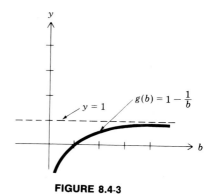

FIGURE 8.4-3

$$\lim_{b \to \infty} \int_1^b \frac{1}{x^2}\,dx = \lim_{b \to \infty} \left(1 - \frac{1}{b}\right) = 1$$

so the area $A = 1$.

Integrals of the form

$$\int_a^\infty f(x)\,dx, \quad \int_{-\infty}^b f(x)\,dx, \quad \text{or} \quad \int_{-\infty}^\infty f(x)\,dx$$

are called *improper integrals*. All three types of integrals occur frequently. Although applications of improper integrals are beyond the range of this text, it is worth remarking that they are invaluable in fields using statistics or probability. The three types of improper integrals mentioned above are defined as follows.

Other types of improper integrals are defined in more advanced texts.

$$\int_a^\infty f(x)\ dx = \lim_{b \to \infty} \int_a^b f(x)\ dx$$

if this limit exists.

$$\int_{-\infty}^b f(x)\ dx = \lim_{a \to -\infty} \int_a^b f(x)\ dx$$

if this limit exists.

$$\int_{-\infty}^\infty f(x)\ dx = \int_{-\infty}^0 f(x)\ dx + \int_0^\infty f(x)\ dx,$$

if each of these limits exists.

When an improper integral exists and is finite, it is called *convergent*; otherwise it is called *divergent*.

EXAMPLE 8.4-1

Evaluate $\int_1^\infty (1/x)\ dx$.

SOLUTION

$$\int_1^b \frac{1}{x}\ dx = \ln|x|\ \Big|_1^b = \ln(b) - \ln(1) = \ln(b)$$

Thus

$$\int_1^\infty \frac{1}{x}\ dx = \lim_{b \to \infty} \int_1^b \frac{1}{x}\ dx = \lim_{b \to \infty} \ln(b) = +\infty$$

This means that $\int_1^\infty (1/x)\ dx$ is divergent, and we say that there is not a finite amount of area under $f(x) = 1/x$ to the right of 1. A comparison of the graphs of $f(x) = 1/x$ and $f(x) = 1/x^2$ makes this result somewhat startling (see Fig. 8.4-4).

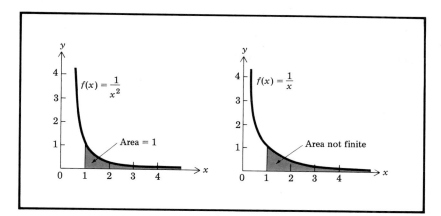

FIGURE 8.4-4

EXAMPLE 8.4-2

Evaluate $\int_{-\infty}^{2} e^x \, dx$.

SOLUTION

$$\int_{a}^{2} e^x \, dx = e^x \Big|_{a}^{2} = e^2 - e^a$$

Thus

$$\int_{-\infty}^{2} e^x \, dx = \lim_{a \to -\infty} \int_{a}^{2} e^x \, dx$$

$$= \lim_{a \to -\infty} (e^2 - e^a)$$

$$= \lim_{a \to -\infty} e^2 - \lim_{a \to -\infty} e^a$$

$$= e^2 - 0 = e^2$$

(see Fig. 8.4-5).

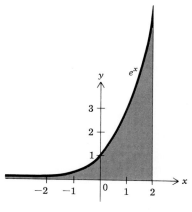

FIGURE 8.4-5

EXAMPLE 8.4-3

Evaluate $\int_{-\infty}^{\infty} xe^{-x^2}\, dx$.

SOLUTION

Letting $t = -x^2$, we have $dt = D_x(-x^2)\, dx = -2x\, dx$. Thus $x\, dx = -\frac{1}{2}\, dt$ and

$$\int xe^{-x^2}\, dx = \int e^t \left(-\tfrac{1}{2} \right) dt$$

$$= -\frac{1}{2} \int e^t\, dt$$

$$= -\frac{1}{2} e^t + C$$

$$= \frac{-e^{-x^2}}{2} + C.$$

Thus

$$\int_{-\infty}^{0} xe^{-x^2}\, dx = \lim_{a \to -\infty} \int_{a}^{0} xe^{-x^2}\, dx$$

$$= \lim_{a \to -\infty} \left. \frac{-e^{-x^2}}{2} \right|_{a}^{0}$$

$$= \lim_{a \to -\infty} \left(-\frac{1}{2} + \frac{e^{-a^2}}{2} \right)$$

$$= -\frac{1}{2}$$

Similarly,

$$\int_{0}^{\infty} xe^{-x^2}\, dx = \lim_{b \to \infty} \int_{0}^{b} xe^{-x^2}\, dx$$

$$= \lim_{b \to \infty} \left. \frac{-e^{-x^2}}{2} \right|_{0}^{b}$$

$$= \lim_{b \to \infty} \left[\frac{-e^{-b^2}}{2} + \frac{1}{2} \right] = \frac{1}{2}$$

Now, since

$$\int_{-\infty}^{\infty} xe^{-x^2}\, dx = \int_{-\infty}^{0} xe^{-x^2}\, dx + \int_{0}^{\infty} xe^{-x^2}\, dx$$

we see that

$$\int_{-\infty}^{\infty} xe^{-x^2}\, dx = -\frac{1}{2} + \frac{1}{2} = 0$$

(see Fig. 8.4-6).

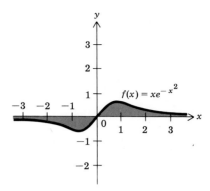

FIGURE 8.4-6 Since the shaded region above the x axis is equal in area to the shaded region below the x axis,

$$\int_{-\infty}^{\infty} xe^{-x^2}\, dx =$$

$$\int_{-\infty}^{0} xe^{-x^2}\, dx + \int_{0}^{\infty} xe^{-x^2}\, dx =$$

$$-\tfrac{1}{2} + \tfrac{1}{2} = 0.$$

Exercises 8.4

In Exercises 1 through 12, evaluate the given improper integral if it converges.

1. $\displaystyle\int_{0}^{\infty} e^{-x}\, dx$

2. $\displaystyle\int_{1}^{\infty} x^{-3}\, dx$

3. $\displaystyle\int_{1}^{\infty} \frac{3}{x^4}\, dx$

4. $\displaystyle\int_{0}^{\infty} e^{x}\, dx$

5. $\displaystyle\int_{1}^{\infty} 7x^{-4}\, dx$

6. $\displaystyle\int_{-\infty}^{0} e^{2x}\, dx$

7. $\displaystyle\int_{-\infty}^{1} xe^{x^2}\, dx$

8. $\displaystyle\int_{1}^{\infty} xe^{x}\, dx$

9. $\displaystyle\int_{-\infty}^{\infty} (x + 1)^{-3}\, dx$

10. $\displaystyle\int_{-\infty}^{\infty} x(x^2 + 4)^{-2}\, dx$

11. $\displaystyle\int_{1}^{\infty} \frac{\ln(x)}{x}\, dx$

12. $\displaystyle\int_{0}^{\infty} \frac{x}{(x^2 + 4)}\, dx$

13. If n is any integer greater than 1, show that $\displaystyle\int x^{-n}\, dx = -1/[(n - 1)x^{n-1}]$. Use this result to show that $\displaystyle\int_{1}^{\infty} x^{-n}\, dx = 1/(n - 1)$.

Self Test   Part IV

1.   Evaluate $\int_0^3 x\sqrt{x^2 + 16}\ dx$

2.   Find $\int x(x + 1)^9\ dx$

3.   Evaluate $\int_{-1}^3 (2x^3 - x + e^{2x})\ dx$

4.   Evaluate $\int_1^3 (2\sqrt{x} - 1)/x\ dx$

5.   Find $\int \dfrac{x}{\sqrt{x^2 + 9}}\ dx$

6.   Find $\int \dfrac{2x^3}{\sqrt{x^2 + 9}}\ dx$. (*Hint*: use the result in question 5.)

7.   Find the mean value m of the function $f(x) = x^2 - 2x$ over the interval $[1, 3]$.

8.   A new packaging process will cut the Good Life Cigarette Company's packaging costs at the rate of $150 - x^2$ thousand dollars per year and increase the delivery costs at the rate of x^2 thousand dollars per year, where x is the number of years after the introduction of the new process. For how many years will it be profitable to use this new process? What are the total savings in packaging costs over this time period?

9.   The price of strawberries varies during the strawberry season. Let $P(t)$ denote the price in dollars per ton t wk after the start of the strawberry season; given that $D_t P(t) = -3t/7 + 15$ and $P(0) = 125$, find $P(t)$.

10. The Rest Well Motel chain anticipates that the amount of money it will spend to replace items stolen from its motels will increase at the rate of $600(1 + x^2)$ dollars per year, where x is the number of years after the first motel in the chain opens. What will be the total loss from theft during the first 3 yr of operation?

Multivariable Calculus

PART V

Functions of Several Variables

9.1. Introduction

Thus far we have been concerned with functions of one variable, which describe the effect of a single given quantity on a second quantity. For example, at one point we discussed the effect of the selling price of paint on the amount of paint that could be sold. In many practical situations, however, life is not so simple. The income tax a person pays depends not only on income, but also on (among other things) the number of dependents and amount of itemized deductions. If a soft drink company markets a lemon-lime drink, a cola, and a root beer, then its revenue depends on the sales of each type of soft drink. In this chapter we provide a brief introduction to the calculus of functions of several variables (quantities that depend on several other quantities) so that we can investigate these more complicated situations and, for example, maximize a profit function P even if several different factors affect P.

9.2. Functions of Several Variables and Their Graphs

It often happens that the value of one quantity depends on the values of several others. For example, the area of a rectangular strip of land depends on the length and width of the strip (Fig. 9.2-1).

$A = xy$

FIGURE 9.2-1

EXAMPLE 9.2-1

The Gusto Soft Drink Company markets a lemon-lime drink, a cola, and a root beer. The company realizes a profit of $1.20 per gallon on the lemon-lime, $2 per gallon on the cola, and $2.80 per gallon on the root beer. Express the total profit P from all three drinks as a function of the number of gallons of each type sold.

SOLUTION

We will use the variables x, y, and z to denote the number of gallons sold of the lemon-lime drink, of the cola, and of the root beer, respectively. Then the profit is

$$P = 1.20x + 2y + 2.80z$$

dollars.

The notation developed for functions of a single variable can be carried over to functions of more than one variable. Thus the profit function above could be written

$$P(x, y, z) = 1.20x + 2y + 2.80z$$

and if 100 gal of lemon-lime, 200 gal of cola, and 70 gal of root beer were sold, the resulting profit would be

$$P(100, 200, 70) = 1.20(100) + 2(200) + 2.80(70)$$

$$= \$716$$

The domain of the function P consists of all triples of real numbers (x, y, z). Of course, if x, y, or z is negative, P cannot be interpreted as the profit of the Gusto Company. The convention about the domain of a function defined by an equation (Section 1.4) also applies to functions of several variables, as illustrated in the following examples.

EXAMPLE 9.2-2

Let $f(x, y) = 2x + y$. Determine the domain of f and compute $f(-1, 0)$.

SOLUTION

Since the right-hand side of $f(x, y) = 2x + y$ is a real number for any pair of numbers (x, y), the domain of f is all pairs of real numbers (x, y). To evaluate $f(-1, 0)$ we substitute -1 for x and 0 for y in $f(x, y) = 2x + y$ to obtain

$$f(-1, 0) = 2(-1) + 0 = -2$$

EXAMPLE 9.2-3

Let $f(x, y, z) = xyz/(x - z)$. Determine the domain of f and compute $f(1, -1, 3)$.

SOLUTION

Since we may never divide by 0, no ordered triple (x, y, z) with $x = z$ can be in the domain of f. However, as long as $x \neq z$, the right-hand side of $f(x, y, z) = xyz/(x - z)$ is a real number, so the domain of f is all triples of real numbers (x, y, z) with $x \neq z$. To compute $f(1, -1, 3)$ we substitute 1 for x, -1 for y, and 3 for z in $f(x, y, z) = xyz/(x - z)$ to obtain

$$f(1, -1, 3) = \frac{(1)(-1)(3)}{1 - 3} = \frac{-3}{-2} = \frac{3}{2}$$

EXAMPLE 9.2-4

Let $f(r, s) = \sqrt{r/s}$. Determine the domain of f and compute $f(-2, -\frac{1}{2})$.

SOLUTION

Since we cannot take the square root of a negative number, we must have $r/s \geq 0$. This means that for $r \neq 0$, r and s must have the same algebraic sign (both positive or both negative). We cannot divide by 0, so we must have $s \neq 0$.

Thus the domain of f is all pairs of real numbers (r, s) with $s \neq 0$ such that either r and s have the same sign or $r = 0$. Setting $r = -2$ and $s = -\frac{1}{2}$, we obtain

$$f\left(-2, -\tfrac{1}{2}\right) = \sqrt{-2/\left(-\tfrac{1}{2}\right)} = \sqrt{2/\tfrac{1}{2}}$$

$$\sqrt{2 \cdot 2} = \sqrt{4} = 2$$

When investigating properties of functions of a single variable, we made extensive use of the graph as the pictorial representation of the function. We shall see that by using a three-dimensional coordinate system, we can represent functions of two variables graphically as surfaces. However, for functions of more than two variables we have no such pictorial devices, so we must rely on other methods to investigate the properties of such functions.

If f is a function of two variables x and y, it is customary to denote the value of f at (x, y) by z, i.e., $z = f(x, y)$.

To construct our three-dimensional coordinate system, we begin with the usual Cartesian plane in a horizontal position (Fig. 9.2-2(a)) and pass a line perpendicular to this plane through the origin as shown. This line is called the z axis, and the upward direction is chosen as the positive one.

Only the positive coordinate axes are shown in Fig. 9.2-2.

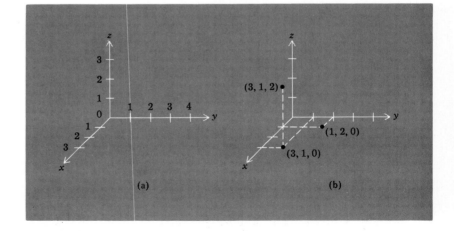

(a)

(b)

FIGURE 9.2-2   Arrows indicate positive direction on the axes

In the triple (x, y, z), x is the first coordinate, y is the second coordinate, and z is the third coordinate.

We can specify the location of any point in three-dimensional space relative to this coordinate system by use of an ordered triple (x, y, z). Ignoring the third coordinate, we locate (x, y) on the horizontal plane in the usual way; then if z is positive, we move upward z units, if z is negative, we move downward $|z|$ units, and if $z = 0$, we remain on the horizontal plane at the point (x, y). Thus the point $(1, 2, 0)$ is on the horizontal plane, whereas (3, 1, 2) is two units above the horizontal plane (Fig. 9.2-2(b)). The point with coordinates (3, 1, − 2) is located two units directly below the point (3, 1) in the horizontal plane. The plane containing the x and y axes is called the xy plane, the plane containing the y and z axes is called the yz plane, and the plane containing the x and z axes is called the xz plane.

Now, if f is a function of two variables, we write $z = f(x, y)$ and think of f as assigning a height z (above or below the xy plane) to each point (x, y) in the xy plane that is also in the domain of f. The *graph* of f is the set of all points (x, y, z) where $z = f(x, y)$. The function f may assign different heights (above or below the xy plane) to various points (x, y) in the domain of f. As shown in Fig. 9.2-3, the graph of f can be visualized as a suitably formed surface or sheet (of some thin material like tin or plastic) over the domain of f.

In previous chapters we referred to the point in the Cartesian plane, on the x axis and 3 units to the right of the origin, either as (3, 0) or simply as 3. Similarly, in a three-dimensional system we refer to a point on the xy plane either as $(x, y, 0)$ or simply as (x, y).

If $f(x, y) = 2x + y$, the point (1, 2, 4) is on the graph of f, and f assigns the height $f(1, 2) = 4$ to the point (1, 2) on the xy plane.

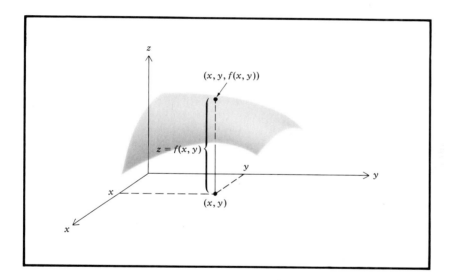

FIGURE 9.2-3

There is no easy method of graphing functions of two variables, but fortunately it is seldom necessary to do so in practical work. The visualization of functions as surfaces may be helpful in understanding the connection between partial derivatives (Section 9.3) and relative maxima and minima (Section 9.4). The notions of limit and continuity can also be interpreted geometrically: A function that is continuous has an uninterrupted graph, i.e., no holes (punctures), gaps (tears), or jumps (sudden changes in altitude). However, we leave the development of these topics to more advanced texts and assume throughout the remainder of this chapter that all functions are continuous and have continuous derivatives at all points of interest. In practice, this does not turn out to be a severe restriction.

Parts of the graphs of $f(x, y) = x^2 + y^2 + 1$ and $f(x, y) = y^2 - x^2$ are shown in Fig. 9.2-4(a) and (b).

The graph in Fig. 9.2-4(b) is known as a saddle surface. The point $(0, 0, 0)$ on this surface is an example of what is termed a "saddle point."

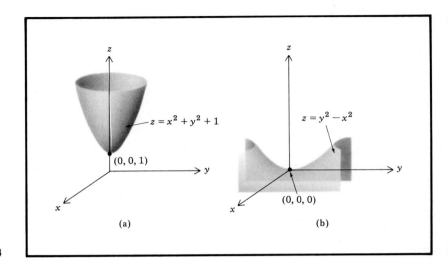

(a)   (b)

FIGURE 9.2-4

Exercises 9.2

In Exercises 1 through 15, determine the domain of the given function and compute the value of the function at the indicated point.

1. $f(x, y) = x + y - 1$, $f(1, -1)$

2. $f(x, y) = \dfrac{y}{2} + 3x$, $f(0, 4)$

3. $f(x, y) = 2e^x - 3y$, $f(0, 1)$

4. $f(x, y) = x(x - y)$, $f(4, -4)$

5. $f(x, y) = \sqrt{x - 1}$, $f(2, 5)$

6. $f(x, y) = \dfrac{x^2 y}{y - 1}$, $f(4, 2)$

7. $f(x, y) = e^{-x} - \ln(y)$, $f(-2, 1)$

8. $f(x, y) = \dfrac{2e^{2x}}{x - y}$, $f(2, 0)$

9. $f(x, y) = \dfrac{\sqrt{y - x}}{y}$, $f(2, 6)$

10. $f(x, y) = \dfrac{4}{xy}$, $f(\tfrac{1}{2}, -4)$

11. $f(x, y) = \dfrac{2\ln(2x)}{y + 4}$, $f(\tfrac{1}{2}, 0)$

12. $f(x, y, z) = 2x + y^3 - \dfrac{1}{(z + 2)}$, $f(1, -1, 2)$

13. $f(x, y, z) = xe^{y-1} - \ln(z - 1)$, $f(4, 2, 2)$

14. $f(x, y, z) = \dfrac{z\sqrt{y - 2}}{\sqrt{x^2 + 2}}$, $f(\sqrt{2}, 6, 2)$

15. $f(x, y, z) = \dfrac{\ln\sqrt{z^2 + 7}}{x - y}$, $f(2, 6, 3)$

In Exercises 16 through 25, construct a three-dimensional coordinate system and sketch the graph of the given function.

16. $f(x, y) = 2$ 17. $f(x, y) = x$

18. $f(x, y) = 2y$ 19. $f(x, y) = -3x$

20. $f(x, y) = x + 2$ 21. $f(x, y) = y - 3$

22. $f(x, y) = e^x + 1$ 23. $f(x, y) = \ln(y) + 2$

24. $f(x, y) = x + y$ 25. $f(x, y) = y - x$

26. The E-Z Open Manufacturing Company makes manual can openers at a production cost of \$2 each and electric can openers at a production cost of \$11 each. Express the total production cost C as a function of the number of units of each type of can opener produced.

27. A Textile Manufacturing Plant employs skilled and unskilled workers. The skilled workers are paid \$11.00 an hour, and the unskilled workers are paid \$5.00 an hour. Express the total hourly wages W paid by the company to its work force as a function of the number of skilled and unskilled workers on the payroll.

9.3.   Partial Derivatives

For functions of more than one variable the concept of the partial derivative plays a role analogous to that of the derivative for functions of one variable. For example, we are often interested in the rate of change of a function with respect to one of its variables while the others remain fixed. In such a case, we want to *differentiate* the function with respect to the one variable, treating the other variables as constants. This process, called *partial differentiation*, results in the *partial derivative* of the function with respect to one particular variable.

EXAMPLE 9.3-1

The temperature of a liquid used in the preparation of a cough syrup, 5 min after being inserted into a preheated oven, is given by the following equation.

$$\text{Temperature of liquid} = T(x, y) = 0.6x + \frac{10{,}000}{y}$$

where x is the temperature of the preheated oven (in degrees Fahrenheit) and y is the amount of liquid (in gallons) inserted in the oven. The equation is valid for $300° \leqslant x \leqslant 550°$ and $100 \leqslant y \leqslant 500$. If the oven is kept at $500°F$, find the rate of change of the temperature of the liquid with respect to the number of gallons of the liquid placed in the oven. What change in the temperature would be caused by placing 201 gal in the oven instead of 200 gal?

SOLUTION

At a fixed temperature of $500°F$, the temperature of y gal of the liquid after 5 min in the oven is

$$T(500, y) = 0.6(500) + \frac{10{,}000}{y} = 300 + 10{,}000y^{-1}$$

degrees. If we now differentiate with respect to y, we get

$$D_y\left[300 + 10{,}000y^{-1}\right] = -10{,}000y^{-2} = -\frac{10{,}000}{y^2}$$

degrees per gallon. The negative value $-10{,}000/y^2$ indicates that the temperature *decreases* by $10{,}000/y^2$ degrees per gallon. Thus when $y = 200$ gal, the temperature changes (decreases) at the rate of $10{,}000/(200)^2 = 10{,}000/40{,}000 = \frac{1}{4}$ of a degree for each additional gallon of liquid placed in the oven. Hence, if we change the amount of liquid inserted by 1 gal from 200 to 201, we expect a change (decrease) of approximately $\frac{1}{4} = 0.25°$ in the temperature of the liquid.

In Example 9.3-1 the actual temperature change asked for is $T(500, 201) - T(500, 200)$, which can be computed directly to be 0.240, correct to three decimal places. The calculus method is often more efficient, however, and offers a good approximation of the desired change.

EXAMPLE 9.3-2

In the situation described in Example 9.3-1, suppose that 100 gal of liquid is placed in an oven preheated to $x°$ ($300° \leqslant x \leqslant 550°$). Find the rate of change of the temperature of the 100 gal of liquid with respect to the temperature of the oven. If the 100 gal of liquid is placed in an oven preheated to $401°$ rather than $400°$, what is the resulting change in the temperature of the liquid?

SOLUTION

The temperature of 100 gal of liquid placed in an oven preheated to $x°$ is

$$T(x, 100) = 0.6x + \frac{10{,}000}{100} = 0.6x + 100$$

Differentiating with respect to x, we obtain the rate of change of the temperature of the 100 gal with respect to the temperature of the oven, i.e., $D_x[0.6x + 100] = 0.6°$ per degree. Thus the temperature of the 100 gal of liquid will rise approximately $0.6°$ for each degree we raise the temperature of the oven. Consequently, we expect a temperature change of $0.6°$ in the liquid if we preheat the oven to $401°$ rather than $400°$.

We can capsulize what took place in the above two examples as follows. We started with a function $T(x, y)$ of two variables x and y. In Example 9.3-1 we replaced the variable x with a *particular* value of x and then differentiated T with respect to y; next, in Example 9.3-2, we replaced the variable y with a *particular* value of y and then differentiated T with respect to x. Now we wish to perform this kind of differentiation without actually specifying a numerical value for the variable that is to be held fixed. We develop a program for doing this in the following examples.

EXAMPLE 9.3-3

Let $f(x, y) = x + 3xy^2$. Substitute $x = 1$ and differentiate f with respect to y.

SOLUTION

Letting $x = 1$ gives $f(1, y) = (1) + 3(1) y^2$. Differentiating with respect to y gives

$$D_y\left[(1) + 3(1) y^2 \right] = 3(1)2y$$

EXAMPLE 9.3-4

Let $f(x, y) = x + 3xy^2$. Substitute $x = 2$ and differentiate with respect to y.

SOLUTION

Letting $x = 2$ gives $f(2, y) = (2) + 3(2) y^2$. Differentiating with respect to y gives

$$D_y\left[(2) + 3(2) y^2 \right] = 3(2)2y$$

EXAMPLE 9.3-5

Let $f(x, y) = x + 3xy^2$. Substitute $x = \pi$ and differentiate with respect to y.

SOLUTION

Letting $x = \pi$ gives $f(\pi, y) = (\pi) + 3(\pi) y^2$. Differentiating with respect to y gives

$$D_y\left[(\pi) + 3(\pi) y^2 \right] = 3\pi 2y$$

Note that the solutions to Examples 9.3-3, 9.3-4, and 9.3-5 can be obtained by substituting $x = 1$, $x = 2$ and $x = \pi$, respectively, into $3x2y$. Moreover, if we *pretend* that x is a constant in $f(x, y) = x + 3xy^2$ and differentiate f with respect to y, we get

$$D_y\left[x + 3xy^2\right] = D_y x + D_y 3xy^2 = 0 + 3xD_y y^2 = 3x2y$$

$D_y x = 0$, since here x is treated as a constant. Similarly, $D_y 3xy^2 = 3xD_y y^2$, since $3x$ is treated as a constant.

This suggests the following procedure. When differentiating a function of two variables x and y with respect to y, treat x as though it were a constant (a fixed numerical value) without actually substituting any value for x, and use the usual rules for differentiation with respect to y.

EXAMPLE 9.3-6

Let $f(x, y) = xy^3 + 1$. Differentiate f with respect to y. Denote the resulting function by f_y and find a rule for $f_y(1, y)$. Compute $f_y(1, 2)$.

SOLUTION

Pretending that x is constant, we differentiate as follows.

$$D_y\left[xy^3 + 1\right] = D_y xy^3 + D_y 1$$

$$= xD_y y^3 + 0 = x \cdot 3y^2 = 3xy^2$$

$D_y xy^3 = xD_y y^3$, since x is treated as a constant.

Denoting this function by f_y, we have $f_y(x, y) = 3xy^2$, so

$$f_y(1, y) = 3(1) y^2 = 3y^2$$

Thus

$$f_y(1, 2) = 3(2)^2 = 12$$

Notice that $f_y(1, y)$ can be obtained by first computing $f(1, y)$ and then differentiating with respect to y. Similarly, $f_y(2, y) = 3(2) y^2 = 6y^2$ can be obtained by first computing $f(2, y)$ and then differentiating with respect to y.

EXAMPLE 9.3-7

Let $f(x, y) = x^2y + x$. Differentiate f with respect to y. Denote the resulting function by f_y and find a rule for $f_y(-2, y)$. Compute $f_y(-2, 7)$.

SOLUTION

Pretending that x is constant, we differentiate as follows.

$$D_y(x^2y + x) = D_y x^2 y + D_y x = x^2 D_y y + 0 = x^2 \cdot 1 = x^2$$

$D_y x = 0$, since x is treated as a constant.

Denoting this function by f_y gives $f_y(x, y) = x^2$, so

$$f_y(-2, y) = (-2)^2 = 4$$

Thus $f_y(-2, 7) = 4$.

Definition of partial derivative

An alternative notation for f_x is $\partial f / \partial x$. An alternative notation for f_y is $\partial f / \partial y$.

The function f_y obtained in Examples 9.3-6 and 9.3-7 is called the *partial derivative of f with respect to y.* In general, if $f(x_1, x_2, \ldots, x_n)$ is a function of n variables, we define the *partial derivative of f with respect to the jth variable x_j* (denoted by f_{x_j}) as the function obtained by differentiating f with respect to x_j, treating all the other variables as constants.

EXAMPLE 9.3-8

Let $f(x, y) = xy^3 + 1$. Differentiate f with respect to x to obtain $f_x(x, y)$. Compute $f_x(1, 2)$.

SOLUTION

$$D_x(xy^3 + 1) = y^3 D_x x + D_x 1$$

$D_x xy^3 = y^3 D_x x$ since y is treated as a constant.

$$= y^3(1) + 0 = y^3$$

Thus $f_x(x, y) = y^3$ and $f_x(1, 2) = 2^3 = 8$.

EXAMPLE 9.3-9

Let $f(x, y) = x^2y + x$. Differentiate f with respect to x to obtain $f_x(x, y)$. Compute $f_x(-2, 7)$.

SOLUTION

$$D_x(x^2y + x) = D_x x^2y + D_x x$$

$$= yD_x x^2 + 1 = 2xy + 1$$

Thus $f_x(x, y) = 2xy + 1$, so $f_x(-2, 7) = 2(-2)(7) + 1 = -27$.

EXAMPLE 9.3-10

Let $f(x, y, z) = 2x^3 + 3x^2z + xyz^2$. Differentiate f with respect to x, y, and z, respectively, to obtain f_x, f_y, and f_z.

SOLUTION

To obtain f_x we differentiate with respect to x. Thus

$$f_x(x, y, z) = D_x(2x^3 + 3x^2z + xyz^2)$$

$$= D_x 2x^3 + D_x 3x^2z + D_x xyz^2$$

$$= 2D_x x^3 + 3zD_x x^2 + yz^2 D_x x$$

$$= 2(3x^2) + 3z(2x) + yz^2(1)$$

$$= 6x^2 + 6xz + yz^2$$

so

$$f_x(x, y, z) = 6x^2 + 6xz + yz^2$$

Differentiating f with respect to y gives

$$D_y(2x^3 + 3x^2z + xyz^2) = D_y 2x^3 + D_y 3x^2z + D_y xyz^2$$

$$= 0 + 0 + xz^2 D_y y$$

$$= xz^2(1) = xz^2$$

Thus

$$f_y(x, y, z) = xz^2$$

Differentiating with respect to z gives

$$D_z\big[2x^3 + 3x^2z + xyz^2\big] = D_z 2x^3 + D_z 3x^2z + D_z xyz^2$$

$$= 0 + 3x^2 D_z z + xy D_z z^2$$

$$= 3x^2(1) + xy(2z)$$

$$= 3x^2 + 2xyz$$

Thus

$$f_z(x, y, z) = 3x^2 + 2xyz$$

EXAMPLE 9.3-11

Let $f(r, s) = e^{rs}/(s + r)$. Differentiate f with respect to r and s to obtain f_r and f_s.

SOLUTION

To obtain f_r we differentiate f with respect to r. Thus

$$f_r(r, s) = D_r\left(\frac{e^{rs}}{s + r}\right)$$

$$= \frac{(s + r)D_r e^{rs} - e^{rs}D_r(s + r)}{(s + r)^2}$$

$$= \frac{(s + r)se^{rs} - e^{rs}(1)}{(s + r)^2}$$

$$= \frac{(s^2 + rs - 1)e^{rs}}{(s + r)^2}$$

so

$$f_r(r, s) = \frac{(s^2 + rs - 1)e^{rs}}{(s + r)^2}$$

The partial derivative of f with respect to s is

$$D_s\left(\frac{e^{rs}}{s+r}\right) = \frac{(s+r)D_s e^{rs} - e^{rs}D_s(s+r)}{(s+r)^2}$$

$$= \frac{(s+r)re^{rs} - e^{rs}(1)}{(s+r)^2}$$

$$= \frac{(r^2 + rs - 1)e^{rs}}{(s+r)^2}$$

so

$$f_s(r,s) = \frac{(r^2 + rs - 1)e^{rs}}{(s+r)^2}$$

The partial derivatives of a function of two variables have a fairly simple geometric interpretation. If $z = f(x, y)$ is a function of two variables and if (x_0, y_0) is a particular point on the xy plane in the domain of f, then the point (x_0, y_0, z_0) is on the graph of f, where $z_0 = f(x_0, y_0)$ (see Fig. 9.3-1). Now, if we intersect the graph of f with

FIGURE 9.3-1

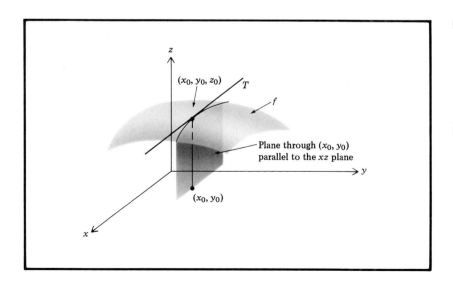

$f_x(x_0, y_0)$ as the slope of a tangent line

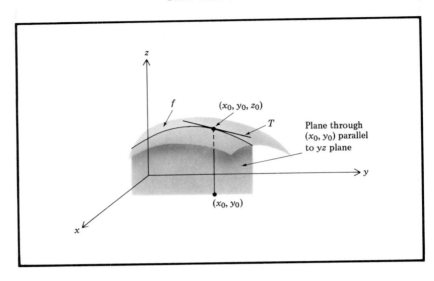

$f_y(x_0, y_0)$ as the slope of a tangent line

FIGURE 9.3-2

the plane through (x_0, y_0) parallel to the xz plane, we obtain a curve in this plane that contains (x_0, y_0, z_0), and the tangent line T to this curve at the point (x_0, y_0, z_0) has slope $f_x(x_0, y_0)$ (see Fig. 9.3-1). Similarly, the intersection of the graph of f with the plane through (x_0, y_0) parallel to the yz plane contains the point (x_0, y_0, z_0) (see Fig. 9.3-2), and the tangent line T to the curve has slope $f_y(x_0, y_0)$.

It should be remarked that in other texts the notations $\partial f / \partial x$ and $\partial f / \partial y$ often replaces f_x and f_y. More generally, if $f(x_1, x_2, \ldots, x_n)$ is a function of n variables, the notation $\partial f / \partial x_j$ has the same meaning as f_{x_j}. Such partial derivatives are themselves functions and can be differentiated. The resulting functions are called *second-order partial derivatives*. The notation for these second-order partial derivatives is as follows. If $z = f(x, y)$, then

When $\partial f / \partial x$ replaces f_x, the notation $f_x(1, 2)$ is often replaced by $\partial f / \partial x|_{(1, 2)}$.

(1) f_{xx} denotes the partial derivative of f_x with respect to x (alternatively, $\partial^2 f / \partial x^2$),

(2) f_{xy} denotes the partial derivative of f_x with respect to y (alternatively, $\partial^2 f / \partial y \partial x$),

(3) f_{yx} denotes the partial derivative of f_y with respect to x (alternatively, $\partial^2 f / \partial x \partial y$),

(4) f_{yy} denotes the partial derivative of f_y with respect to y (alternatively, $\partial^2 f / \partial y^2$).

EXAMPLE 9.3-12

Let $f(x, y) = y^2 e^{2x} + x^2 y$. Compute f_{xx}, f_{xy}, f_{yx}, and f_{yy}.

SOLUTION

$$f_x(x, y) = D_x\left(y^2 e^{2x} + x^2 y\right)$$

$$= y^2 D_x e^{2x} + y D_x x^2 = 2y^2 e^{2x} + 2yx$$

so

$$f_{xx}(x, y) = D_x f_x(x, y)$$

$$= D_x\left(2y^2 e^{2x} + 2yx\right)$$

$$= 2y^2 D_x e^{2x} + 2y D_x x$$

$$= 4y^2 e^{2x} + 2y$$

and

$$f_{xy}(x, y) = D_y f_x(x, y)$$

$$= D_y\left(2y^2 e^{2x} + 2yx\right)$$

$$= 2e^{2x} D_y y^2 + 2x D_y y$$

$$= 4y e^{2x} + 2x$$

Also,

$$f_y(x, y) = D_y\left(y^2 e^{2x} + x^2 y\right)$$

$$= e^{2x} D_y y^2 + x^2 D_y y$$

$$= e^{2x}(2y) + x^2$$

$$= 2y e^{2x} + x^2$$

so

$$f_{yy}(x, y) = D_y f_y(x, y)$$

$$= D_y\left(2y e^{2x} + x^2\right)$$

$$= 2e^{2x} D_y y + D_y x^2 = 2e^{2x}$$

and

$$f_{yx}(x, y) = D_x f_y(x, y)$$

$$= D_x\left(2y e^{2x} + x^2\right)$$

$$= 2y D_x e^{2x} + D_x x^2 = 4y e^{2x} + 2x$$

EXAMPLE 9.3-13

Let $f(x, y) = xye^x + 3x^2 \ln(y)$. Compute f_{xy} and f_{yx}.

SOLUTION

$$f_x(x, y) = D_x\left[xye^x + 3x^2 \ln(y)\right]$$

$$= D_x(xye^x) + D_x\left[3x^2 \ln(y)\right]$$

$$= yD_x(xe^x) + \ln(y)D_x(3x^2)$$

$$= y\left[1(e^x) + x(e^x)\right] + 6x \ln(y)$$

$$= ye^x + yxe^x + 6x \ln(y)$$

Thus

$$f_{xy}(x, y) = D_y\left[ye^x + yxe^x + 6x \ln(y)\right]$$

$$= D_y(ye^x) + D_y(yxe^x) + D_y\left[6x \ln(y)\right]$$

$$= e^xD_yy + xe^xD_yy + 6xD_y \ln(y)$$

$$= e^x + xe^x + 6x\left(\frac{1}{y}\right) = e^x + xe^x + \frac{6x}{y}$$

Now,

$$f_y(x, y) = D_y\left[xye^x + 3x^2 \ln(y)\right]$$

$$= D_y(xye^x) + D_y\left[3x^2 \ln(y)\right]$$

$$= xe^xD_yy + 3x^2D_y \ln(y)$$

$$= xe^x + 3x^2\left(\frac{1}{y}\right)$$

so,

$$f_{yx}(x, y) = D_x\left[xe^x + 3x^2\left(\frac{1}{y}\right)\right]$$

$$= D_x(xe^x) + D_x\left[3x^2\left(\frac{1}{y}\right)\right]$$

$$= D_x(xe^x) + \left(\frac{1}{y}\right)D_x3x^2$$

$$= 1(e^x) + xe^x + \frac{1}{y}(6x)$$

$$= e^x + xe^x + \frac{6x}{y}$$

Notice in Examples 9.3-12 and 9.3-13 that $f_{xy}(x, y) = f_{yx}(x, y)$. This is always the case provided that f_x, f_y, f_{xy}, and f_{yx} are each continuous. The two derivatives f_{xy} and f_{yx} are usually referred to as the *mixed partials* of f.

Equality of mixed partials

We close this section with an application of partial derivatives to economics.

EXAMPLE 9.3-14

The number of boxes of cigars produced in a day by the Heaven Scent Cigar Company depends on the number x of workers employed that day and the number y of packing machines in use on that day. The company has determined that the number N of cigars produced in a day is given by $N(x, y) = x^2 + 10xy + 3y^2$ for $10 \leqslant x \leqslant 100$ and $2 \leqslant y \leqslant 20$. Thus if 10 workers are present and three machines are working on a particular day, then $N(10, 3) = (10)^2 + 10(10)(3) + 3(3)^2 = 427$ boxes of cigars are manufactured on that day. The partial derivative N_x is called the *marginal productivity of labor*, and N_y is called the *marginal productivity of machines*. Find $N_x(10, 3)$ and $N_y(10, 3)$ and interpret these results.

SOLUTION

From $N(x, y) = x^2 + 10xy + 3y^2$ we find

$$N_x(x, y) = D_x(x^2 + 10xy + 3y^2)$$

$$= D_x x^2 + 10y D_x x + D_x 3y^2$$

$$= 2x + 10y$$

and

$$N_y(x, y) = D_y(x^2 + 10xy + 3y^2)$$

$$= D_y x^2 + 10x D_y y + 3 D_y y^2$$

$$= 10x + 3(2y) = 10x + 6y$$

Now $N_x(x, 3)$ is the rate of change of production with respect to the number of workers employed with exactly three machines in use. Thus we interpret $N_x(10, 3) = 2(10) + 10(3) = 50$ as an approximation to the number of additional boxes of cigars that would be produced on a day when three machines were in use if we increased the work force from 10 to 11 workers. Similarly, $N_y(10, y)$ is the rate of change of production with respect to the number of machines in use on a day when exactly 10 workers are present. Thus $N_y(10, 3) = 10(10) + 6(3) = 118$ approximates the number of additional boxes of cigars that would be produced on a day when 10 workers were present if we increased the number of machines in use from three to four.

Exercises 9.3

In Exercises 1 through 12, find the partial derivatives f_x and f_y of the given function f and evaluate f_x and f_y at the given point P.

1. $f(x, y) = 2x - 6y$, $P = (-1, 2)$

2. $f(x, y) = xy + x^2$, $P = (2, 4)$

3. $f(x, y) = x^2 y^2 + 2x$, $P = (-2, 1)$

4. $f(x, y) = 4x^2 y + 2x^3 - 8xy^3$, $P = (1, 1)$

5. $f(x, y) = 2x^4 y - 8x^2 y + \sqrt{x}$, $P = (4, 2)$

6. $f(x, y) = \sqrt{x^2 y + 2}$, $P = (1, 2)$

7. $f(x, y) = \dfrac{xy}{\sqrt{x^2 + y^2}}$, $P = (2, 0)$

8. $f(x, y) = e^{2xy} + xy^3$, $P = (0, 3)$

9. $f(x, y) = \ln(x^2 + y) - 8xy$, $P = (-1, 2)$

10. $f(x, y) = (xe^y + 1)[\ln(xy) - 2]$, $P = (-1, -2)$

11. $f(x, y) = xe^{y^2} - y \ln(x^3)$, $P = (1, 2)$

12. $f(x, y) = \ln(xe^{xy} + y^2)$, $P = (0, 1)$

In Exercises 13 through 20, find the second-order partial derivatives $f_{xx}, f_{yy}, f_{xy},$ and f_{yx} of the given function f and evaluate each at the given point P.

13. $f(x, y) = xy^3 - 4x^2y^2, \quad P = (1, -1)$

14. $f(x, y) = 2xy - 8x^2y + 4y^3, \quad P = (2, 0)$

15. $f(x, y) = 3x^2e^y - 8ye^x, \quad P = (0, 2)$

16. $f(x, y) = 2x \ln(3y) - 8xy^2, \quad P = (-1, 2)$

17. $f(x, y) = 5 \ln(x^2y) - 8e^{2xy}, \quad P = (1, 1)$

18. $f(x, y) = 2x^2ye^x - 8xy^2, \quad P = (0, 2)$

19. $f(x, y) = \dfrac{e^{xy}}{x^2y}, \quad P = (1, 2)$

20. $f(x, y) = \dfrac{2x^2 - xy}{e^{2y} + x}, \quad P = (1, 0)$

In Exercises 21 through 30, find all the first-order partial derivatives of the given function.

21. $f(x, y, z) = 2x^3y - 3y^2x + 4xz$

22. $f(x, y, z) = xe^{yz} - z \ln(xy)$

23. $f(x, y, z) = \sqrt{xz + y^2} - x^2ye^z$

24. $f(t, u, v) = e^{tv} - 8tuv^2 + \ln(uv)$

25. $f(r, s, t) = r^2ste^{st^2}$

26. $g(p, q, r) = \dfrac{p^2qr}{r^2 + q^2}$

27. $w(x, u, v) = (2x^2v + u)(x^3 + 3uv - u^2)$

28. $z(r, s, t) = r\sqrt{t} - \dfrac{st}{r} + \ln(r^t)$

29. $v(p, q, r, s) = (e^{pqs} + rps)[\ln(p + q) - r^2s]$

30. $z(u, v, w, y) = y\sqrt{v^2w + u} + 2vwe^{uy^2} - \ln(v^2wy)$

31. The Frizzy-Free Company manufactures and sells electric hair dryers. The company finds that the number N of dryers it can sell per month depends upon the selling price x of each dryer and the amount y of money spent each month on advertising, as given by the equation $N(x, y) = 1000(25 - x) + (y/10)$ for $10 \le x \le 15$ and $500 \le y \le 1000$. Estimate the effect on the number N of dryers sold per month if the price per dryer is raised from \$13 to \$14 and \$1000 is spent per month on advertising (use an appropriate partial derivative).

32. Eltrec Electronics, Inc. estimates that the production cost C per unit for its best quality stereo receiver is given by $C(x, y) = 10x^2 + 15y^2 - 10xy + 20$, where x is the cost of transistors (in dollars) and y is the cost of labor (in dollars per hour). Compute and interpret $C_x(4, 5)$ and $C_y(5, 4)$.

9.4   Maxima and Minima

The concepts of maximum and minimum values for functions of several variables are virtually identical with these notions for functions of one variable. We will restrict our attention to functions of two variables, however, for the sake of simplicity and because in this case we can visualize our results via the geometric representation of the graphs of such functions as surfaces. In determining the maximum and minimum values of a function of two variables, the partial derivatives will play a role analogous to that of the derivative for functions of one variable. As with functions of one variable, a point on the graph of a function f (of two variables) is called a *relative maximum* if it is higher than any nearby point on the graph (Fig. 9.4-1(a)); a point on the graph of f is called a *relative minimum* if it is lower than any nearby point (Fig. 9.4-1(b)).

Definition of relative maximum

Definition of relative minimum

Geometrically, a relative maximum point corresponds to a mountain cap or peak, whereas a relative minimum point corresponds to the bottom of a cup or a valley. The graphs in Fig. 9.4-2 suggest that there is again a connection between horizontal tangent lines and relative maximum and minimum points of a function of two variables (see also Figs. 9.3-1 and 9.3-2).

FIGURE 9.4-1

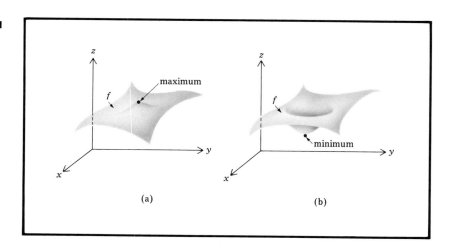

FIGURE 9.4-2

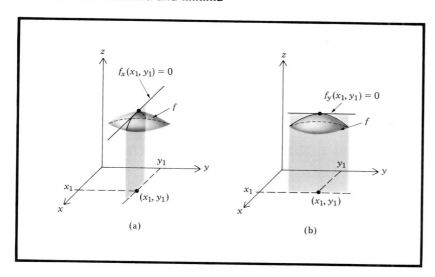

(a)

(b)

Recall from Section 9.3 that partial derivatives give the slopes of tangent lines to the graph of f lying in certain vertical planes.

Let us consider briefly the function f given by $f(x, y) = x^2 + y^2 + 1$ (Fig. 9.2-4(a)). Since the square of any number is greater than or equal to 0, $x^2 + y^2 \geqslant 0$ and so $x^2 + y^2 + 1 \geqslant 1$. Thus $f(x, y) = x^2 + y^2 + 1 \geqslant 1$, and since $f(x, y)$ gives the height above the xy plane of the point $(x, y, f(x, y))$ on the graph of f, we see that every point on the graph of f is at least one unit above the xy plane. Since $f(0, 0) = 0^2 + 0^2 + 1 = 1$, we have that $(0, 0, f(0, 0)) = (0, 0, 1)$ is exactly one unit above the xy plane and consequently is lower than any other point on the graph of f; therefore, it is a minimum point of f. Now

$$f_x(x, y) = D_x(x^2 + y^2 + 1) = 2x$$

and

$$f_y(x, y) = D_y(x^2 + y^2 + 1) = 2y$$

so

$$f_x(0, 0) = 2 \cdot 0 = 0 \quad \text{and} \quad f_y(0, 0) = 2 \cdot 0 = 0$$

Thus at the minimum point $(0, 0, 1)$ we have two horizontal tangent lines, one in the x direction (parallel to the x axis) and one in the y direction (parallel to the y axis); see Fig. 9.2-4(a). In general, if the first-order partial derivatives of a function f are defined in some region of the xy plane, then the relative maximum and minimum values of f in the region occur only at points (a, b), where both $f_x(a, b) = 0$ and $f_y(a, b) = 0$. Such a point (a, b), for which both $f_x(a, b) = 0$ and $f_y(a, b) = 0$, is called a *critical point* of f. It is important to recognize that $f(a, b)$ may be neither a relative maximum value nor a relative minimum value of f even if (a, b) is a critical point. For example, the

If (x, y, z) is a relative maximum point of f, then $z = f(x, y)$ is a relative maximum value of f. If (x, y, z) is a relative minimum point of f, then $z = f(x, y)$ is a relative minimum value of f.

Definition of a critical point of a function

point $(0, 0)$ is a critical point of the function $f(x, y) = y^2 - x^2$ (see Fig. 9.2-4(b)), but $f(0, 0)$ is neither a relative maximum nor a relative minimum value of f; rather, the point $(0, 0, f(0, 0)) = (0, 0, 0)$ is a saddle point.

Second derivative test for functions of two variables

There is a procedure analogous to the second derivative test (Section 6.3) that usually enables us to decide whether a critical point yields a relative maximum value, a relative minimum value, or neither. Given a function f of two variables and a critical point (a, b), we set $A = f_{xx}(a, b)$, $B = f_{xy}(a, b)$, $C = f_{yy}(a, b)$, and $D = AC - B^2$. Then

(1)   if $D > 0$ and $A < 0$, then $(a, b, f(a, b))$ is a relative maximum point of f (so $f(a, b)$ is a relative maximum value of f),

(2)   if $D > 0$ and $A > 0$, then $(a, b, f(a, b))$ is a relative minimum point of f (so $f(a, b)$ is a relative minimum value of f),

(3)   if $D < 0$, then $(a, b, f(a, b))$ is a saddle point of f (neither a relative maximum point nor a relative minimum point),

(4)   if $D = 0$, no conclusion can be reached about the point $(a, b, f(a, b))$.

EXAMPLE 9.4-1

Use the test above to find the relative maximum and minimum points of $f(x, y) = x^2 + y^2 + 1$.

SOLUTION

Although earlier in this section we showed that $(0, 0, 1)$ is a minimum point of this function, it will be instructive to use the test to verify this result. First, we see that

$$f_x(x, y) = D_x(x^2 + y^2 + 1) = 2x$$

and

$$f_y(x, y) = D_y(x^2 + y^2 + 1) = 2y$$

Now, setting these partial derivatives equal to 0 yields $2x = 0$ and $2y = 0$, so $x = y = 0$. Thus $(0, 0)$ is the only critical point of f and $(0, 0, f(0, 0)) = (0, 0, 1)$ is the only candidate for a relative maximum or minimum point of f. Now

$$f_{xx}(x, y) = D_x f_x(x, y) = D_x(2x) = 2$$

hence

$$A = f_{xx}(0, 0) = 2 > 0$$

Also

$$f_{xy}(x, y) = D_y f_x(x, y) = D_y(2x) = 0$$

so

$$B = f_{xy}(0, 0) = 0$$

Moreover,

$$f_{yy}(x, y) = D_y f_y(x, y) = D_y[2y] = 2$$

so

$$C = f_{yy}(0, 0) = 2$$

Finally,

$$D = AC - B^2 = 2 \cdot 2 - 0^2 = 4 > 0$$

and since $A = 2 > 0$, by the second derivative test $(0, 0, 1)$ is a relative minimum point of f. Furthermore, it is the only point that is a relative maximum or minimum point of f.

As in Example 9.4-1, the preliminary steps in applying the second derivative test to find the relative maximum and minimum values of f are

Preliminary steps in using the second derivative test

(1) compute f_x and f_y,
(2) solve $f_x(x, y) = 0$ and $f_y(x, y) = 0$ simultaneously to obtain the critical points of f,
(3) compute f_{xx}, f_{xy}, and f_{yy},
(4) compute A, B, C, and D for *each* critical point and apply the test to determine whether the critical point yields a relative maximum value, a relative minimum value, or neither.

EXAMPLE 9.4-2

Let $f(x, y) = 2x^3 - 2xy + y^2 - 20y$. Find the relative maximum and minimum points of f.

SOLUTION

We proceed as outlined above.

$$f_x(x, y) = D_x(2x^3 - 2xy + y^2 - 20y)$$

$$= 2D_x x^3 - 2yD_x x + D_x y^2 - D_x 20y$$

$$= 6x^2 - 2y$$

$$f_y(x, y) = D_y(2x^3 - 2xy + y^2 - 20y)$$

$$= D_y 2x^3 - 2xD_y y + D_y y^2 - D_y 20y$$

$$= -2x + 2y - 20$$

Thus we need to solve simultaneously

$$\begin{cases} f_x(x, y) = 0 \\ f_y(x, y) = 0 \end{cases} \quad \text{or} \quad \begin{cases} 6x^2 - 2y = 0 \\ -2x + 2y - 20 = 0 \end{cases}$$

From $6x^2 - 2y = 0$ we obtain

$$6x^2 = 2y \quad \text{or} \quad y = 3x^2$$

Substituting this value for y in $-2x + 2y - 20 = 0$ yields

$$-2x + 2(3x^2) - 20 = 0$$

or

$$6x^2 - 2x - 20 = 0$$

Thus we have

$$3x^2 - x - 10 = 0$$

i.e.,

$$(3x + 5)(x - 2) = 0$$

so

$$3x + 5 = 0 \quad \text{or} \quad x - 2 = 0$$

which means $x = -\frac{5}{3}$ or $x = 2$. Thus we have two values for x, $-\frac{5}{3}$ and 2; if $x = -\frac{5}{3}$, then $y = 3x^2 = 3(-\frac{5}{3})^2 = \frac{25}{3}$, so $(-\frac{5}{3}, \frac{25}{3})$ is a critical point of f. For $x = 2$, we see that $y = 3x^2 = 3(2)^2 = 12$, so $(2, 12)$ is also a critical point of f. Now,

$$f_{xx}(x, y) = D_x f_x(x, y) = D_x(6x^2 - 2y) = 12x$$

$$f_{xy}(x, y) = D_y f_x(x, y) = D_y(6x^2 - 2y) = -2$$

and

$$f_{yy}(x, y) = D_y f_y(x, y) = D_y(-2x + 2y - 20) = 2$$

Thus for the critical point $(-\frac{5}{3}, \frac{25}{3})$, we find that

$$A = f_{xx}\left(-\tfrac{5}{3}, \tfrac{25}{3}\right) = 12\left(-\tfrac{5}{3}\right) = -20 < 0$$

$$B = f_{xy}\left(-\tfrac{5}{3}, \tfrac{25}{3}\right) = -2$$

$$C = f_{yy}\left(-\tfrac{5}{3}, \tfrac{25}{3}\right) = 2$$

and

$$D = AC - B^2 = (-20)(2) - (-2)^2 = -36 < 0$$

This means that $(-\frac{5}{3}, \frac{25}{3}, f(-\frac{5}{3}, \frac{25}{3}))$ is a saddle point of f, since $D < 0$. For the critical point $(2, 12)$ we have

$$A = f_{xx}(2, 12) = 12(2) = 24 > 0$$

$$B = f_{xy}(2, 12) = -2$$

$$C = f_{yy}(2, 12) = 2$$

and

$$D = AC - B^2 = (24)(2) - (-2)^2 = 44 > 0$$

Thus $(2, 12, f(2, 12))$ is a relative minimum point of f, since $D > 0$ and $A > 0$. The relative minimum value of f is

$$f(2, 12) = 2(2)^3 - 2(2)(12) + (12)^2 - 20(12) = -128$$

The theory of absolute maximum and absolute minimum values for functions of two variables is quite similar to that for functions of one variable. However, in most cases of *practical* importance, the *absolute* maximum and *absolute* minimum value of a given function of two variables and the relative maximum and relative minimum value of the function coincide. This is the assumption under which we will operate when trying to determine absolute maximum and minimum values in this book.

EXAMPLE 9.4-3

The Stay Fresh Deodorant Company promotes its product through direct television advertisement and by sponsoring sports events such as golf and tennis tournaments. The company statisticians estimate that if the company spends x thousand dollars on television advertisements and y thousand dollars sponsoring sports events for the year, the company can anticipate selling $N(x, y) = 2500 - 0.2x^2 + 50x - 0.03y^2 + 15y$ cans of deodorant per day. Determine the values of x and y that will maximize daily sales.

SOLUTION

Since

$$N(x, y) = 2500 - 0.2x^2 + 50x - 0.03y^2 + 15y$$

we see that

$$N_x(x, y) = -0.4x + 50$$

and

$$N_y(x, y) = -0.06y + 15$$

Thus, setting $N_x(x, y) = 0$ and $N_y(x, y) = 0$, we have

$$-0.4x + 50 = 0 \quad \text{and} \quad -0.06y + 15 = 0$$

so $x = 125$ and $y = 250$; this means that $(125, 250)$ is the only critical point of f. Now

$$N_{xx}(x, y) = -0.4, \quad N_{yy}(x, y) = -0.06, \quad \text{and} \quad N_{xy}(x, y) = 0$$

so

$$A = N_{xx}(125, 250) = -0.4 < 0$$

$$B = N_{xy}(125, 250) = 0$$

$$C = N_{yy}(125, 250) = -0.06 < 0$$

and

$$D = AC - B^2 = (-0.4)(-0.06) - 0 = 0.024 > 0$$

Thus since $D > 0$ and $A < 0$, the point $(125, 250, N(125, 250))$ is the maximum point of N, i.e., $x = 125$ thou-

sand dollars and $y = 250$ thousand dollars should be spent per
year to maximize daily sales. The maximum daily sales are

$$N(125, 250) = 2500 - 0.2(125)^2 + 50(125)$$

$$- 0.03(250)^2 + 15(250) = 7500$$

cans of deodorant.

Exercises 9.4

In Exercises 1 through 15, find the relative maximum, relative minimum,
and saddle points of the given function.

1. $f(x, y) = x^2 + y^2 - 1$

2. $f(x, y) = 2 - x^2 - y^2$

3. $f(x, y) = (x - 1)^2 + (y - 1)^2$

4. $f(x, y) = x^2 + 2x + y^2$

5. $f(x, y) = 10 + x^2 + y^2 + 2x - 4y$

6. $f(x, y) = -x^2 - y^2 + 4x + 4y - 4$

7. $f(x, y) = 4x^2 - 2y^2$

8. $f(x, y) = xy + \dfrac{4}{x}$

9. $f(x, y) = x^2 + y^4 - 4y^2$

10. $f(x, y) = x^3 + 2xy - y^3$

11. $f(x, y) = x^3 + y^3 - 4xy$

12. $f(x, y) = 4x^2y^2 - x - y$

13. $g(s, t) = 2s^2 - 8st + 6t^2$

14. $z(v, w) = 2v^3 - 2w^2 - 2v - 4$

15. $h(u, v) = \dfrac{1}{u} - \dfrac{1}{v} + 2u - v$

16. The monthly profit P of the Cumberland Sugar Refinery is given
by $P(x, y) = 12{,}000 + 20x - 2x^2 + 100y - y^2$, where x is the
cost of labor (in dollars per hour) and y is the cost of sugar cane (in
dollars per ton). Find the values of x and y that yield maximum
profit. What is the maximum profit?

17. A building contractor estimates that the labor cost C for the construction of a building is given by $C(x, y) = x^2 + y^2 - 80x - 40y + 22{,}000$, where x is the number of days of skilled labor and y is the number of days of semiskilled labor required to construct the building. Find the values of x and y that will minimize the labor cost C. What is the minimum labor cost?

18. The True-Touch Company manufactures and sells manual and electric typewriters. The company finds that its monthly profit P (in hundreds of dollars) is given by $P(x, y) = 16{,}220 + 16x + 20y - (x^2/100) - (y^2/100)$, where x is the number of electric typewriters produced per month and y is the number of manual typewriters produced. How many typewriters of each kind should the company produce per month in order to maximize the profit P? What is the maximum monthly profit?

19. An Architectural Engineering Company pays its architects $6 per hour and its designers $5 per hour. The company estimates that a particular project will cost $x^2 + y^2 - 6x - 10y + 534$ dollars, where x is the number of architects and y is the number of designers used on the project. Find x and y such that the cost of the project will be minimized. What is the minimum cost?

20. Find three numbers x, y, z whose sum is 40 and whose product is a maximum.

21. The Friendly Fish Supply Company sells two types of 50-gal aquariums, one rectangular and the other circular. If the rectangular aquarium is priced at $$x$ and the circular aquarium at $$y$, the number of rectangular aquariums sold per month is $600 - 40x + 30y$ and the number of circular aquariums sold per month is $800 + 50x - 60y$. How much should the company charge for each type of aquarium in order to maximize its monthly sales revenue R from the aquariums? What is the maximum monthly sales revenue?

22. The Crown Container Corporation receives an order for 500,000 rectangular cardboard boxes. Each box is to be constructed from 96 square ft of material and will include a top. Find the dimensions of the boxes such that the volume of each box is maximum.

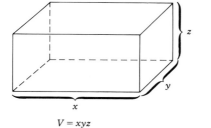

$V = xyz$

9.5　Lagrange Multipliers

Frequently we want to find the maximum and minimum values of a function of several variables where some special restrictions are placed on the variables. For a function of two variables, these restrictions (or constraints) can usually be expressed in the form $g(x, y) = 0$. In such a situation the powerful method of Lagrange multipliers offers a useful alternative to the techniques we have already developed. For simplicity, we describe the Lagrange method for a function of two variables, but the method is applicable to functions of more than two variables as well (see Example 9.5-2).

Lagrange Multipliers. If $f(x, y)$ is a relative maximum or minimum of a function f subject to the restriction $g(x, y) = 0$, then the values of x and y must simultaneously satisfy the equations

$$f_x(x, y) + tg_x(x, y) = 0$$

$$f_y(x, y) + tg_y(x, y) = 0$$

$$g(x, y) = 0$$

provided that all the indicated partial derivatives exist. The variable t is called the *Lagrange multiplier.*

EXAMPLE 9.5-1

A firm has agreed to supply 8000 gallons of a pineapple-orange juice mix to a resort each month. The total cost of producing a monthly batch composed of x thousand gallons of pineapple juice and y thousand gallons of orange juice is $f(x, y) = x^2 + 4y^2 - xy - 5$ (in hundreds of dollars). What mix is least expensive to produce?

SOLUTION

(1) Write the restriction $x + y = 8$ as $x + y - 8 = 0$; thus $g(x, y) = x + y - 8$.
(2) Compute $f_x(x, y) + tg_x(x, y) = 2x - y + t$ and $f_y(x, y) + tg_y(x, y) = 8y - x + t$.
(3) Solve the simultaneous equations

(a) $2x - y + t = 0$
(b) $8y - x + t = 0$
(c) $x + y - 8 = 0$

Subtracting equation (b) from equation (a) yields $3x - 9y = 0$, or $x = 3y$.

Next, substitute $3y$ for x in equation (c) to obtain $3y + y - 8 = 0$, or $4y = 8$, or $y = 2$. Now substitute $y = 2$ into equation (c) to obtain $x + 2 - 8 = 0$, or $x = 6$. Thus the point $(x, y) = (6, 2)$ is the only candidate for yielding a relative minimum subject to the constraint $x + y = 8$, so the mix should contain 6000 gallons of pineapple juice and 2000 gallons of orange juice. To see that $f(6, 2)$ is actually a relative minimum and not a relative maximum, we note that $f(5, 3) = 41 > 35 = f(6, 2)$.

It is usually difficult to determine whether a solution obtained by the Lagrange method is a relative maximum or a relative minimum. We will rely on context and computation of sample values to aid in this determination.

EXAMPLE 9.5-2

Contraband crossing the Mexican border into the U. S. is intercepted at three major check points, X, Y, and Z. The border patrol must make a total of 52 "stop and search" checks per day at these three points, including at least one such check at each point. The daily value of the contraband intercepted is given by the formula $f(x, y, z) = x^2yz + x + y + z$ dollars, where x, y, and z denote the number of checks made at the points X, Y, and Z, respectively. What values for x, y, and z will maximize the value of the intercepted contraband?

SOLUTION

(1) Write the restriction $x + y + z = 52$ as $x + y + z - 52 = 0$; thus $g(x, y, z) = x + y + z - 52$.

(2) Compute

$$f_x(x, y, z) + tg_x(x, y, z) = 2xyz + 1 + t;$$

$$f_y(x, y, z) + tg_y(x, y, z) = x^2z + 1 + t;$$

$$f_z(x, y, z) + tg_z(x, y, z) = x^2y + 1 + t.$$

(3) Solve the simultaneous equations

(a) $2xyz + 1 + t = 0$

(b) $x^2z + 1 + t = 0$

(c) $x^2y + 1 + t = 0$

(d) $x + y + z - 52 = 0$

Subtracting $1 + t$ from each of equations (a), (b), and (c) yields $-1 - t = 2xyz = x^2z = x^2y$.

From $x \geqslant 1$ and $x^2z = x^2y$ we obtain $z = y$. Thus $2xyz = x^2z$ becomes $2xz^2 = x^2z$. Then from $z \geqslant 1$ we conclude $2z = x$. Finally, $x + y + z - 52 = 0$ yields $2z + z + z = 52$, or $4z = 52$, or $z = 13$. So $y = 13$ and $x = 26$. To rule out the possibility that $x = 26$, $y = 13$, $z = 13$ produces a relative minimum for f, we note that $f(25, 14, 13) = 113,802 < 114,296 = f(26, 13, 13)$. Thus 26 checks at X and 13 at each Y and Z allow the patrol to intercept the maximum of \$114,296 worth of contraband per day.

Exercises 9.5

In Exercises 1 through 4, use the method of Lagrange multipliers to find the relative maximum and minimum values of the given function f subject to the given constraint.

1. $f(x, y) = x^2 + y^2$ subject to the constraint $x + y = 1$.

2. $f(x, y) = 2x^2 + 6y^2 + 4xy + 10$ subject to the constraint $x + y = 20$.

3. $f(x, y) = x^2 + 2y^2 - xy$ subject to the constraint $x + y = 8$.

4. $f(x, y, z) = x^2 + y^2 + 2z^2$ subject to the constraint $x + y + z = 30$.

5. Use the method of Lagrange multipliers to maximize the product of two numbers whose sum is 24.

6. Kane's kennel buys two types of dog food, Brand X and Brand Y. Brand X costs \$3 per bag, and Brand Y costs \$2 per bag. If the kennel spends \$44 per week on food and $f(x, y) = 6x^2 + 15x + y^2 + 10y$ gives the number of kilocalories obtained from x bags of Brand X and y bags of Brand Y, find the minimum number of kilocalories that can be obtained from a week's supply of dog food.

7. Find the dimensions of the rectangular box of maximum volume that can be produced from 6 square feet of material.

1. Determine the domain of the function $f(x, y) = 2x/\sqrt{x - y}$ and compute $f(2, 1)$. **Self-Test   Part V**

2. Determine the domain of the function $f(x, y, z) = ze^x/x\sqrt{y}$ and compute $f(3, \pi, 0)$.

3. Find the partial derivatives f_x and f_y of the function $f(x, y) = ye^{x^2} - xy^3$ and compute $f_x(0, 1)$ and $f_y(0, 1)$.

4. Find the second-order partial derivatives f_{xx}, f_{yy}, f_{xy}, and f_{yx} of the function $f(x, y) = e^{2y}/x^2 - \ln(xy)$.

5. Find the relative maximum, relative minimum, and saddle points for the function $f(x, y) = 1/(x^2 + y^2 + 1)$.

6. Find the relative maximum, relative minimum, and saddle points for the function $f(x, y) = 4 - x^2 - 2y^2 + x$.

7. Find the relative maximum, relative minimum, and saddle points for the function $f(x, y) = (1/x) + xy - (8/y)$.

8. A surgical supply company has determined that its monthly profit (in thousands of dollars) is approximated by the formula $P(x, y) = 4x + 2y - x^2 + xy - y^2$, where x is its monthly salary expense and y is the inventory value at the beginning of the month. Here x and y are given in units of \$10,000. Find values of x and y that will maximize the profit P.

Trigonometric Functions

Appendix A

A.1. Degree and Radian Measure

In attempting to measure angles we face the same problem encountered in measuring length and area, namely, the selection of a unit of measure. One commonly used unit, called a *degree*, is defined as $\frac{1}{360}$ of a circle (see Fig. A.1-1). To get some idea of the size of a degree, imagine dividing a pie into 360 equal servings; each serving represents a one-degree angle.

Definition of a degree

The notation 1° is read *one degree*, and $x°$ is read x *degrees*. We further subdivide a degree as follows: *one minute* (1′) is one sixtieth of

Definition of a minute and a second

FIGURE A.1-1

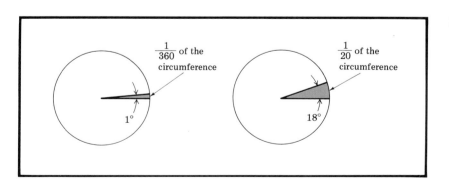

$\frac{1}{360}$ of the circumference

$\frac{1}{20}$ of the circumference

1°

18°

FIGURE A.1-2

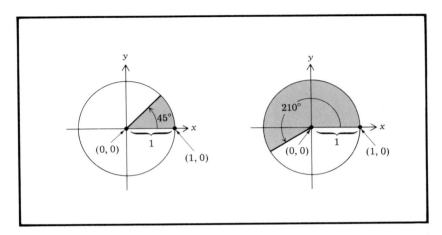

one degree, and *one second* (1″) is one sixtieth of one minute. In discussing angles it is convenient to represent them as drawn in a circle of radius 1, with center at the origin of a Cartesian plane. The angle is said to be in *standard position* if one side of the angle is on the positive *x* axis (see Fig. A.1-2). The side of an angle in standard position on the positive *x* axis is called the *initial side* of the angle, and the other side is called the *terminal side*. Since there are 360° in a circle, a quarter-circle measures $\frac{1}{4}(360°) = 90°$, a half-circle measures $\frac{1}{2}(360°) = 180°$, and three quarters of a circle measures $\frac{3}{4}(360°) = 270°$ (see Fig. A.1-3).

A circle of radius one is called a *unit circle*.

Initial and terminal sides of an angle in standard position

FIGURE A.1-3

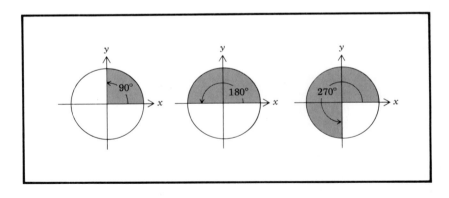

Another unit of measure for angles is called the *radian*, and it turns out that radian measure is preferable to degree measure when we are employing the techniques of calculus. Imagine a first quadrant angle in standard position such that the arc length on the unit circle from the initial side of the angle to the terminal side is one unit (see Fig. A.1-4(a)); such an angle has a measure of 1 *radian*. In fact, if we construct a circle of radius *r* and mark off an arc on the circle equal in length to *r*, then

Definition of a radian

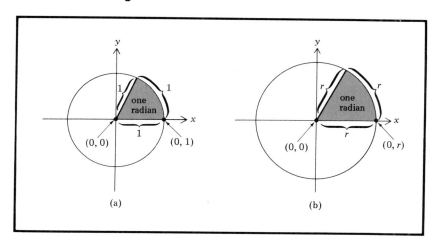

the angle whose sides intersect the endpoints of this arc has measure one radian (see Fig. A.1-4(b)). Recall that a circle of radius r has a circumference of length $2\pi r$ and that π is approximated (to two decimal places) by 3.14. Consequently, the length of the circumference of a circle of radius one is approximately $2(3.14)(1) = 6.28$, and the length of the circumference of a circle of radius r is approximately $2(3.14)r = 6.28r$. Thus, an arc of length r can be marked off on the circumference of a circle of radius r six times, with some of the circumference left over (see Fig. A.1-5). More precisely, an arc of length r can be marked off 2π (approximately 6.28) times, which means there are 2π radians in a circle. Since a circle contains $360°$, it is clear that a radian is much larger than a degree. It is reasonable to ask how many degrees there are in a radian, that is, what is the degree measure of an angle of measure one radian?

There are roughly six radians in a circle.

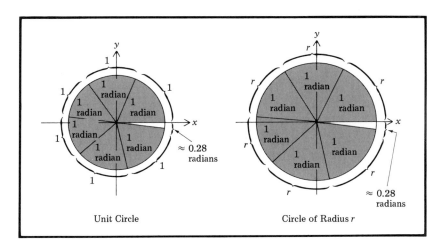

1 radian is approximately 57°17′45″.

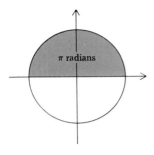

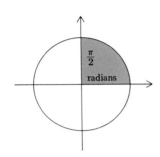

To answer this question, we observe that there are 2π radians and 360 degrees in any circle, so 2π radians = 360 degrees. Thus π radians = 180 degrees, and 1 radian = $180°/\pi \approx 57\frac{1}{3}°$.

The formula π radians = 180 degrees tells us that

$$\frac{\pi}{180} \text{ radians} = 1°$$

and if we now multiply both sides of this equation by x we have a formula for converting x degrees to radians, namely

$$x \text{ degrees} = x(\pi/180) \text{ radians}$$

Thus the radian measures of the "standard" angles of 30°, 45°, 60°, and 90° are

$$30° = 30\left(\frac{\pi}{180}\right) \text{ radians} = \frac{\pi}{6} \text{ radians}$$

$$45° = 45\left(\frac{\pi}{180}\right) \text{ radians} = \frac{\pi}{4} \text{ radians}$$

$$60° = 60\left(\frac{\pi}{180}\right) \text{ radians} = \frac{\pi}{3} \text{ radians}$$

$$90° = 90\left(\frac{\pi}{180}\right) \text{ radians} = \frac{\pi}{2} \text{ radians}$$

Now, multiplying both sides of the equation 1 radian = $180/\pi$ degrees by x produces a formula for converting x radians to degrees, namely, x radians = $x(180/\pi)$ degrees. For instance, $\pi/10$ radians = $(\pi/10)(180/\pi)$ degrees = 18 degrees, and $2\pi/3$ radians = $(2\pi/3)(180/\pi)$ degrees = 120 degrees (see Fig. A.1-6).

FIGURE A.1-6

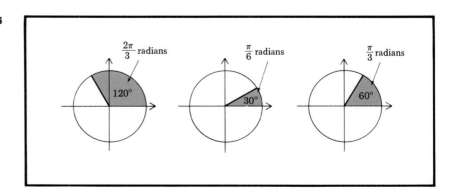

Exercises A.1

In Exercises 1 through 12, draw a unit circle with center at the origin of a Cartesian plane. Then draw an angle of the given size in standard position.

1. $135°$ 2. $315°$
3. $60°$ 4. $225°$
5. $300°$ 6. $120°$
7. $\pi/2$ radians 8. $\pi/4$ radians
9. $3\pi/2$ radians 10. $5\pi/4$ radians
11. $3\pi/4$ radians 12. $5\pi/3$ radians

In Exercises 13 through 20, convert the given degree measures to radian measure.

13. $20°$ 14. $5°$
15. $37°$ 16. $118°$
17. $340°$ 18. $272°$
19. $45\pi°$ 20. $100\pi°$

In Exercises 21 through 28, convert the given radian measures to degree measure.

21. $\pi/3$ radians   22. $2\pi/5$ radians
23. $2\pi/15$ radians 24. $17\pi/30$ radians
25. $9\pi/12$ radians 26. $35/36$ radians
27. 3 radians   28. 1.5 radians

A.2   The Trigonometric Functions

Originally, the trigonometric functions were defined for acute angles (angles between $0°$ and $90°$) by using the sides of right triangles. For example, if θ is an angle in a right triangle (other than the $90°$ angle), then (see Fig. A.2-1)

$$\text{sine } \theta = \frac{\text{length of the side opposite the angle } \theta}{\text{length of the hypotenuse}}$$

$$\text{cosine } \theta = \frac{\text{length of the side adjacent to the angle } \theta}{\text{length of hypotenuse}}$$

Definition of the sine and cosine of an acute angle

This definition is satisfactory for the sine and cosine of angles between $0°$ and $90°$ (or 0 to $\pi/2$ radians). However, we would like to change the domains of these trigonometric functions from angles to numbers and, if

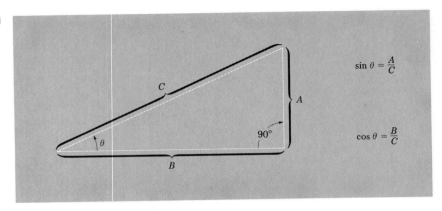

$$\sin \theta = \frac{A}{C}$$

$$\cos \theta = \frac{B}{C}$$

possible, have the domains consist of all real numbers. This would allow us to treat these functions with the methods of the calculus. To accomplish this, we first extend the notion of an angle of x radians to include negative values of x; then we define the sine and cosine of an angle of x radians (for any real number x), and finally, we define the *cosine of the number x* to be the *cosine of the angle of x radians* and the *sine of the number x* to be the *sine of the angle of x radians*.

Sine and cosine of a number

Construction of an angle having radian measure x

We construct an angle (in standard position) having radian measure x, for any real number x, as follows:

(1) If $x \geq 0$, measure an arc of length x on the unit circle in the counterclockwise direction from the point $(1, 0)$. The endpoints of this arc determine an angle of radian measure x (see Fig. A.2-2).

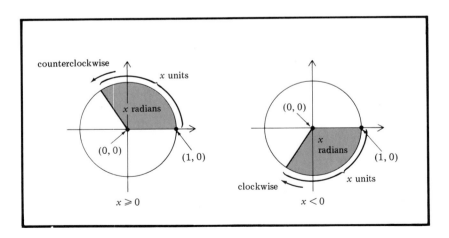

(2) If $x < 0$, measure an arc of length $|x|$ on the unit circle in a clockwise direction from the point $(1, 0)$. The endpoints of this arc determine an angle of radian measure x (see Fig. A.2-2).

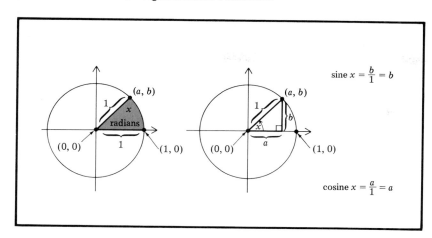

Now we can define the sine and cosine of any real number x. Let (a, b) be the point where the terminal side of the angle of radian measure x (as described above) intersects the unit circle (see Fig. A.2-3). Then we define sine $x = b$ and cosine $x = a$.

Definition of the sine and cosine of any number x

As suggested by Fig. A.2-3, when x denotes an angle of radian measure between 0 and $\pi/2$ these definitions are the same as the definitions made in terms of the sides of right triangles. From this point on we adopt the abbreviations

$$\sin x \quad \text{for} \quad \text{sine } x$$

and

$$\cos x \quad \text{for} \quad \text{cosine } x$$

EXAMPLE A.2-1

If $x = 0$, then $(a, b) = (1, 0)$; hence $\sin 0 = 0$ and $\cos 0 = 1$. If $x = \pi/2$, then $(a, b) = (0, 1)$; hence $\sin(\pi/2) = 1$ and $\cos(\pi/2) = 0$.

The definitions of $\sin x$ and $\cos x$ are quite simple, but these functions share a difficulty found with the functions $\ln x$ and e^x, namely, that for a specified numerical value of x it is not a simple matter to determine a numerical value for $\sin x$ or $\cos x$. For example, given the number $x = 0.5$, how can we determine $\sin 0.5$ and $\cos 0.5$? A crude estimate can be obtained if we construct a unit circle on finely ruled graph paper, measure a length of thread 0.5 units long, lay the thread out in a counterclockwise direction on the circle from the point $(1, 0)$, and

FIGURE A.2-4

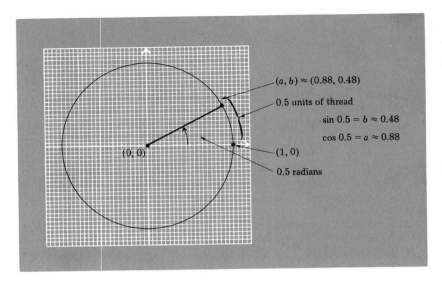

attempt to read off the coordinates a and b of the point (a, b) at the end of the thread (see Fig. A.2-4). However, this procedure is highly impractical, and, in fact, there is no easy way to compute the values of sin 0.5 and cos 0.5 directly. Fortunately, tables such as Tables B and C in the back of the book are readily accessible to provide values of the trigonometric functions. Although it is difficult to compute the values of the sin and cos functions directly, we can learn much about the properties of these functions, as we were able to do for the logarithm and exponential functions.

From Table C, we find that sin 0.5 = 0.4794 and cos 0.5 = 0.8776.

Perhaps the most immediate fundamental fact about the functions sin and cos is that for any real number x, $\sin^2 x + \cos^2 x = 1$. This follows from the fact that the point $(\cos x, \sin x) = (a, b)$ is a point on the unit circle (see Fig. A.2-3), so its distance from $(0, 0)$ is 1. Thus

Sin^2 x means $(\sin x)^2$, and $\cos^2 x$ means $(\cos x)^2$.

$$\sqrt{(a - 0)^2 + (b - 0)^2} = 1, \text{ which means that } \sqrt{a^2 + b^2} = 1 \text{ or}$$
$a^2 + b^2 = 1$, so $\sin^2 x + \cos^2 x = 1$. We can use this fact, together with

FIGURE A.2.5

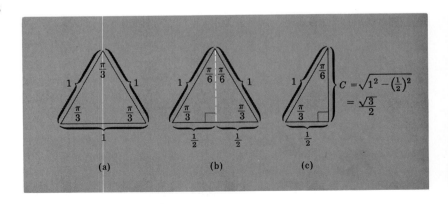

the definition of sin and cos in terms of the lengths of sides of triangles, to determine $\sin x$ and $\cos x$ for $x = \pi/6$, $\pi/3$, and $\pi/4$. Consider the isosceles triangle in Fig. A.2-5(a). Passing a line through the upper vertex of this triangle so as to bisect the upper angle produces a pair of right triangles (Fig. A.2-5(b)). The lengths of two sides of each of these right triangles are known immediately, and the length of their common side, c, can be computed by the Pythagorean theorem as

$$\sqrt{1^2 - \left(\tfrac{1}{2}\right)^2} = \sqrt{1 - \tfrac{1}{4}} = \sqrt{\tfrac{3}{4}} = \frac{\sqrt{3}}{2}$$

Thus from Fig. A.2-5(c) we have

$$\sin \frac{\pi}{3} = \frac{\text{opposite}}{\text{hypotenuse}} = \frac{\sqrt{3}/2}{1} = \frac{\sqrt{3}}{2}$$

$$\cos \frac{\pi}{3} = \frac{\text{adjacent}}{\text{hypotenuse}} = \frac{1/2}{1} = \frac{1}{2}$$

$$\sin \frac{\pi}{6} = \frac{\text{opposite}}{\text{hypotenuse}} = \frac{1/2}{1} = \frac{1}{2}$$

and

$$\cos \frac{\pi}{6} = \frac{\text{adjacent}}{\text{hypotenuse}} = \frac{\sqrt{3}/2}{1} = \frac{\sqrt{3}}{2}$$

To determine $\sin(\pi/4)$ and $\cos(\pi/4)$, note from Fig. A.2-6 that $\cos(\pi/4) = \sin(\pi/4)$. Thus since $\sin^2(\pi/4) + \cos^2(\pi/4) = 1$, we have

$$\cos^2 \frac{\pi}{4} + \cos^2 \frac{\pi}{4} = 1$$

or

$$2 \cos^2 \frac{\pi}{4} = 1$$

Thus

$$\cos^2 \frac{\pi}{4} = \frac{1}{2}$$

and

$$\cos \frac{\pi}{4} = \frac{1}{\sqrt{2}}$$

(since $\cos(\pi/4) > 0$). Finally, $\sin(\pi/4) = \cos(\pi/4) = 1/\sqrt{2}$. We summarize these results in Table A.2-1.

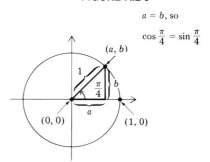

FIGURE A.2-6

$a = b$, so

$\cos \dfrac{\pi}{4} = \sin \dfrac{\pi}{4}$

TABLE A.2-1

x	$\cos x$	$\sin x$
0	1	0
$\pi/6$	$\sqrt{3}/2$	$1/2$
$\pi/4$	$1/\sqrt{2}$	$1/\sqrt{2}$
$\pi/3$	$1/2$	$\sqrt{3}/2$
$\pi/2$	0	1

In addition to the sin and cos functions, we define four more trigono-metric functions as follows.

$$\text{tangent } x = \frac{\sin x}{\cos x} \quad \text{(abbreviation: tan } x)$$

$$\text{secant } x = \frac{1}{\cos x} \quad \text{(abbreviation: sec } x)$$

$$\text{cotangent } x = \frac{\cos x}{\sin x} \quad \text{(abbreviation: ctn } x)$$

$$\text{cosecant } x = \frac{1}{\sin x} \quad \text{(abbreviation: csc } x)$$

Using Table A.2-1, we can easily determine the values of these additional trigonometric functions for $x = 0, \pi/6, \pi/4, \pi/3$, and $\pi/2$. These values are given in Table A.2-2.

TABLE A.2-2

x	$\tan x$	$\sec x$	$\text{ctn } x$	$\csc x$
0	0	1	undefined	undefined
$\pi/6$	$1/\sqrt{3}$	$2/\sqrt{3}$	$\sqrt{3}$	2
$\pi/4$	1	$\sqrt{2}$	1	$\sqrt{2}$
$\pi/3$	$\sqrt{3}$	2	$1/\sqrt{3}$	$2/\sqrt{3}$
$\pi/2$	undefined	undefined	0	1

The quadrant in which the terminal side of an angle of measure x radians lies determines the sign of each of the functions $\sin x$, $\cos x$ and $\tan x$. This can be seen from Fig. A.2-3 and the fact that $\tan x = \sin x/\cos x$. Table A.2-3 summarizes the results.

TABLE A.2-3

Quadrant	sin x	cos x	tan x
I	+	+	+
II	+	−	−
III	−	−	+
IV	−	+	−

EXAMPLE A.2-2

Find cos x and tan x if sin $x = \frac{1}{2}$ and if the terminal side of an angle (in standard position) of x radians lies in quadrant II.

SOLUTION

Since $\sin^2 x + \cos^2 x = 1$ and since sin $x = \frac{1}{2}$, we have $(\frac{1}{2})^2 + \cos^2 x = 1$; this means that $\cos^2 x = 1 - \frac{1}{4} = \frac{3}{4}$, so cos $x = \pm \sqrt{3}/2$. Now, from Table A.2-3 we see that cos x is negative, since the terminal side of an angle of x radians (in standard position) lies in quadrant II. Thus cos $x = -\sqrt{3}/2$. Finally,

$$\tan x = \frac{\sin x}{\cos x} = \frac{1/2}{-\sqrt{3}/2} = \frac{-1}{\sqrt{3}}$$

Exercises A.2

In Exercises 1 through 25, sketch an angle of the given size in standard position on a unit circle in a Cartesian plane. Then find the values of the six trigonometric functions at the given angle (without using tables).

1. $5\pi/4$ radians
2. $2\pi/3$ radians
3. $4\pi/3$ radians
4. $7\pi/6$ radians
5. $5\pi/6$ radians
6. $11\pi/6$ radians
7. $7\pi/4$ radians
8. $5\pi/4$ radians
9. $-\pi/2$ radians
10. $-\pi/6$ radians
11. $-5\pi/6$ radians
12. $-4\pi/3$ radians

13.	$-7\pi/6$ radians	14.	$-5\pi/4$ radians
15.	$-11\pi/6$ radians	16.	3π radians
17.	$7\pi/2$ radians	18.	$-13\pi/6$ radians
19.	$10\pi/3$ radians	20.	$-7\pi/3$ radians
21.	$450°$	22.	$-720°$
23.	$-405°$	24.	$390°$
25.	$-765°$		

In Exercises 26 through 40, use Table C in the back of the book to evaluate the six trigonometric functions at the given angle. (Convert the angle to radian measure if necessary.)

26.	1 radian	27.	0.67 radians
28.	$13°$	29.	$62°$
30.	$72°14'$	31.	$36°45'$
32.	-1.12 radians	33.	3.27 radians
34.	-4.56 radians	35.	7.18 radians
36.	$385°37'$	37.	$-145°12'$
38.	-3.85 radians	39.	8.62 radians
40.	$-642°27'$		

41. Show that $1 + \tan^2 x = \sec^2 x$

42. Show that $\cos x \cdot \tan x \cdot \csc x = 1$

43. Show that $1 + \operatorname{ctn}^2 x = \csc^2 x$

44. Show that $\dfrac{\sin x}{1 + \cos x} + \dfrac{1 + \cos x}{\sin x} = 2 \csc x$

45. Find $\cos x$ and $\tan x$ if $\sin x = \frac{1}{3}$ and the terminal side of an angle of x radians (in standard position) lies in quadrant III.

46. Find $\sin x$ and $\csc x$ if $\cos x = \frac{1}{4}$ and the terminal side of an angle of x radians (in standard position) lies in quadrant IV.

A.3　The Graphs of the Sine, Cosine, and Tangent Functions

In order to graph the function $\sin x$ we will use the following table of values together with the definition of the sine function. The first three values in Table A.3-1 have been reproduced from Table A.2-1, and the remaining three values are easily obtained from Fig. A.3-1(a). The chart in Fig. A.3-1(b) indicates that as x varies from 0 to $\pi/2$, $\sin x$ (the y coordinate of the endpoint of the indicated arc of length x) varies from 0 to 1; as x varies from $\pi/2$ to π, $\sin x$ varies from 1 to 0; and so on. It is clear that when x exceeds 2π the values previously obtained for $\sin x$ are repeated, so that for any number x, $\sin(x + 2\pi) = \sin x$. Figure A.3-2 suggests another important fact about the sine function, namely, that $\sin(-x) = -\sin x$ (e.g., $\sin(-\pi/6) = -\sin \pi/6 = \frac{-1}{2}$).

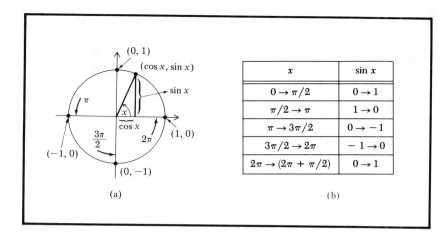

(a)   (b)

TABLE A.3-1

x	0	$\pi/6$	$\pi/2$	π	$3\pi/2$	2π
$\sin x$	0	$\frac{1}{2}$	1	0	-1	0

This information leads us to sketch the graph of sin x as shown in Fig. A.3-3.

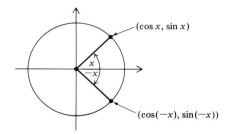

FIGURE A.3-2. $\sin(-x) = -\sin x.$

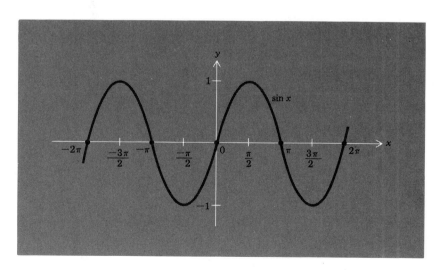

The graph of sin x

FIGURE A.3-3

Now, referring again to Fig. A.3-1(a), we see that as x varies from 0 t $\pi/2$, cos x (the first coordinate of the endpoint of the arc of length varies from 1 to 0; as x varies from $\pi/2$ to π, cos x varies from 0 to $-$;

FIGURE A.3-4

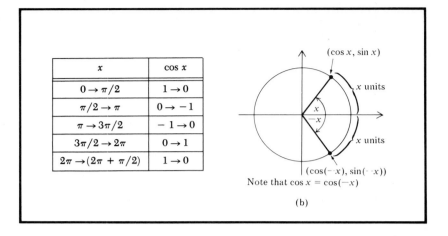

x	$\cos x$
$0 \rightarrow \pi/2$	$1 \rightarrow 0$
$\pi/2 \rightarrow \pi$	$0 \rightarrow -1$
$\pi \rightarrow 3\pi/2$	$-1 \rightarrow 0$
$3\pi/2 \rightarrow 2\pi$	$0 \rightarrow 1$
$2\pi \rightarrow (2\pi + \pi/2)$	$1 \rightarrow 0$

Note that $\cos x = \cos(-x)$

(b)

and so on. Furthermore, as with the sine function, it is clear that $\cos(x + 2\pi) = \cos x$. This information is summarized in the chart in Fig. A.3-4(a). Also, from Fig. A.3-4(b), we see that $\cos(-x) = \cos x$ (e.g., $\cos(-\pi/6) = \cos(\pi/6) = \sqrt{3}/2$). Using this information, we sketch the graph of $\cos x$ as shown in Fig. A.3-5. Since the sine and cosine functions

FIGURE A.3-5

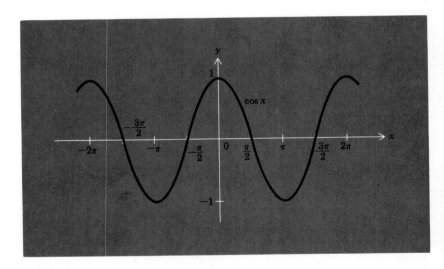

The graph of $\cos x$

repeat the same set of values every 2π radians, we say that the sine and cosine functions have *period* 2π. This is expressed by the two formulas $\sin(x + 2\pi) = \sin x$ and $\cos(x + 2\pi) = \cos x$. The tangent function, however, repeats the same set of values every π radians (and is said to have period π), since

$$\tan(x + \pi) = \frac{\sin(x + \pi)}{\cos(x + \pi)} = \frac{-\sin x}{-\cos x} = \frac{\sin x}{\cos x} = \tan x$$

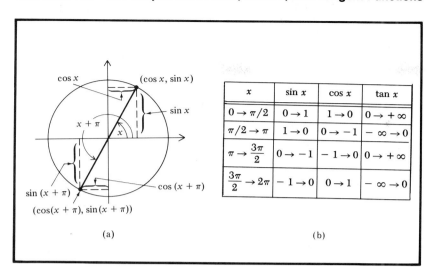

x	$\sin x$	$\cos x$	$\tan x$
$0 \to \pi/2$	$0 \to 1$	$1 \to 0$	$0 \to +\infty$
$\pi/2 \to \pi$	$1 \to 0$	$0 \to -1$	$-\infty \to 0$
$\pi \to \dfrac{3\pi}{2}$	$0 \to -1$	$-1 \to 0$	$0 \to +\infty$
$\dfrac{3\pi}{2} \to 2\pi$	$-1 \to 0$	$0 \to 1$	$-\infty \to 0$

(a)   (b)

(The validity of the formulas $\sin(x + \pi) = -\sin x$ and $\cos(x + \pi) = -\cos x$ is suggested by Fig. A.3-6(a).) Some sample values for $\tan x$ are easily obtained from values of $\sin x$ and $\cos x$ (since $\tan x = \sin x / \cos x$), and these values together with the chart in Fig. A.3-6(b) lead us to sketch the graph of $\tan x$ as shown in Fig. A.3-7.

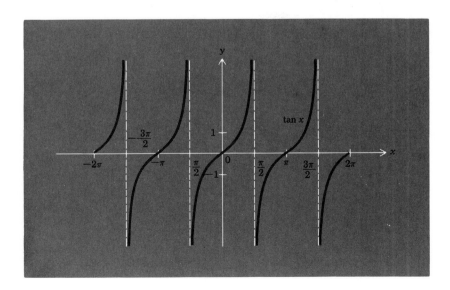

The graph of $\tan x$

Exercises A.3

In Exercises 1 through 12, sketch the graphs of the given functions.

1. $\sin 2x$ 2. $\cos 4x$
3. $\sin(x/2)$ 4. $\cos(x/4)$
5. $\tan 2x$ 6. $4 \sin 2x$
7. $-3 \cos 3x$ 8. $\operatorname{ctn} x$
9. $\csc x$ 10. $\sec x$
11. $\sin(x + \pi/2)$ 12. $\cos(x - \pi/2)$

A.4   Derivatives and Integrals of the Trigonometric Functions

The graphs of $\sin x$ and $\cos x$ lead one to suspect that it is possible to draw a tangent line at each point of each graph, and this is indeed the case. However, to prove that $\sin x$ and $\cos x$ are differentiable at each point x (that the graphs have a tangent line at each point) is somewhat difficult, so we present the following formulas without proof.

The derivatives of the sine and cosine functions

$(1)\quad D_x \sin x = \cos x$

$(2)\quad D_x \cos x = -\sin x$

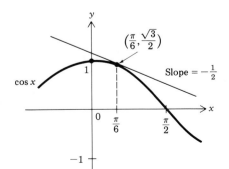

FIGURE A.4-1

EXAMPLE A.4-1

Find the slope of the tangent line to the graph of $\cos x$ at the point $(\pi/6, \sqrt{3}/2)$. Sketch the tangent line at this point.

SOLUTION

$D_x \cos x = -\sin x$, so $D_x \cos(\pi/6) = -\sin(\pi/6) = -\frac{1}{2}$. The tangent line is sketched in Fig. A.4-1.

EXAMPLE A.4-2

Find $D_x \cos(x^2)$.

SOLUTION

The notation $\cos(x^2)$ denotes the composition $f \circ g$, where $f(x) = \cos x$ and $g(x) = x^2$. Thus, using the chain rule, we have $D_x \cos(x^2) = -\sin(x^2) \cdot D_x(x^2) = -2x \sin(x^2)$.

EXAMPLE A.4-3

Find $D_x \cos^2 x$.

SOLUTION

Using the chain rule, we have $D_x \cos^2 x = D_x[\cos x]^2 = 2 \cos x \, D_x \cos x = -2 \cos x \sin x$.

EXAMPLE A.4-4

Find $D_x \tan x$.

The derivatives of $\tan x$, $\operatorname{ctn} x$, $\sec x$, and $\csc x$

SOLUTION

By the quotient rule,

$$D_x \tan x = D_x \left(\frac{\sin x}{\cos x} \right)$$

$$= \frac{\cos x \cdot D_x \sin x - \sin x \cdot D_x \cos x}{(\cos x)^2}$$

$$= \frac{\cos x (\cos x) - \sin x (-\sin x)}{\cos^2 x}$$

$$= \frac{\cos^2 x + \sin^2 x}{\cos^2 x}$$

$$= \frac{1}{\cos^2 x} = \left[\frac{1}{\cos x} \right]^2 = \sec^2 x$$

In much the same way as in Example A.4-4, we can verify the following formulas:

$$D_x \text{ ctn } x = -\csc^2 x$$

$$D_x \sec x = \sec x \tan x$$

$$D_x \csc x = -\csc x \text{ ctn } x$$

Table A.4-1 gives the derivatives of the composite of each of the six trignometric functions with a differentiable function $f(x)$.

TABLE A.4-1

(1)	$D_x \sin(f(x)) = \cos(f(x)) \cdot D_x f(x)$
(2)	$D_x \cos(f(x)) = -\sin(f(x)) \cdot D_x f(x)$
(3)	$D_x \tan(f(x)) = \sec^2(f(x)) \cdot D_x f(x)$
(4)	$D_x \text{ ctn}(f(x)) = -\csc^2(f(x)) \cdot D_x f(x)$
(5)	$D_x \sec(f(x)) = \sec(f(x)) \cdot \tan(f(x)) \cdot D_x f(x)$
(6)	$D_x \csc(f(x)) = -\csc(f(x)) \cdot \text{ctn}(f(x)) \cdot D_x f(x)$

EXAMPLE A.4-5

Find $D_x \sin(\cos x)$

SOLUTION

$$D_x \sin(\cos x) = \cos(\cos x) \cdot D_x \cos x = -\sin x \cdot \cos(\cos x).$$

EXAMPLE A.4-6

Sketch the graph of $f(x) = \sin 2x$ over the interval $[0, 2\pi]$, using the derivative to aid accuracy (see Exercise A.3-1).

SOLUTION

First we compute $D_x f$ as $D_x \sin 2x = \cos(2x) \cdot D_x(2x)$ $= 2 \cos 2x$. Next we set $D_x f(x) = 0$ to obtain $2 \cos 2x = 0$ or $\cos 2x = 0$. Now $\cos 2x = 0$ providing $2x = \pm \pi/2, \pm 3\pi/2,$ $\pm 5\pi/2, \pm 7\pi/2, \pm 9\pi/2$, etc., so $x = \pm \pi/4, \pm 3\pi/4,$ $\pm 5\pi/4, \pm 7\pi/4, \pm 9\pi/4$, etc. Of these critical values for f only $\pi/4, 3\pi/4, 5\pi/4,$ and $7\pi/4$ are in the interval $[0, 2\pi]$. Now if $0 < x < \pi/4$, we have $0 < 2x < \pi/2$; hence $\cos 2x$ is positive, so $D_x f(x) = 2 \cos 2x$ is positive. If $\pi/4 < x < 3\pi/4$, then $\pi/2 < 2x < 3\pi/2$, so $\cos 2x$ is negative and $D_x f(x) = 2 \cos 2x$ is negative. Similarly, $3\pi/4 < x < 5\pi/4$ assures that $3\pi/2 < 2x < 5\pi/2$ and that $D_x f(x) = 2 \cos 2x$ is positive, whereas $5\pi/4 < x < 7\pi/4$ yields that $5\pi/2 < 2x < 7\pi/2$ and that $D_x f(x) = 2 \cos 2x$ is negative. Thus $f(x) = \sin 2x$ has relative maximum values at $x = \pi/4$ and $x = 5\pi/4$ and relative minimum values at $x = 3\pi/4$ and $x = 7\pi/4$. The graph of $f(x) = \sin 2x$ is shown in Fig. A.4-2, and the above information is displayed in Table A.4-2.

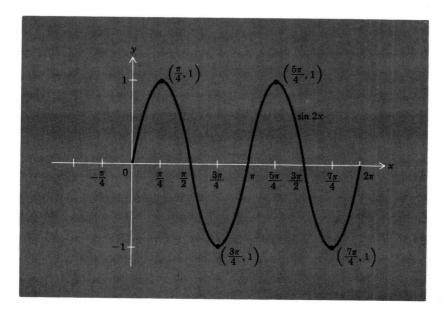

FIGURE A.4-2

Note: $\sin(2 \cdot 5\pi/4) = \sin(5\pi/2)$ $= \sin(\pi/2 + 2\pi) = \sin(\pi/2) = 1.$

TABLE A.4-2

Critical Point	Interval	$D_x f(x)$	f	Results
	$\left(0, \dfrac{\pi}{4}\right)$	positive	increasing	
$x = \dfrac{\pi}{4}$		0	$f\left(\dfrac{\pi}{4}\right) = 1$	Relative maximum value at $\pi/4$
	$\left(\dfrac{\pi}{4}, \dfrac{3\pi}{4}\right)$	negative	decreasing	
$x = \dfrac{3\pi}{4}$		0	$f\left(\dfrac{3\pi}{4}\right) = -1$	Relative minimum value at $3\pi/4$
	$\left(\dfrac{3\pi}{4}, \dfrac{5\pi}{4}\right)$	positive	increasing	
$x = \dfrac{5\pi}{4}$		0	$f\left(\dfrac{5\pi}{4}\right) = 1$	Relative maximum value at $5\pi/4$
	$\left(\dfrac{5\pi}{4}, \dfrac{7\pi}{4}\right)$	negative	decreasing	
$x = \dfrac{7\pi}{4}$		0	$f\left(\dfrac{7\pi}{4}\right) = -1$	Relative minimum value at $7\pi/4$
	$\left(\dfrac{7\pi}{4}, 2\pi\right)$	positive	increasing	

Another formulation of $D_x \sin x = \cos x$ is $\displaystyle\int \cos x \, dx = \sin x + C.$ Similarly, $D_x \cos x = -\sin x$ shows that $\displaystyle\int \sin x \, dx = -\cos x + C.$ Continuing in this way, using Table A.4-1, we produce the following table of antiderivatives for the trigonometric functions.

TABLE A.4-3

(1) $\int \sin x \, dx = -\cos x + C$	(4) $\int \csc x \operatorname{ctn} x \, dx = -\csc x + C$
(2) $\int \cos x \, dx = \sin x + C$	(5) $\int \sec x \tan x \, dx = \sec x + C$
(3) $\int \sec^2 x \, dx = \tan x + C$	(6) $\int \csc^2 x \, dx = -\operatorname{ctn} x + C$

Integration formulas for trigonometric functions

EXAMPLE A.4-7

Find $\int_0^{\pi} \sin x \, dx$.

SOLUTION

$\int_0^{\pi} \sin x \, dx = -\cos x \Big|_0^{\pi} = [-\cos \pi] - [-\cos 0] = [-(-1)]$
$- [-(1)] = 2$. Thus the area of the shaded region in Fig. A.4-3 is 2 square units.

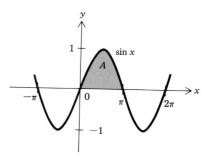

FIGURE A.4-3

EXAMPLE A.4-8

Use the integration technique of substitution to evaluate $\int \tan x \, dx$.

SOLUTION

Writing $\tan x = \sin x / \cos x$, we have

$$\int \tan x \, dx = \int \frac{\sin x}{\cos x} \, dx = \int \frac{1}{\cos x} \sin x \, dx$$

Letting $t = \cos x$, we get $dt = -\sin x \, dx$, so

$$\int \frac{1}{\cos x} \sin x \, dx = -\int \frac{1}{\cos x} (-\sin x) \, dx$$

$$= -\int \frac{1}{t} \, dt = -\ln|t| + C$$

$$= -\ln|\cos x| + C$$

EXAMPLE A.4-9

Use integration by parts to evaluate $\int x \cos x \, dx$.

SOLUTION

Let $f(x) = x$ and $D_x g(x) = \cos x$. Then $g(x) = \int D_x g(x) \, dx = \int \cos x \, dx = \sin x + C$ and $D_x f(x) = 1$. Thus (choosing $C = 0$),

$$\int x \cos x \, dx = \int f(x) D_x g(x) \, dx$$

$$= f(x) g(x) - \int g(x) D_x f(x) \, dx$$

$$= x \sin x - \int \sin x \, dx$$

$$= x \sin x + \cos x \, dx$$

EXAMPLE A.4-10

Evaluate $\int \sin x \cos x \, dx$.

SOLUTION

Let $t = \sin x$; then $dt = \cos x \, dx$, hence

$$\int \sin x \cos x \, dx = \int t \, dt = \frac{t^2}{2} + C = \tfrac{1}{2} \sin^2 x + C$$

Exercises A.4

In Exercises 1 through 12, find the derivative of the given function.

1. $f(x) = \cos(3x)$

2. $f(x) = \cos(\tan x)$

3. $f(x) = \tan(x^3)$

4. $f(x) = \sin^2 x$

5. $f(x) = \sin(2x)\cos(4x)$

6. $f(x) = e^{2x}\tan x$

7. $f(x) = \operatorname{ctn}\left(\dfrac{x}{2}\right)$

8. $f(x) = \dfrac{\sin x}{\tan x}$

9. $f(x) = 3\sec(x^2)$

10. $f(x) = \csc(\cos x)$

11. $f(x) = \tan(\ln x)$

12. $f(x) = \csc(\operatorname{ctn} x)$

13. Verify that $D_x \operatorname{ctn} x = -\csc^2 x$

14. Verify that $D_x \sec x = \sec x \tan x$

15. Verify that $D_x \csc x = -\csc x \operatorname{ctn} x$

In Exercises 16 through 20, find the first and second derivatives of the given function. Then find the critical points, relative maximum points, and relative minimum points and sketch the graph.

16. $f(x) = \cos(4x)$

17. $f(x) = \sin\left(\dfrac{x}{2}\right)$

18. $f(x) = \sin\left(x + \dfrac{\pi}{2}\right)$

19. $f(x) = 4\sin(2x)$

20. $f(x) = \sin x + \cos x$

In Exercises 21 through 30, find the given antiderivatives.

21. $\displaystyle\int \sin(2x)\,dx$

22. $\displaystyle\int \cos\left(\dfrac{x}{2}\right)dx$

23. $\displaystyle\int \tan(4x)\,dx$

24. $\displaystyle\int \operatorname{ctn} x\,dx$

25. $\displaystyle\int x\sin x\,dx$

26. $\displaystyle\int \dfrac{\sin x}{\cos^2 x}\,dx$

27. $\displaystyle\int x\sin(x^2)\,dx$

28. $\displaystyle\int \sec\left(x + \dfrac{\pi}{2}\right)\tan\left(x + \dfrac{\pi}{2}\right)dx$

29. $\displaystyle\int x^3\cos(x^4)\,dx$

30. $\displaystyle\int (1 + \sin^2 x\,\cos x)\,dx$

In Exercises 31 through 36, evaluate the given definite integral.

31. $\displaystyle\int_0^{\pi/2} \cos x\,dx$

32. $\displaystyle\int_0^{\pi} \sin x\,\cos^2 x\,dx$

33. $\displaystyle\int_0^{\pi/4} \sec^2 x\,\tan x\,dx$

34. $\displaystyle\int_0^{\pi/4} \sec^2 x\,\tan^2 x\,dx$

35. $\displaystyle\int_0^{\pi/2} x^2\sin x\,dx$

36. $\displaystyle\int_0^{\pi/4} (2 + \tan x)\,dx$

37. Find the area of the region bounded by $f(x) = \sin x$ and the x axis between $x = 0$ and $x = \pi/4$.

38. Find the area of the region bounded by $f(x) = \tan x$ and the x axis between $x = 0$ and $x = \pi/4$.

Exponential and Logarithmic Functions

Appendix B

B.1. Exponential Functions

For each real number a and positive integer n, a^n is defined as the product $a \cdot a \cdot a \cdots a$, where the factor a occurs n times. The number a is called the *base* and n the *exponent*. Thus $a^5 = a \cdot a \cdot a \cdot a \cdot a$ and $3^4 = 3 \cdot 3 \cdot 3 \cdot 3 = 81$. Directly from this definition come the basic rules for exponents:

(1) $a^m \cdot a^n = a^{m+n}$ and

(2) $(a^m)^n = a^{m \cdot n}$.

For example, $3^2 \cdot 3^4 = (3 \cdot 3)(3 \cdot 3 \cdot 3 \cdot 3) = 3 \cdot 3 \cdot 3 \cdot 3 \cdot 3 \cdot 3 = 3^6 = 3^{2+4}$ and $(3^2)^3 = 3^2 \cdot 3^2 \cdot 3^2 = (3 \cdot 3)(3 \cdot 3)(3 \cdot 3) = 3 \cdot 3 \cdot 3 \cdot 3 \cdot 3 \cdot 3 = 3^6 = 3^{2 \cdot 3}$. If $a \neq 0$, then *negative exponents* are defined by

$$a^{-n} = \frac{1}{a^n}$$

(where n is a positive integer), and *zero exponents* are defined by $a^0 = 1$. It is not hard to demonstrate that the rules $a^m \cdot a^n = a^{m+n}$ and $(a^m)^n = a^{m \cdot n}$ hold even if negative or 0 exponents are used (for $a \neq 0$). Thus

$$a^{-3} \cdot a^2 = \frac{1}{a^3} a^2 = \frac{a^2}{a^3} = \frac{a \cdot a}{a \cdot a \cdot a} = \frac{1}{a} = a^{-1} = a^{(-3)+2}$$

and

$$(a^{-2})^3 = a^{-2} \cdot a^{-2} \cdot a^{-2}$$

$$= \frac{1}{a^2} \cdot \frac{1}{a^2} \cdot \frac{1}{a^2}$$

$$= \frac{1}{a^6} = a^{-6} = a^{(-2) \cdot 3}$$

When defining rational exponents, such as $a^{1/3}$, we also wish to have rules (1) and (2) above hold. Thus we want $(a^{1/3})^3 = a^{(1/3) \cdot 3} = a^1 = a$, so $a^{1/3}$ must be a number that equals a when cubed. Similarly, we want $(a^{1/2})^2 = a^{(1/2) \cdot 2} = a^1 = a$, so $a^{1/2}$ should be a number that when squared equals a. This presents a problem if $a < 0$, since no real number squared could equal a negative number a. Consequently, we will define rational exponents only for numbers $a > 0$. The definition of $a^{1/n}$ for $a > 0$ and n a positive integer is this: $a^{1/n}$ is the positive number b such that $b^n = a$. For example, $8^{1/3} = 2$ because $2^3 = 2 \cdot 2 \cdot 2 = 8$, and $81^{1/4} = 3$ because $3^4 = 3 \cdot 3 \cdot 3 \cdot 3 = 81$. If m and n are positive integers and $a > 0$, we define

$$a^{m/n} = \left[a^{1/n}\right]^m \quad \text{and} \quad a^{-m/n} = \frac{1}{a^{m/n}}$$

In advanced courses in mathematics it is shown that a^x can be defined for every real number x (for $a > 0$). Thus it is sensible to consider such expressions as $5^{\sqrt{2}}$ and 2^{π}. Once a^x has been defined for every real number $x (a > 0)$, it can be shown that the following laws of exponents hold (where $a > 0$, $b > 0$).

(1) $a^x a^y = a^{x+y}$ (2) $(a^x)^y = a^{xy}$

(3) $\dfrac{a^x}{a^y} = a^{x-y}$ (4) $(ab)^x = a^x b^x$

(5) $a^{-x} = \dfrac{1}{a^x}$ (6) $\left(\dfrac{a}{b}\right)^x = \dfrac{a^x}{b^x}$

Definition of an exponential function

The function f defined by $f(x) = a^x (a > 0)$ is called an *exponential function*.

EXAMPLE B.1-1

Sketch the graph of the exponential function $f(x) = 2^x$.

SOLUTION

We first compute the following table of values.

x	0	1	-1	2	-2	3	-3	4	-4	5	-5	6	-6
2^x	1	2	$\frac{1}{2}$	4	$\frac{1}{4}$	8	$\frac{1}{8}$	16	$\frac{1}{16}$	32	$\frac{1}{32}$	64	$\frac{1}{64}$

The trend of these values is clear; as x increases indefinitely so does 2^x, and as x decreases indefinitely 2^x becomes smaller and smaller. We plot a few points and connect them with a smooth curve to obtain the graph of $f(x) = 2^x$ (see Fig. B.1-1(a)).

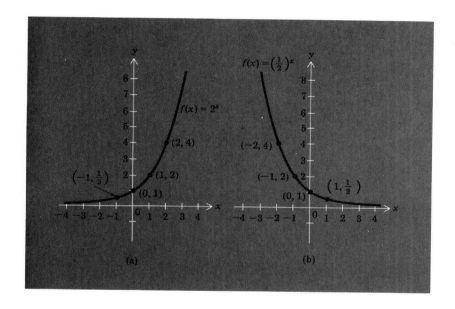

FIGURE B.1-1. (a) Graph of $f(x)$ $= 2^x$. (b) Graph of $f(x) = (\frac{1}{2})^x$.

EXAMPLE B.1-2

Sketch the graph of $f(x) = (\frac{1}{2})^x$.

SOLUTION

First we compute a table of values, as follows.

x	0	1	-1	2	-2	3	-3	4	-4	5	-5	6	-6
$(\frac{1}{2})^x$	1	$\frac{1}{2}$	2	$\frac{1}{4}$	4	$\frac{1}{8}$	8	$\frac{1}{16}$	16	$\frac{1}{32}$	32	$\frac{1}{64}$	64

The values of $(\frac{1}{2})^x$ become smaller and smaller as x increases indefinitely, whereas the values of $(\frac{1}{2})^x$ increase indefinitely as x decreases. The graph is sketched in Fig. B.1-1(b). The graphs of $f(x) = 2^x$ and $f(x) = (\frac{1}{2})^x$ are typical of the graphs of $f(x) = a^x$ for $a > 0$ $(a \neq 1)$. If $a > 1$, the graph is similar to that in Fig. B.1-1(a); if $a < 1$, the graph is similar to that in Fig. B.1-1(b).

Exercises B.1

In Exercises 1 through 10, evaluate the given quantity.

1. 2^3	2. 3^{-3}
3. $(2^2)^2$	4. $2^2 \cdot 2^{-3}$
5. $16^{1/4}$	6. $(2^{-1})^3$
7. $(1/8)^{1/3}$	8. $(27)^{-1/3}$
9. $(8)^{2/3}$	10. $(81)^{-3/4}$

In Exercises 11 through 20, simplify the given expression and write it without negative exponents.

11. $a^2 \cdot a^4$	12. a^6/a^2
13. $(a^{-2})^3$	14. $(8a^3)^{1/3}$
15. $(36y^4x^0)^{-1/2}$	16. $(x^2)^{-2} \cdot (x^{-1})^2$
17. $(16x^4y^4)^{3/4}$	18. $(8x^6y^9)^{-2/3}$
19. $-(81x^{-8}y^4)^{-1/2}$	20. $-(27x^{-3}y^{-8})^{-2/3}$

In Exercises 21 through 24, sketch the graph of the given function.

21. $f(x) = (\frac{1}{3})^x$	22. $f(x) = 4^x$
23. $f(x) = 5^x$	24. $f(x) = (\frac{1}{5})^x$

B.2. Logarithmic Functions

Logarithmic functions can be obtained from exponential functions by first reformulating the equation $x = a^y$, using what is referred to as logarithmic notation. For $a > 0$ and $a \neq 1$, $x = a^y$ is also indicated by

If $a = e$, then we write $\ln x$ instead of $\log_e x$.

$$y = \log_a x$$

The equation $y = \log_a x$ is read y *is the logarithm of x to the base a.* Thus a logarithm is nothing more than an exponent. For example, since $2^3 = 8$, we can write

$$3 = \log_2 8$$

Similarly, since $10^3 = 1000$, we can write

$$3 = \log_{10} 1000$$

Obviously, since $a^1 = a$, we always have $\log_a a = 1$; and $a^0 = 1$ yields $\log_a 1 = 0$. A function $g(x) = \log_a x$ is called a logarithmic function and is closely related to the exponential function $f(x) = a^x$. For these two functions we have $g[f(x)] = x$ for all x and $f[g(x)] = x$ for $x > 0$. The domain of a logarithmic function is $(0, \infty)$, since for x to be in the domain of $g(x) = \log_a x$ we must have $x = a^y$, and a^y is positive.

Definition of a logarithmic function

Originally, logarithms were used to simplify numerical calculations. Since our number system uses base 10, logarithms to this base were the most convenient for such computations. Because of this widespread use of the base 10 logarithms, $\log_{10} x$ was abbreviated $\log x$, and $\log_{10} x$ was termed the *common logarithm of x.*

Table B.2-1 illustrates the use of logarithmic notation.

TABLE B.2-1

Exponential form	Logarithmic form
$10^2 = 100$	$\log_{10} 100 = 2$
$3^2 = 9$	$\log_3 9 = 2$
$10^{-2} = \frac{1}{100}$	$\log_{10} \frac{1}{100} = -2$
$(\frac{1}{2})^3 = \frac{1}{8}$	$\log_{1/2} \frac{1}{8} = 3$
$5^{-3} = \frac{1}{125}$	$\log_5 \frac{1}{125} = -3$
$10^{-4} = 0.0001$	$\log_{10} 0.0001 = -4$
$\pi^0 = 1$	$\log_\pi 1 = 0$

The basic rules governing the use of logarithms can be derived from the rules for exponents. These rules are given below, where a, x, and y are positive real numbers, $a \neq 1$, and r is any real number.

(1) $\log_a xy = \log_a x + \log_a y$

(2) $\log_a x^r = r \log_a x$

(3) $\log_a(x/y) = \log_a x - \log_a y$

(4) $\log_a a = 1$

(5) $\log_a 1 = 0$

EXAMPLE B.2-1

Sketch the graph of the logarithmic function $f(x) = \log_2 x$.

SOLUTION

We first compute the following table of values.

x	1	2	$\frac{1}{2}$	4	$\frac{1}{4}$	8	$\frac{1}{8}$	16	$\frac{1}{16}$
$\log_2 x$	0	1	-1	2	-2	3	-3	4	-4

Now, we see that as the values of x are taken closer and closer to 0, the values of $\log_2 x$ decrease indefinitely; as the values of x increase indefinitely, so do the values of $\log_2 x$. These observations together with the given table of values allow us to sketch the graph of $\log_2 x$ as in Fig. B.2-1(a).

FIGURE B.2-1.   (a)Graph of $f(x)$ = $\log_2 x$. (b) Graph of $f(x)$ = $\log_{1/2}(x)$.

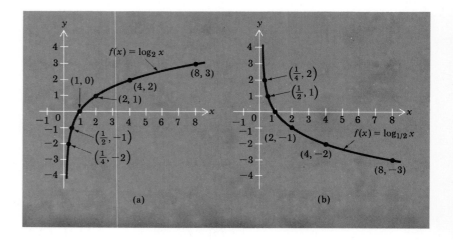

The graph of $\log_{1/2}(x)$ shown in Fig. B.2-1(b) can be produced by constructing a table of values as in Example B.2-1 and noting that as values of x are chosen closer and closer to 0, $\log_{1/2} x$ increases indefinitely; as x increases, the values of $\log_{1/2} x$ decrease. The graph of $\log_2 x$ is typical of logarithmic functions to base a with $a > 1$, and the graph of $\log_{1/2} x$ is typical of the graphs of logarithmic functions to base a with $a < 1$.

The following differentiation and integration formulas are frequently useful when working with exponential and logarithmic functions.

Differentiation and integration formulas for exponential and logarithmic functions

(1) $D_x a^x = a^x \ln a$

(2) $D_x \log_a x = \dfrac{\log_a e}{x}$

(3) $\displaystyle\int a^x \, dx = \dfrac{a^x}{\ln a} + C$

(4) $\displaystyle\int \ln x \, dx = x \ln x - x + C$

(5) $\displaystyle\int \log_a x \, dx = x \log_a x - x \log_a e + C$

Exercises B.2

In Exercises 1 through 6, write the given equation using logarithmic notation.

1. $5^2 = 25$

2. $8^{1/3} = 2$

3. $16^{1/2} = 4$

4. $2^{-3} = \frac{1}{8}$

5. $64^{1/2} = 8$

6. $2^0 = 1$

In Exercises 7 through 12, write the given equation using exponential notation.

7. $\log_2 16 = 4$

8. $\log_3 27 = 3$

9. $\log_{10} 1000 = 3$

10. $\log_4 1 = 0$

11. $\log_{50} 50 = 1$

12. $\log_{1/2}(\frac{1}{4}) = 2$

In Exercises 13 through 18, write the expression as a sum or difference of logarithms of single terms.

13. $\log_a 2x^3 y$

14. $\log_5 8zw$

15. $\log_a \dfrac{2x^3}{4y^2}$

16. $\log_a \left(\dfrac{(x^4 y)^{1/2}}{z^2} \right)$

17. $\log_a \left(\dfrac{x^3}{y^2 z} \right)^{1/4}$

18. $\log_a \left(\dfrac{(x^2 y)^{1/2}}{z^2} \right)^{1/4}$

In Exercises 19 through 22, sketch the graph of the given function.

19. $f(x) = \log_4 x$

20. $f(x) = \log_{1/4} x$

21. $f(x) = \log_{10} x$

22. $f(x) = \log_{1/3} x$

In Exercises 23 through 28, find the derivative of the given function.

23. $f(x) = 2^x$

24. $f(x) = 4^{x^2}$

25. $f(x) = \log_4 x$

26. $f(x) = \log_5 (3x^2)$

27. $f(x) = 10^{(x^2 + 2x)}$

28. $f(x) = \ln(4x^2 + 2x)$

In Exercises 29 through 34, find the indefinite integral.

29. $\displaystyle\int 2^x \, dx$

30. $\displaystyle\int \log_4 x \, dx$

31. $\displaystyle\int x5^{x^2} \, dx$

32. $\displaystyle\int x \ln (x^2 + 1) \, dx$

33. $\displaystyle\int x^3 \log_5 (x^4 + 2) \, dx$

34. $\displaystyle\int x^2 \log_{1/2}\left(\frac{x^3}{3} + 2 \right) dx$

Tables of Exponential, Logarithmic, and Trigonometric Functions

TABLE A
Exponential and Natural Logarithm Functions

x	e^x	e^{-x}	$\ln(x)$	x	e^x	e^{-x}	$\ln(x)$
0.00	1.0000	1.0000	—	0.56	1.7507	0.5712	−0.5798
0.01	1.0101	0.9900	−4.6052	0.57	1.7683	0.5655	−0.5621
0.02	1.0202	0.9802	−3.9120	0.58	1.7860	0.5599	−0.5447
0.03	1.0305	0.9704	−3.5066	0.59	1.8040	0.5543	−0.5276
0.04	1.0408	0.9608	−3.2189	0.60	1.8221	0.5488	−0.5108
0.05	1.0513	0.9512	−2.9957	0.61	1.8404	0.5434	−0.4943
0.06	1.0618	0.9418	−2.8134	0.62	1.8589	0.5379	−0.4780
0.07	1.0725	0.9324	−2.6593	0.63	1.8776	0.5326	−0.4620
0.08	1.0833	0.9231	−2.5257	0.64	1.8965	0.5273	−0.4463
0.09	1.0942	0.9139	−2.4079	0.65	1.9155	0.5220	−0.4308
0.10	1.1052	0.9048	−2.3026	0.66	1.9348	0.5169	−0.4155
0.11	1.1163	0.8958	−2.2073	0.67	1.9542	0.5117	−0.4005
0.12	1.1275	0.8869	−2.1203	0.68	1.9739	0.5066	−0.3857
0.13	1.1388	0.8781	−2.0402	0.69	1.9937	0.5016	−0.3711
0.14	1.1503	0.8694	−1.9661	0.70	2.0138	0.4966	−0.3567
0.15	1.1618	0.8607	−1.8971	0.71	2.0340	0.4916	−0.3425
0.16	1.1735	0.8521	−1.8326	0.72	2.0544	0.4868	−0.3285
0.17	1.1853	0.8437	−1.7720	0.73	2.0751	0.4819	−0.3147
0.18	1.1972	0.8353	−1.7148	0.74	2.0959	0.4771	−0.3011
0.19	1.2092	0.8270	−1.6607	0.75	2.1170	0.4724	−0.2877
0.20	1.2214	0.8187	−1.6094	0.76	2.1383	0.4677	−0.2744
0.21	1.2337	0.8106	−1.5606	0.77	2.1598	0.4630	−0.2614
0.22	1.2461	0.8025	−1.5141	0.78	2.1815	0.4584	−0.2485
0.23	1.2586	0.7945	−1.4697	0.79	2.2034	0.4538	−0.2357
0.24	1.2712	0.7866	−1.4271	0.80	2.2255	0.4493	−0.2231
0.25	1.2840	0.7788	−1.3863	0.81	2.2479	0.4449	−0.2107
0.26	1.2969	0.7711	−1.3471	0.82	2.2705	0.4404	−0.1985
0.27	1.3100	0.7634	−1.3093	0.83	2.2933	0.4360	−0.1863
0.28	1.3231	0.7558	−1.2730	0.84	2.3164	0.4317	−0.1744
0.29	1.3364	0.7483	−1.2379	0.85	2.3396	0.4274	−0.1625
0.30	1.3499	0.7408	−1.2040	0.86	2.3632	0.4232	−0.1508
0.31	1.3634	0.7334	−1.1712	0.87	2.3869	0.4190	−0.1393
0.32	1.3771	0.7261	−1.1394	0.88	2.4109	0.4148	−0.1278
0.33	1.3910	0.7189	−1.1087	0.89	2.4351	0.4107	−0.1165
0.34	1.4049	0.7118	−1.0788	0.90	2.4596	0.4066	−0.1054
0.35	1.4191	0.7047	−1.0498	0.91	2.4843	0.4025	−0.0943
0.36	1.4333	0.6977	−1.0217	0.92	2.5093	0.3985	−0.0834
0.37	1.4477	0.6907	−0.9943	0.93	2.5345	0.3946	−0.0726
0.38	1.4623	0.6839	−0.9676	0.94	2.5600	0.3906	−0.0619
0.39	1.4770	0.6771	−0.9416	0.95	2.5857	0.3867	−0.0513
0.40	1.4918	0.6703	−0.9163	0.96	2.6117	0.3829	−0.0408
0.41	1.5068	0.6637	−0.8916	0.97	2.6379	0.3791	−0.0305
0.42	1.5220	0.6570	−0.8675	0.98	2.6645	0.3753	−0.0202
0.43	1.5373	0.6505	−0.8440	0.99	2.6912	0.3716	−0.0101
0.44	1.5527	0.6440	−0.8210	1.00	2.7183	0.3679	0.0000
0.45	1.5683	0.6376	−0.7985	1.01	2.7456	0.3642	0.0100
0.46	1.5841	0.6313	−0.7765	1.02	2.7732	0.3606	0.0198
0.47	1.6000	0.6250	−0.7550	1.03	2.8011	0.3570	0.0296
0.48	1.6161	0.6188	−0.7340	1.04	2.8292	0.3535	0.0392
0.49	1.6323	0.6126	−0.7133	1.05	2.8577	0.3499	0.0488
0.50	1.6487	0.6065	−0.6931	1.06	2.8864	0.3465	0.0583
0.51	1.6653	0.6005	−0.6733	1.07	2.9154	0.3430	0.0677
0.52	1.6820	0.5945	−0.6539	1.08	2.9447	0.3396	0.0770
0.53	1.6989	0.5886	−0.6349	1.09	2.9743	0.3362	0.0862
0.54	1.7160	0.5827	−0.6162	1.10	3.0042	0.3329	0.0953
0.55	1.7333	0.5769	−0.5978				

TABLE A
Exponential and Natural Logarithm Functions

x	e^x	e^{-x}	$\ln(x)$	x	e^x	e^{-x}	$\ln(x)$
1.11	3.0344	0.3296	0.1044	1.66	5.2593	0.1901	0.5068
1.12	3.0649	0.3263	0.1133	1.67	5.3122	0.1882	0.5128
1.13	3.0957	0.3230	0.1222	1.68	5.3656	0.1864	0.5188
1.14	3.1268	0.3198	0.1310	1.69	5.4195	0.1845	0.5247
1.15	3.1582	0.3166	0.1398	1.70	5.4739	0.1827	0.5306
1.16	3.1899	0.3135	0.1484	1.71	5.5290	0.1809	0.5365
1.17	3.2220	0.3104	0.1570	1.72	5.5845	0.1791	0.5423
1.18	3.2544	0.3073	0.1655	1.73	5.6407	0.1773	0.5481
1.19	3.2871	0.3042	0.1740	1.74	5.6973	0.1755	0.5539
1.20	3.3201	0.3012	0.1823	1.75	5.7546	0.1738	0.5596
1.21	3.3535	0.2982	0.1906	1.76	5.8124	0.1720	0.5653
1.22	3.3872	0.2952	0.1989	1.77	5.8709	0.1703	0.5710
1.23	3.4212	0.2923	0.2070	1.78	5.9299	0.1686	0.5766
1.24	3.4556	0.2894	0.2151	1.79	5.9895	0.1670	0.5822
1.25	3.4903	0.2865	0.2231	1.80	6.0496	0.1653	0.5878
1.26	3.5254	0.2837	0.2311	1.81	6.1104	0.1637	0.5933
1.27	3.5609	0.2808	0.2390	1.82	6.1719	0.1620	0.5988
1.28	3.5966	0.2780	0.2469	1.83	6.2339	0.1604	0.6043
1.29	3.6328	0.2753	0.2546	1.84	6.2965	0.1588	0.6098
1.30	3.6693	0.2725	0.2624	1.85	6.3598	0.1572	0.6152
1.31	3.7062	0.2698	0.2700	1.86	6.4237	0.1557	0.6206
1.32	3.7434	0.2671	0.2776	1.87	6.4883	0.1541	0.6259
1.33	3.7810	0.2645	0.2852	1.88	6.5535	0.1526	0.6313
1.34	3.8190	0.2618	0.2927	1.89	6.6194	0.1511	0.6366
1.35	3.8574	0.2592	0.3001	1.90	6.6859	0.1496	0.6419
1.36	3.8962	0.2567	0.3075	1.91	6.7531	0.1481	0.6471
1.37	3.9354	0.2541	0.3148	1.92	6.8210	0.1466	0.6523
1.38	3.9749	0.2516	0.3221	1.93	6.8895	0.1451	0.6575
1.39	4.0149	0.2491	0.3293	1.94	6.9588	0.1437	0.6627
1.40	4.0552	0.2466	0.3365	1.95	7.0287	0.1423	0.6678
1.41	4.0960	0.2441	0.3436	1.96	7.0993	0.1409	0.6729
1.42	4.1371	0.2417	0.3507	1.97	7.1707	0.1395	0.6780
1.43	4.1787	0.2393	0.3577	1.98	7.2427	0.1381	0.6831
1.44	4.2207	0.2369	0.3646	1.99	7.3155	0.1367	0.6881
1.45	4.2631	0.2346	0.3716	2.00	7.3891	0.1353	0.6931
1.46	4.3060	0.2322	0.3784	2.01	7.4633	0.1340	0.6981
1.47	4.3492	0.2299	0.3853	2.02	7.5383	0.1327	0.7031
1.48	4.3929	0.2276	0.3920	2.03	7.6141	0.1313	0.7080
1.49	4.4371	0.2254	0.3988	2.04	7.6906	0.1300	0.7129
1.50	4.4817	0.2231	0.4055	2.05	7.7679	0.1287	0.7178
1.51	4.5267	0.2209	0.4121	2.06	7.8460	0.1275	0.7227
1.52	4.5722	0.2187	0.4187	2.07	7.9248	0.1262	0.7275
1.53	4.6182	0.2165	0.4253	2.08	8.0045	0.1249	0.7324
1.54	4.6646	0.2144	0.4318	2.09	8.0849	0.1237	0.7372
1.55	4.7115	0.2122	0.4383	2.10	8.1662	0.1225	0.7419
1.56	4.7588	0.2101	0.4447	2.11	8.2482	0.1212	0.7467
1.57	4.8066	0.2080	0.4511	2.12	8.3311	0.1200	0.7514
1.58	4.8550	0.2060	0.4574	2.13	8.4149	0.1188	0.7561
1.59	4.9037	0.2039	0.4637	2.14	8.4994	0.1177	0.7608
1.60	4.9530	0.2019	0.4700	2.15	8.5849	0.1165	0.7655
1.61	5.0028	0.1999	0.4762	2.16	8.6711	0.1153	0.7701
1.62	5.0531	0.1979	0.4824	2.17	8.7583	0.1142	0.7747
1.63	5.1039	0.1959	0.4886	2.18	8.8463	0.1130	0.7793
1.64	5.1552	0.1940	0.4947	2.19	8.9352	0.1119	0.7839
1.65	5.2070	0.1920	0.5008	2.20	9.0250	0.1108	0.7885

TABLE A
Exponential and Natural Logarithm Functions

x	e^x	e^{-x}	$\ln(x)$	x	e^x	e^{-x}	$\ln(x)$
2.21	9.1157	0.1097	0.7930	2.76	15.800	0.0633	1.0152
2.22	9.2073	0.1086	0.7975	2.77	15.959	0.0627	1.0188
2.23	9.2999	0.1075	0.8020	2.78	16.119	0.0620	1.0225
2.24	9.3933	0.1065	0.8065	2.79	16.281	0.0614	1.0260
2.25	9.4877	0.1054	0.8109	2.80	16.445	0.0608	1.0296
2.26	9.5831	0.1044	0.8154	2.81	16.610	0.0602	1.0332
2.27	9.6794	0.1033	0.8198	2.82	16.777	0.0596	1.0367
2.28	9.7767	0.1023	0.8242	2.83	16.945	0.0590	1.0403
2.29	9.8749	0.1013	0.8286	2.84	17.116	0.0584	1.0438
2.30	9.9742	0.1003	0.8329	2.85	17.288	0.0578	1.0473
2.31	10.074	0.0993	0.8372	2.86	17.462	0.0573	1.0508
2.32	10.176	0.0983	0.8416	2.87	17.637	0.0567	1.0543
2.33	10.278	0.0973	0.8459	2.88	17.814	0.0561	1.0578
2.34	10.381	0.0963	0.8502	2.89	17.993	0.0556	1.0613
2.35	10.486	0.0954	0.8544	2.90	18.174	0.0550	1.0647
2.36	10.591	0.0944	0.8587	2.91	18.357	0.0545	1.0682
2.37	10.697	0.0935	0.8629	2.92	18.541	0.0539	1.0716
2.38	10.805	0.0926	0.8671	2.93	18.728	0.0534	1.0750
2.39	10.913	0.0916	0.8713	2.94	18.916	0.0529	1.0784
2.40	11.023	0.0907	0.8755	2.95	19.106	0.0523	1.0818
2.41	11.134	0.0898	0.8796	2.96	19.298	0.0518	1.0852
2.42	11.246	0.0889	0.8838	2.97	19.492	0.0513	1.0886
2.43	11.359	0.0880	0.8879	2.98	19.688	0.0508	1.0919
2.44	11.473	0.0872	0.8920	2.99	19.886	0.0503	1.0953
2.45	11.588	0.0863	0.8961	3.00	20.086	0.0498	1.0986
2.46	11.705	0.0854	0.9002	3.01	20.287	0.0493	1.1019
2.47	11.822	0.0846	0.9042	3.02	20.491	0.0488	1.1053
2.48	11.941	0.0837	0.9083	3.03	20.697	0.0483	1.1086
2.49	12.061	0.0829	0.9123	3.04	20.905	0.0478	1.1119
2.50	12.182	0.0821	0.9163	3.05	21.115	0.0474	1.1151
2.51	12.305	0.0813	0.9203	3.06	21.328	0.0469	1.1184
2.52	12.429	0.0805	0.9243	3.07	21.542	0.0464	1.1217
2.53	12.554	0.0797	0.9282	3.08	21.758	0.0460	1.1249
2.54	12.680	0.0789	0.9322	3.09	21.977	0.0455	1.1282
2.55	12.807	0.0781	0.9361	3.10	22.198	0.0450	1.1314
2.56	12.936	0.0773	0.9400	3.11	22.421	0.0446	1.1346
2.57	13.066	0.0765	0.9439	3.12	22.646	0.0442	1.1378
2.58	13.197	0.0758	0.9478	3.13	22.874	0.0437	1.1410
2.59	13.330	0.0750	0.9517	3.14	23.104	0.0433	1.1442
2.60	13.464	0.0743	0.9555	3.15	23.336	0.0429	1.1474
2.61	13.599	0.0735	0.9594	3.16	23.571	0.0424	1.1506
2.62	13.736	0.0728	0.9632	3.17	23.807	0.0420	1.1537
2.63	13.874	0.0721	0.9670	3.18	24.047	0.0416	1.1569
2.64	14.013	0.0714	0.9708	3.19	24.288	0.0412	1.1600
2.65	14.154	0.0707	0.9746	3.20	24.533	0.0408	1.1632
2.66	14.296	0.0699	0.9783	3.21	24.779	0.0404	1.1663
2.67	14.440	0.0693	0.9821	3.22	25.028	0.0400	1.1694
2.68	14.585	0.0686	0.9858	3.23	25.280	0.0396	1.1725
2.69	14.732	0.0679	0.9895	3.24	25.534	0.0392	1.1756
2.70	14.880	0.0672	0.9933	3.25	25.790	0.0388	1.1787
2.71	15.029	0.0665	0.9969	3.26	26.050	0.0384	1.1817
2.72	15.180	0.0659	1.0006	3.27	26.311	0.0380	1.1848
2.73	15.333	0.0652	1.0043	3.28	26.576	0.0376	1.1878
2.74	15.487	0.0646	1.0080	3.29	26.843	0.0373	1.1909
2.75	15.643	0.0639	1.0116	3.30	27.113	0.0369	1.1939

TABLE A
Exponential and Natural Logarithm Functions

x	e^x	e^{-x}	$\ln(x)$	x	e^x	e^{-x}	$\ln(x)$
3.31	27.385	0.0365	1.1969	3.86	47.465	0.0211	1.3507
3.32	27.660	0.0362	1.2000	3.87	47.942	0.0209	1.3533
3.33	27.938	0.0358	1.2030	3.88	48.424	0.0207	1.3558
3.34	28.219	0.0354	1.2060	3.89	48.911	0.0204	1.3584
3.35	28.503	0.0351	1.2090	3.90	49.402	0.0202	1.3610
3.36	28.789	0.0347	1.2119	3.91	49.899	0.0200	1.3635
3.37	29.079	0.0344	1.2149	3.92	50.400	0.0198	1.3661
3.38	29.371	0.0340	1.2179	3.93	50.907	0.0196	1.3686
3.39	29.666	0.0337	1.2208	3.94	51.419	0.0194	1.3712
3.40	29.964	0.0334	1.2238	3.95	51.935	0.0193	1.3737
3.41	30.265	0.0330	1.2267	3.96	52.457	0.0191	1.3762
3.42	30.569	0.0327	1.2296	3.97	52.985	0.0189	1.3788
3.43	30.877	0.0324	1.2326	3.98	53.517	0.0187	1.3813
3.44	31.187	0.0321	1.2355	3.99	54.055	0.0185	1.3838
3.45	31.500	0.0317	1.2384	4.00	54.598	0.0183	1.3863
3.46	31.817	0.0314	1.2413	4.01	55.147	0.0181	1.3888
3.47	32.137	0.0311	1.2442	4.02	55.701	0.0180	1.3913
3.48	32.460	0.0308	1.2470	4.03	56.261	0.0178	1.3938
3.49	32.786	0.0305	1.2499	4.04	56.826	0.0176	1.3962
3.50	33.115	0.0302	1.2528	4.05	57.397	0.0174	1.3987
3.51	33.448	0.0299	1.2556	4.06	57.974	0.0172	1.4012
3.52	33.784	0.0296	1.2585	4.07	58.557	0.0171	1.4036
3.53	34.124	0.0293	1.2613	4.08	59.145	0.0169	1.4061
3.54	34.467	0.0290	1.2641	4.09	59.740	0.0167	1.4085
3.55	34.813	0.0287	1.2669	4.10	60.340	0.0166	1.4110
3.56	35.163	0.0284	1.2698	4.11	60.947	0.0164	1.4134
3.57	35.517	0.0282	1.2726	4.12	61.559	0.0162	1.4159
3.58	35.874	0.0279	1.2754	4.13	62.178	0.0161	1.4183
3.59	36.234	0.0276	1.2782	4.14	62.803	0.0159	1.4207
3.60	36.598	0.0273	1.2809	4.15	63.434	0.0158	1.4231
3.61	36.966	0.0271	1.2837	4.16	64.072	0.0156	1.4255
3.62	37.338	0.0268	1.2865	4.17	64.715	0.0155	1.4279
3.63	37.713	0.0265	1.2892	4.18	65.366	0.0153	1.4303
3.64	38.092	0.0263	1.2920	4.19	66.023	0.0151	1.4327
3.65	38.475	0.0260	1.2947	4.20	66.686	0.0150	1.4351
3.66	38.861	0.0257	1.2975	4.21	67.357	0.0148	1.4375
3.67	39.252	0.0255	1.3002	4.22	68.033	0.0147	1.4398
3.68	39.646	0.0252	1.3029	4.23	68.717	0.0146	1.4422
3.69	40.045	0.0250	1.3056	4.24	69.408	0.0144	1.4446
3.70	40.447	0.0247	1.3083	4.25	70.105	0.0143	1.4469
3.71	40.854	0.0245	1.3110	4.26	70.810	0.0141	1.4493
3.72	41.264	0.0242	1.3137	4.27	71.522	0.0140	1.4516
3.73	41.679	0.0240	1.3164	4.28	72.240	0.0138	1.4540
3.74	42.098	0.0238	1.3191	4.29	72.966	0.0137	1.4563
3.75	42.521	0.0235	1.3218	4.30	73.700	0.0136	1.4586
3.76	42.948	0.0233	1.3244	4.31	74.440	0.0134	1.4609
3.77	43.380	0.0231	1.3271	4.32	75.189	0.0133	1.4633
3.78	43.816	0.0228	1.3297	4.33	75.944	0.0132	1.4656
3.79	44.256	0.0226	1.3324	4.34	76.708	0.0130	1.4679
3.80	44.701	0.0224	1.3350	4.35	77.478	0.0129	1.4702
3.81	45.150	0.0221	1.3376	4.36	78.257	0.0128	1.4725
3.82	45.604	0.0219	1.3403	4.37	79.044	0.0127	1.4748
3.83	46.063	0.0217	1.3429	4.38	79.838	0.0125	1.4770
3.84	46.525	0.0215	1.3455	4.39	80.640	0.0124	1.4793
3.85	46.993	0.0213	1.3481	4.40	81.451	0.0123	1.4816

TABLE A
Exponential and Natural Logarithm Functions

x	e^x	e^{-x}	$\ln(x)$	x	e^x	e^{-x}	$\ln(x)$
4.41	82.269	0.0122	1.4839	4.96	142.594	0.0070	1.6014
4.42	83.096	0.0120	1.4861	4.97	144.027	0.0069	1.6034
4.43	83.931	0.0119	1.4884	4.98	145.474	0.0069	1.6054
4.44	84.775	0.0118	1.4907	4.99	146.936	0.0068	1.6074
4.45	85.627	0.0117	1.4929	5.00	148.413	0.0067	1.6094
4.46	86.488	0.0116	1.4951	5.01	149.905	0.0067	1.6114
4.47	87.357	0.0114	1.4974	5.02	151.411	0.0066	1.6134
4.48	88.235	0.0113	1.4996	5.03	152.933	0.0065	1.6154
4.49	89.121	0.0112	1.5019	5.04	154.470	0.0065	1.6174
4.50	90.017	0.0111	1.5041	5.05	156.022	0.0064	1.6194
4.51	90.922	0.0110	1.5063	5.06	157.591	0.0063	1.6214
4.52	91.836	0.0109	1.5085	5.07	159.174	0.0063	1.6233
4.53	92.759	0.0108	1.5107	5.08	160.774	0.0062	1.6253
4.54	93.691	0.0107	1.5129	5.09	162.390	0.0062	1.6273
4.55	94.632	0.0106	1.5151	5.10	164.022	0.0061	1.6292
4.56	95.583	0.0105	1.5173	5.11	165.670	0.0060	1.6312
4.57	96.544	0.0104	1.5195	5.12	167.335	0.0060	1.6332
4.58	97.514	0.0103	1.5217	5.13	169.017	0.0059	1.6351
4.59	98.494	0.0102	1.5239	5.14	170.716	0.0059	1.6371
4.60	99.484	0.0101	1.5261	5.15	172.431	0.0058	1.6390
4.61	100.484	0.0100	1.5282	5.16	174.164	0.0057	1.6409
4.62	101.494	0.0099	1.5304	5.17	175.915	0.0057	1.6429
4.63	102.514	0.0098	1.5326	5.18	177.683	0.0056	1.6448
4.64	103.544	0.0097	1.5347	5.19	179.469	0.0056	1.6467
4.65	104.585	0.0096	1.5369	5.20	181.272	0.0055	1.6487
4.66	105.636	0.0095	1.5390	5.21	183.094	0.0055	1.6506
4.67	106.698	0.0094	1.5412	5.22	184.934	0.0054	1.6525
4.68	107.770	0.0093	1.5433	5.23	186.793	0.0054	1.6544
4.69	108.853	0.0092	1.5454	5.24	188.670	0.0053	1.6563
4.70	109.947	0.0091	1.5476	5.25	190.566	0.0052	1.6582
4.71	111.052	0.0090	1.5497	5.26	192.481	0.0052	1.6601
4.72	112.168	0.0089	1.5518	5.27	194.416	0.0051	1.6620
4.73	113.296	0.0088	1.5539	5.28	196.370	0.0051	1.6639
4.74	114.434	0.0087	1.5560	5.29	198.343	0.0050	1.6658
4.75	115.584	0.0087	1.5581	5.30	200.337	0.0050	1.6677
4.76	116.746	0.0086	1.5602	5.31	202.350	0.0049	1.6696
4.77	117.919	0.0085	1.5623	5.32	204.384	0.0049	1.6715
4.78	119.104	0.0084	1.5644	5.33	206.438	0.0048	1.6734
4.79	120.301	0.0083	1.5665	5.34	208.513	0.0048	1.6752
4.80	121.510	0.0082	1.5686	5.35	210.608	0.0047	1.6771
4.81	122.732	0.0081	1.5707	5.36	212.725	0.0047	1.6790
4.82	123.965	0.0081	1.5728	5.37	214.863	0.0047	1.6808
4.83	125.211	0.0080	1.5748	5.38	217.022	0.0046	1.6827
4.84	126.469	0.0079	1.5769	5.39	219.203	0.0046	1.6845
4.85	127.740	0.0078	1.5790	5.40	221.406	0.0045	1.6864
4.86	129.024	0.0078	1.5810	5.41	223.632	0.0045	1.6882
4.87	130.321	0.0077	1.5831	5.42	225.879	0.0044	1.6901
4.88	131.631	0.0076	1.5851	5.43	228.149	0.0044	1.6919
4.89	132.954	0.0075	1.5872	5.44	230.442	0.0043	1.6938
4.90	134.290	0.0074	1.5892	5.45	232.758	0.0043	1.6956
4.91	135.639	0.0074	1.5913	5.46	235.097	0.0043	1.6974
4.92	137.003	0.0073	1.5933	5.47	237.460	0.0042	1.6993
4.93	138.380	0.0072	1.5953	5.48	239.847	0.0042	1.7011
4.94	139.770	0.0072	1.5974	5.49	242.257	0.0041	1.7029
4.95	141.175	0.0071	1.5994	5.50	244.692	0.0041	1.7047

TABLE A
Exponential and Natural Logarithm Functions

x	e^x	e^{-x}	$\ln(x)$	x	e^x	e^{-x}	$\ln(x)$
5.51	247.151	0.0040	1.7066	6.6	735.095	0.00136	1.8871
5.52	249.635	0.0040	1.7084	6.7	812.406	0.00123	1.9021
5.53	252.144	0.0040	1.7102	6.8	897.847	0.00111	1.9169
5.54	254.678	0.0039	1.7120	6.9	992.275	0.00101	1.9315
5.55	257.238	0.0039	1.7138	7.0	1096.63	0.00091	1.9459
5.56	259.823	0.0038	1.7156	7.1	1211.97	0.00083	1.9601
5.57	262.434	0.0038	1.7174	7.2	1339.43	0.00075	1.9741
5.58	265.072	0.0038	1.7192	7.3	1480.30	0.00068	1.9879
5.59	267.736	0.0037	1.7210	7.4	1635.98	0.00061	2.0015
5.60	270.426	0.0037	1.7228	7.5	1808.04	0.00055	2.0149
5.61	273.144	0.0037	1.7246	7.6	1998.20	0.00050	2.0281
5.62	275.889	0.0036	1.7263	7.7	2208.35	0.00045	2.0412
5.63	278.662	0.0036	1.7281	7.8	2440.60	0.00041	2.0541
5.64	281.463	0.0036	1.7299	7.9	2697.28	0.00037	2.0669
5.65	284.291	0.0035	1.7317	8.0	2980.96	0.00034	2.0794
5.66	287.149	0.0035	1.7334	8.5	4914.77	0.00020	2.1401
5.67	290.035	0.0034	1.7352	9.0	8103.08	0.00012	2.1972
5.68	292.949	0.0034	1.7370	9.5	13359.73	0.00007	2.2513
5.69	295.894	0.0034	1.7387	10.0	22026.47	0.00005	2.3026
5.70	298.867	0.0033	1.7405				
5.71	301.871	0.0033	1.7422				
5.72	304.905	0.0033	1.7440				
5.73	307.969	0.0032	1.7457				
5.74	311.064	0.0032	1.7475				
5.75	314.191	0.0032	1.7492				
5.76	317.348	0.0032	1.7509				
5.77	320.538	0.0031	1.7527				
5.78	323.759	0.0031	1.7544				
5.79	327.013	0.0031	1.7561				
5.80	330.300	0.0030	1.7579				
5.81	333.619	0.0030	1.7596				
5.82	336.972	0.0030	1.7613				
5.83	340.359	0.0029	1.7630				
5.84	343.779	0.0029	1.7647				
5.85	347.234	0.0029	1.7664				
5.86	350.724	0.0029	1.7681				
5.87	354.249	0.0028	1.7699				
5.88	357.809	0.0028	1.7716				
5.89	361.405	0.0028	1.7733				
5.90	365.037	0.0027	1.7750				
5.91	368.706	0.0027	1.7766				
5.92	372.412	0.0027	1.7783				
5.93	376.155	0.0027	1.7800				
5.94	379.935	0.0026	1.7817				
5.95	383.753	0.0026	1.7834				
5.96	387.610	0.0026	1.7851				
5.97	391.506	0.0026	1.7867				
5.98	395.440	0.0025	1.7884				
5.99	399.415	0.0025	1.7901				
6.0	403.429	0.00248	1.7918				
6.1	445.858	0.00224	1.8083				
6.2	492.749	0.00203	1.8245				
6.3	544.572	0.00184	1.8405				
6.4	601.845	0.00166	1.8563				
6.5	665.142	0.00150	1.8718				

TABLE B
Trigonometric Functions
x in degrees

x (degrees)	Sin(x)	Cos(x)	Tan(x)
0	0.0000	1.0000	0.0000
1	0.0175	0.9998	0.0175
2	0.0349	0.9994	0.0349
3	0.0523	0.9986	0.0524
4	0.0698	0.9976	0.0699
5	0.0872	0.9962	0.0875
6	0.1045	0.9945	0.1051
7	0.1219	0.9925	0.1228
8	0.1392	0.9903	0.1405
9	0.1564	0.9877	0.1584
10	0.1736	0.9848	0.1763
11	0.1908	0.9816	0.1944
12	0.2079	0.9781	0.2126
13	0.2250	0.9744	0.2309
14	0.2419	0.9703	0.2493
15	0.2588	0.9659	0.2679
16	0.2756	0.9613	0.2867
17	0.2924	0.9563	0.3057
18	0.3090	0.9511	0.3249
19	0.3256	0.9455	0.3443
20	0.3420	0.9397	0.3640
21	0.3584	0.9336	0.3839
22	0.3746	0.9272	0.4040
23	0.3907	0.9205	0.4245
24	0.4067	0.9135	0.4452
25	0.4226	0.9063	0.4663
26	0.4384	0.8988	0.4877
27	0.4540	0.8910	0.5095
28	0.4695	0.8829	0.5317
29	0.4848	0.8746	0.5543
30	0.5000	0.8660	0.5774
31	0.5150	0.8572	0.6009
32	0.5299	0.8480	0.6249
33	0.5446	0.8387	0.6494
34	0.5592	0.8290	0.6745
35	0.5736	0.8192	0.7002
36	0.5878	0.8090	0.7265
37	0.6018	0.7986	0.7536
38	0.6157	0.7880	0.7813
39	0.6293	0.7771	0.8098
40	0.6428	0.7660	0.8391
41	0.6561	0.7547	0.8693
42	0.6691	0.7431	0.9004
43	0.6820	0.7314	0.9325
44	0.6947	0.7193	0.9657
45	0.7071	0.7071	1.0000

TABLE C
Trigonometric Functions
x in radians

x (radians)	Sin(x)	Cos(x)	Tan(x)
0.00	0.0000	1.0000	0.0000
0.01	0.0100	1.0000	0.0100
0.02	0.0200	0.9998	0.0200
0.03	0.0300	0.9996	0.0300
0.04	0.0400	0.9992	0.0400
0.05	0.0500	0.9988	0.0500
0.06	0.0600	0.9982	0.0601
0.07	0.0699	0.9976	0.0701
0.08	0.0799	0.9968	0.0802
0.09	0.0899	0.9960	0.0902
0.10	0.0998	0.9950	0.1003
0.11	0.1098	0.9940	0.1104
0.12	0.1197	0.9928	0.1206
0.13	0.1296	0.9916	0.1307
0.14	0.1395	0.9902	0.1409
0.15	0.1494	0.9888	0.1511
0.16	0.1593	0.9872	0.1614
0.17	0.1692	0.9856	0.1717
0.18	0.1790	0.9838	0.1820
0.19	0.1889	0.9820	0.1923
0.20	0.1987	0.9801	0.2027
0.21	0.2085	0.9780	0.2131
0.22	0.2182	0.9759	0.2236
0.23	0.2280	0.9737	0.2341
0.24	0.2377	0.9713	0.2447
0.25	0.2474	0.9689	0.2553
0.26	0.2571	0.9664	0.2660
0.27	0.2667	0.9638	0.2768
0.28	0.2764	0.9611	0.2876
0.29	0.2860	0.9582	0.2984
0.30	0.2955	0.9553	0.3093
0.31	0.3051	0.9523	0.3203
0.32	0.3146	0.9492	0.3314
0.33	0.3240	0.9460	0.3425
0.34	0.3335	0.9428	0.3537
0.35	0.3429	0.9394	0.3650
0.36	0.3523	0.9359	0.3764
0.37	0.3616	0.9323	0.3879
0.38	0.3709	0.9287	0.3994
0.39	0.3802	0.9249	0.4111
0.40	0.3894	0.9211	0.4228
0.41	0.3986	0.9171	0.4346
0.42	0.4078	0.9131	0.4466
0.43	0.4169	0.9090	0.4586
0.44	0.4259	0.9048	0.4708
0.45	0.4350	0.9004	0.4831
0.46	0.4439	0.8961	0.4954
0.47	0.4529	0.8916	0.5080
0.48	0.4618	0.8870	0.5206
0.49	0.4706	0.8823	0.5334
0.50	0.4794	0.8776	0.5463
0.51	0.4882	0.8727	0.5594
0.52	0.4969	0.8678	0.5726
0.53	0.5055	0.8628	0.5859
0.54	0.5141	0.8577	0.5994
0.55	0.5227	0.8525	0.6131

TABLE C
Trigonometric Functions
x in radians

x (radians)	Sin(x)	Cos(x)	Tan(x)	x (radians)	Sin(x)	Cos(x)	Tan(x)
0.56	0.5312	0.8473	0.6269	1.11	0.8957	0.4447	2.0143
0.57	0.5396	0.8419	0.6410	1.12	0.9001	0.4357	2.0660
0.58	0.5480	0.8365	0.6552	1.13	0.9044	0.4267	2.1198
0.59	0.5564	0.8309	0.6696	1.14	0.9086	0.4176	2.1759
0.60	0.5646	0.8253	0.6841	1.15	0.9128	0.4085	2.2345
0.61	0.5729	0.8196	0.6989	1.16	0.9168	0.3993	2.2958
0.62	0.5810	0.8139	0.7139	1.17	0.9208	0.3902	2.3600
0.63	0.5891	0.8080	0.7291	1.18	0.9246	0.3809	2.4273
0.64	0.5972	0.8021	0.7445	1.19	0.9284	0.3717	2.4979
0.65	0.6052	0.7961	0.7602	1.20	0.9320	0.3624	2.5722
0.66	0.6131	0.7900	0.7761	1.21	0.9356	0.3530	2.6503
0.67	0.6210	0.7838	0.7923	1.22	0.9391	0.3436	2.7328
0.68	0.6288	0.7776	0.8087	1.23	0.9425	0.3342	2.8198
0.69	0.6365	0.7712	0.8253	1.24	0.9458	0.3248	2.9119
0.70	0.6442	0.7648	0.8423	1.25	0.9490	0.3153	3.0096
0.71	0.6518	0.7584	0.8595	1.26	0.9521	0.3058	3.1133
0.72	0.6594	0.7518	0.8771	1.27	0.9551	0.2963	3.2236
0.73	0.6669	0.7452	0.8949	1.28	0.9580	0.2867	3.3413
0.74	0.6743	0.7385	0.9131	1.29	0.9608	0.2771	3.4672
0.75	0.6816	0.7317	0.9316	1.30	0.9636	0.2675	3.6021
0.76	0.6889	0.7248	0.9505	1.31	0.9662	0.2579	3.7471
0.77	0.6961	0.7179	0.9697	1.32	0.9687	0.2482	3.9033
0.78	0.7033	0.7109	0.9893	1.33	0.9711	0.2385	4.0723
0.79	0.7104	0.7038	1.0092	1.34	0.9735	0.2288	4.2556
0.80	0.7174	0.6967	1.0296	1.35	0.9757	0.2190	4.4552
0.81	0.7243	0.6895	1.0505	1.36	0.9779	0.2092	4.6734
0.82	0.7311	0.6822	1.0717	1.37	0.9799	0.1994	4.9131
0.83	0.7379	0.6749	1.0934	1.38	0.9819	0.1896	5.1774
0.84	0.7446	0.6675	1.1156	1.39	0.9837	0.1798	5.4707
0.85	0.7513	0.6600	1.1383	1.40	0.9854	0.1700	5.7979
0.86	0.7578	0.6524	1.1616	1.41	0.9871	0.1601	6.1654
0.87	0.7643	0.6448	1.1853	1.42	0.9887	0.1502	6.5811
0.88	0.7707	0.6372	1.2097	1.43	0.9901	0.1403	7.0555
0.89	0.7771	0.6294	1.2346	1.44	0.9915	0.1304	7.6018
0.90	0.7833	0.6216	1.2602	1.45	0.9927	0.1205	8.2381
0.91	0.7895	0.6137	1.2864	1.46	0.9939	0.1106	8.9886
0.92	0.7956	0.6058	1.3133	1.47	0.9949	0.1006	9.8874
0.93	0.8076	0.5978	1.3409	1.48	0.9959	0.0907	10.9834
0.94	0.8076	0.5898	1.3692	1.49	0.9967	0.0807	12.3499
0.95	0.8134	0.5817	1.3984	1.50	0.9975	0.0707	14.1014
0.96	0.8192	0.5735	1.4284	1.51	0.9982	0.0608	16.4281
0.97	0.8249	0.5653	1.4592	1.52	0.9987	0.0508	19.6695
0.98	0.8305	0.5570	1.4910	1.53	0.9992	0.0408	24.4984
0.99	0.8360	0.5487	1.5237	1.54	0.9995	0.0308	32.4611
1.00	0.8415	0.5403	1.5574	1.55	0.9998	0.0208	48.0785
1.01	0.8468	0.5319	1.5922	1.56	0.9999	0.0108	92.6205
1.02	0.8521	0.5234	1.6281	1.57	1.0000	0.0008	1255.77
1.03	0.8573	0.5148	1.6652	$\pi/2$	1.0000	0.0000	∞
1.04	0.8624	0.5062	1.7036				
1.05	0.8674	0.4976	1.7433				
1.06	0.8724	0.4889	1.7844				
1.07	0.8772	0.4801	1.8270				
1.08	0.8820	0.4713	1.8712				
1.09	0.8866	0.4625	1.9171				
1.10	0.8912	0.4536	1.9648				

Answers to Selected Exercises

Exercises 1.2

1. smallest $= -5$; largest $= 6$
3. smallest $= -3\frac{1}{2}$; largest $= 4$
5. smallest $= -3.25$; largest $= 3$
7. smallest $= -2$; largest $= 2$
9. smallest $= -2\pi$; largest $= 2\pi + 1$
11. $2.01 < 2.05 < 2.1$
13. $0.02 < 0.08 < 0.17$
15. $0.12 < 0.13 < \pi - 3$
17. 0
19. 3
21. -1
29. 0.375
31. $0.\overline{428571}$
37. $-2.\overline{09}$
39. $135/90$
41. $248/9$
49. rational
51. irrational
57. $1.213 + (0.0001)\pi$
59. No; if $x = \sqrt{2}$ and $y = -\sqrt{2}$, then $x + y = 0$, a rational number

Exercises 1.3

1.

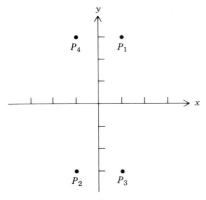

3.

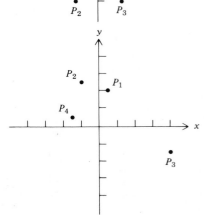

11. slope $= \frac{4}{3}$; length $= 5$

13. slope $= -\frac{2}{3}$; length $= \sqrt{13}$

15. slope undefined; length $= 4$

17. slope $= \frac{2}{5}$; length $= 2\sqrt{29}$

21. a. $|P_2P_3| = \sqrt{(-6+5)^2 + (5-1)^2}$
 $= \sqrt{1 + 16} = \sqrt{17}$

 $|P_3P_1| = \sqrt{(-2+6)^2 + (4-5)^2}$
 $= \sqrt{16 + 1} = \sqrt{17}$

 b. $|P_1P_2| = \sqrt{(-1+3)^2 + (0-0)^2}$
 $= \sqrt{4} = 2$

 $|P_2P_3| = \sqrt{(-2+1)^2 + (\sqrt{3}-0)^2}$
 $= \sqrt{1 + 3} = 2$

 $|P_3P_1| = \sqrt{(-3+2)^2 + (0-\sqrt{3})^2}$
 $= \sqrt{1 + 3} = 2$

23. highest at 30 min after admission; after 45 min, pulse rate $= 120$ beats/min; $P_4 = (45,120)$; admission pulse rate $= 120$ beats/min

25. After being in the box $1\frac{1}{2}$ hr, the rat began to learn.

27. After 24 hr in the hospital, the patient began to recover.

Exercises 1.4

1. $-2, 10, -6, 4$

3. $3, 11, 9, 5, -1$

5. $-5, -6, -4, -7, -2$

7. $13, 4, 8, 53$

9. $2, 0, 3, 5, 10$

11. $0, \frac{4}{3}, \frac{2}{3}, \frac{5}{6}, \frac{7}{6}$

13. $0, \frac{3}{11}, -\frac{3}{11}, \frac{10}{102}, -\frac{5}{27}$

15. $0, 25, 9, 4, 16$

17. $f(x) = 2x$; domain: all real x

19. $f(x) = x - 2$; domain: all real x

21. $f(x) = x^2$; domain: all real x

23. $f(x) = 2x + 1$; domain: all real x

25. $f(x) = 1/x^2$; domain: all real x except 0

27.

Integers x, $0 \leqslant x \leqslant 100$	x	10	50	75	100
$S(x)$	$3000 + 100x$	4000	8000	10,500	13,000

$S(0) = 3000$

29. the following Tuesday

33. $S(x) = 0.1x + 125$
good week: $249

35. $P(t) = 10t + 40$
$P(4.5) = 85$

Exercises 2.1

1. $f(1) = 3; f(2) = 6; f(3) = 9; f(4) = 12; f(5) = 15$

3. $f(-4) = 12; f(-2) = 2; f(0) = 0; f(2) = 6; f(3) = 20$

5. $f(-2) = -4; f(0) = 2; f(1) = 5; f(2) = 8; f(3) = 11$

7. $f(-4) = 7; f(-2) = 6; f(0) = 5; f(2) = 4; f(4) = 3$

9. $f(-3) = 3; f(-1) = 1; f(0) = 0; f(1) = 1; f(3) = 3$

11. $f(-3) = -7; f(-1) = -3; f(0) = -1; f(1) = 1; f(3) = 5$

13. $f(-2) = 7; f(-1) = 4; f(0) = 3; f(1) = 4; f(2) = 7$

15. $f(-3) = 5; f(-1) = 3; f(0) = 2; f(1) = 3; f(3) = 5$

17. $t(-3) = 1; t(-2) = 0; t(-1) = 1; t(0) = 2; t(1) = 3; t(3) = 5$

19. $t(-3) = 5; t(-1) = 3; t(0) = 2; t(1) = 3; t(3) = 5$

21. $S(160) = 182$

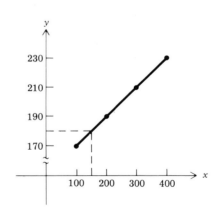

23. 10,000

25. 4.2, 6.6

27. 620,500

29. 40, 60

31. twice as fast; 2.3

33. 160 mm; no, because only months are in the domain

Exercises 2.2

1.

x	1	2	3	4	5	6
$(f + g)(x)$	-13	3	0	-8	8	3
$(f - g)(x)$	-7	1	8	-4	2	-1
$(fg)(x)$	30	2	-16	12	15	2
$(f/g)(x)$	$\frac{10}{3}$	2	-1	3	$\frac{5}{3}$	$\frac{1}{2}$

7.

all real x	x	-3	-1	0	2	4
$(f + g)(x)$	$x^2 + x + 3$	9	3	3	9	23
$(f - g)(x)$	$-x^2 + x + 1$	-11	-1	1	-1	-11
$(fg)(x)$	$(x + 2)(x^2 + 1)$	-10	2	2	20	102
$(f/g)(x)$	$(x + 2)/(x^2 + 1)$	$-\frac{1}{10}$	$\frac{1}{2}$	2	$\frac{4}{5}$	$\frac{6}{17}$

9. $(f + g)(x) = \sqrt{x - 2} + 7$; $(f - g)(x) = \sqrt{x - 2} - 7$; $(fg)(x) = 7\sqrt{x - 2}$; $(f/g)(x) = \sqrt{x - 2}/7$; domains: all real $x \geqslant 2$.

15. $(f + g)(x) = x^2 + 5x + 1 + \sqrt{x - 3}$; $(f - g)(x) = x^2 + 5x + 1 - \sqrt{x - 3}$; $(fg)(x) = (x^2 + 5x + 1)\sqrt{x - 3}$; $(f/g)(x) = (x^2 + 5x + 1)/\sqrt{x - 3}$; domain of $f + g$, $f - g$, and fg is all real numbers $\geqslant 3$; domain of f/g is all real numbers > 3.

17. $(f + g)(x) = x^2 + x - 12$; $(f - g)(x) = -x^2 + x + 6$; $(fg)(x) = (x - 3)(x^2 - 9)$; $(f/g)(x) = (x - 3)/(x^2 - 9)$; domain of $f + g$, $f - g$, and fg is all real numbers; domain of f/g is all real numbers $\neq \pm 3$.

19. $(f + g)(x) = |x + 1| + |x| + 1$; $(f - g)(x) = |x + 1| - |x| - 1$; $(fg)(x) = (|x + 1|)(|x| + 1)$; $(f/g)(x) = |x + 1|/(|x| + 1)$; domains: all real x.

21. $(f + g)(x) = 2\sqrt{x}$; $(f - g)(x) = 0$; $(fg)(x) = (\sqrt{x})^2$; $(f/g)(x) = 1$; domain of $f + g$, $f - g$, and fg is all real $x \geqslant 0$; domain of f/g is all real $x > 0$.

23.

all real x	x	-2	-1	0	1	2
$(f + g)(x)$	$x^2 + 2$	6	3	2	3	6
$(f - g)(x)$	$x^2 - 2$	2	-1	-2	-1	2
$f(x)$	x^2	4	1	0	1	4

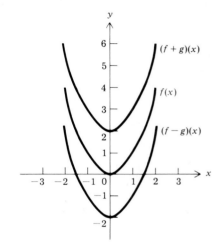

25. $2f(x) = 2x^3$; $(-1)f(x) = -x^3$

27. $(f + g)(x) = |x| + x$; $(f - g)(x) = |x| - x$

29. $S(x) = 0.79x$; $T(x) = 1.50 + 0.01x$; $(S + T)(x) = 1.50 + 0.80x$; $(S + T)(x) =$ total cost, delivered, of x lbs of seed; \$81.50

31. $(E + U + M)(x) = 1200 + 1.20x + x^2/100,000$; total daily cost to produce x units; \$1560.90

Exercises 2.3

1. $a = 1$, $b = 0$, $m = 1$

3. $a = -1$, $b = 0$, $m = -1$

5. $a = -6$, $b = -2$, $m = -6$

7. $a = 0$, $b = \frac{3}{2}$, $m = 0$

9. $a = -1$, $b = -\frac{1}{12}$, $m = -1$

11. $m = -2$

13. $m = -2$

15. $m = -3$

17. $m = 0$

19. $m = \frac{4}{5}$

21. $y = 2x - 6$

23. $y = 4x + 16$

25. $y = -7x + 38$

27. $y = -6x - 29\frac{1}{4}$

29. $y = -\frac{3}{4}x - \frac{3}{4}$

31. $y = -x + 17$

33. $y = -2x + 4$

35. $y = x + 4$

37. $y = -\frac{4}{31}x - 2\frac{3}{31}$

39. $y = -\frac{44}{21}x + \frac{169}{21}$

43.

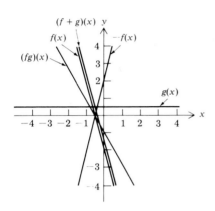

45.

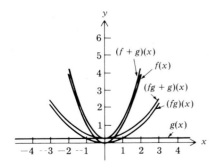

51. $C(x) = 12x + 1000;\ \$4600$

52. $P(t) = 5t + 30;\ 80$

57. $R(t) = 6t + 102;\ 156$ cubic ft per min

Exercises 2.4

1. $p(x) = x^3 + x^2$; degree = 3; leading coefficient = 1

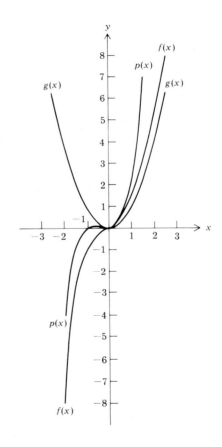

3. $p(x) = x^3 + x - 2$; degree = 3; leading coefficient = 1

5. $p(x) = 2x^7 + 2x^5 - 1$; degree = 7; leading coefficient = 2

7. $p(x) = 4x^4 + 4x^3 - 2$; degree = 4; leading coefficient = 4

9. $p(x) = 5x^6 + 5x + \frac{1}{2}$; degree = 6; leading coefficient = 5

11. $f(x)$: not polynomial; $g(x)$: polynomial, 1, $\sqrt{2}$; $h(x)$: polynomial, 0, 7.5; $q(x)$ polynomial, 1, $\frac{3}{2}$

13. $f(x)$: not polynomial; $g(x)$: polynomial, 1, $-\frac{1}{2}$; $h(x)$: polynomial, 7, -1; $q(x)$: not polynomial

15. $f(x)$: polynomial, 19, -7; $g(x)$: not polynomial; $h(x)$: not polynomial; $q(x)$: not polynomial

17. Vertex: $(\frac{3}{2}, -6\frac{1}{4})$; x intercepts: $-1, 4$

19. Vertex: $(1, 0)$; x intercept: 1

21. Vertex: $(\frac{1}{3}, 3\frac{2}{3})$; no x intercepts

23. Vertex: $(\frac{3}{8}, \frac{17}{16})$; x intercepts: $(3 \pm \sqrt{17})/8$

27. $P(x) = -x^2/1000 + 12x - 6000$; 6000 openers; \$30,000

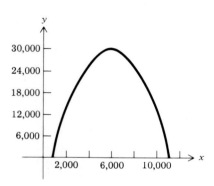

29. 30 min, 35

Exercises 2.5

1.

x	1	3	5	7	9
$(g \circ f)(x)$	60	200	-23	-4	0

5.

x	-3	-1	0	1	3
$(g \circ f)(x)$	2	2	2	2	2

7.

x	2	4	6	8	10
$(g \circ f)(x)$	0	0	0	0	0

11. $(g \circ f)(x) = 6x - 1$; domain: all real x; $(f \circ g)(x) = 6x + 2$; domain: all real x

13. $(g \circ f)(x) = x^3$; domain: all real x; $(f \circ g)(x) = x^3$; domain: all real x

15. $(g \circ f)(x) = x^2 - 2x$; domain: all real x; $(f \circ g)(x) = x^2 - 4x + 2$; domain: all real x

17. $(g \circ f)(x) = 4$; domain: all real x; $(f \circ g)(x) = -3$; domain: all real x

23. $(g \circ f)(x) = |x - 3|^2 + 1$; domain: all real x; $(f \circ g)(x) = |x^2 - 2|$; domain: all real x

25. $h(x) = (g \circ f)(x)$ where $f(x) = x^2 + 4$, $g(x) = \sqrt{x}$

27. $h(x) = (g \circ f)(x)$ where $f(x) = x^4 - 2x$, $g(x) = x^4 + 10$

29. $r(x) = (g \circ f)(x)$ where $f(x) = 2x^2 - 1$, $g(x) = 1/\sqrt{x}$

31. $w(t) = (g \circ f)(t)$ where $f(t) = t^2 + 1$, $g(t) = t^3 - 4/t$

33. $(g \circ f)(c) = g(f(c)) = c$; also $(f \circ g)(c) = f(g(c)) = f(c)$

35. $(D \circ p)(d) = 3[(65 - 2d)^2/400] + 4$; 26.7

37. $(S \circ C)(x) = 3.45x/1000) + 4.6 + (172.5/x)$; \$9.89

39. $(R \circ P)(t) = \frac{1}{2}(-t^2/600 + 13t/60 - 1)^2 + 2(-t^2/600 + 13t/60 - 1)$; 16

PART I   Self Test

The bold number in parentheses after each answer indicates the chapter and section from which the problem is taken.

1. 502/495; (**1.2**)

2. 2, -2; (**1.2**)

3. $\sqrt{5}$; (**1.3**)

4.

All real numbers x where $x > -1$	x	0	1	3	8
$f(x)$	$2/\sqrt{x+1}$	2	$\sqrt{2}$	1	$\frac{2}{3}$

(1.4)

5. $y = -\frac{3}{4}x + \frac{5}{4}$; **(2.3)**

6. $(f + g)(x) = x - 1$; all real x
$(f - g)(x) = 1$; all real x
$(fg)(x) = (\frac{1}{2}x)(\frac{1}{2}x - 1)$; all real x
$(f/g)(x) = (\frac{1}{2})/(\frac{1}{2}x - 1)$; all real x except $x = 2$
$(f \circ g)(x) = \frac{1}{4}x = \frac{1}{2}$; all real x
$(g \circ f)(x) = \frac{1}{4}x - 1$; all real x; **(2.2, 2.5)**

7. domain: all real $x \geqslant 0$

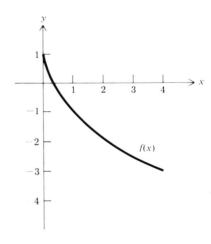

domain: all real x

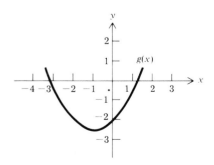

domain: all real $x \neq 0$

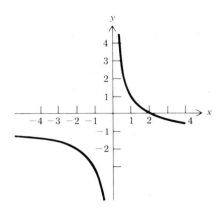

(2.1, 2.4)

8. -2, $(1, 1)$ $(0, 3)$; **(2.3)**

9. 1970

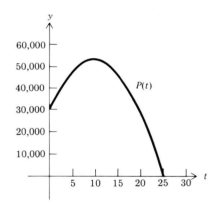

(2.4)

10. $C(x) = 0.14x + 2$

All real $x > 0$	x	50	100	200	500
$C(x)$	$0.14x + 2$	9	16	30	72

$89.50; **(1.4)**

11. $D(x) = -15/2x + 2225$; 1137; **(2.3)**

Exercises 3.2

1. $s(n) = (-1)^n$; $-1, 1, -1, 1, -1, 1$

3. $s(n) = 1 - (1/n)$; $0, \frac{1}{2}, \frac{2}{3}, \frac{3}{4}, \frac{4}{5}, \frac{5}{6}$

5. $s(n) = 3 + (-1)^n/n$; $2, \frac{7}{2}, \frac{8}{3}, \frac{13}{4}, \frac{14}{5}, \frac{19}{6}$

7. $s(n) = [1 + (-1)^n] + (1/n)$; $1, \frac{5}{2}, \frac{1}{3}, \frac{9}{4}, \frac{1}{5}, \frac{13}{6}$

9. $s(n) = (-1)^{n+1}(4/2^n)$; $2, -1, \frac{1}{2}, -\frac{1}{4}, \frac{1}{8}, -\frac{1}{16}$

11. $s(n) = 2$

12. $s(n) = n^2$

15. $s(n) = 1/n^2$

17. $s(n) = 1/2n$

19. $s(n) = (-1)^{n+1}[1/2(n + 1)]$

21. $\frac{3}{4}(15), \frac{9}{16}(15), \frac{27}{64}(15), \frac{81}{256}(15)$. This sequence has the property that $s(n) = (\frac{3}{4})s(n - 1)$ for $n > 1$, and to obtain $s(1)$ we multiply 15 by $\frac{3}{4}$. Thus to get from the first term of this sequence to $s(n)$ we must multiply 15 by $\frac{3}{4}$ n times.

23. 168,000; 176,400; 185,220; 194,481

25. 760,000; 722,000; 685,900; 651,605

Exercises 3.3

1. $2, 3, \frac{7}{2}, \frac{15}{4}$; 4

3. $2, \frac{8}{3}, \frac{26}{9}, \frac{80}{27}$; 3

5. $3, \frac{15}{4}, \frac{63}{16}, \frac{255}{64}$; 4

7. $8, \frac{48}{5}, \frac{248}{25}, \frac{1248}{125}$; 10

9. $\frac{2}{3}, \frac{8}{9}, \frac{26}{27}, \frac{80}{81}$; 1

11. $\frac{3}{2}, \frac{9}{4}, \frac{21}{8}, \frac{45}{16}$; 3

13. $s(n) = \frac{1}{3} + \frac{1}{(3)^2} + \cdots + \frac{1}{3^n} = \frac{1}{3}(1 + \frac{1}{3} + \cdots + \frac{1}{3^n})$ $= \frac{1}{2}[1 - (\frac{1}{3})^n]$; $\frac{1}{2}$ gal

15. $\frac{1}{3}$ million dollars

17. 32 ft

Exercises 3.4

1. $s(n) = \frac{1}{4}n$; $\frac{1}{4}, \frac{1}{8}, \frac{1}{12}, \frac{1}{16}, \frac{1}{20}, \frac{1}{24}$; 0

3. $s(n) = \frac{(-1)^n}{3n}$, $-\frac{1}{3}, \frac{1}{6}, -\frac{1}{9}, \frac{1}{12}, -\frac{1}{15}, \frac{1}{18}$; 0

5. $s(n) = 1 + \frac{1}{n^2}$, $2, \frac{5}{4}, \frac{10}{9}, \frac{17}{16}, \frac{26}{25}, \frac{37}{36}$; 1

7. $s(n) = 7.25 - (1)^n$, 6.25, 6.25, 6.25, 6.25, 6.25, 6.25; 6.25

9. $s(n) = 3 + (-1)^n$; 2, 4, 2, 4, 2, 4; no limit

11. $s(n) = -1 - \frac{(-1)^n}{n}$; $0, -\frac{3}{2}, -\frac{2}{3}, -\frac{5}{4}, -\frac{4}{5}, -\frac{7}{6}$; -1

13. no limit

15. 0

17. $\frac{2}{3}$

19. $s(n) = (\frac{4}{2}n + 1)$; 0

21. $s(n) = 2 - \frac{(1/n)}{2 + n}$; 0

23. $-\frac{1}{2}$

25. $\frac{3}{5}$

27. $\frac{1}{2}$

29. -5

31. $\frac{3}{7}$

33. $1, \frac{1}{16}, \frac{9}{9}, \frac{1}{36}, \frac{1}{25}, \frac{1}{64}$; 0

35. $2, 4, \frac{26}{9}, \frac{18}{5}, \frac{74}{25}, \frac{24}{7}$; 3

37. $-1, \frac{3}{2}, \frac{3}{14}, \frac{7}{4}, \frac{5}{21}, \frac{11}{6}$; no limit

39. 0; value of \$800,000 will dwindle toward 0

41. 4000

43. $s(n) = 60 - 50(\frac{5}{6})^{n-1}$; 60

Exercises 3.5

1. $s(n) = 2 + n$; 3, 4, 5, 6, 7, 8; $+\infty$

3. $s(n) = 2n + \frac{1}{n}$, 3, $4\frac{1}{2}$, $6\frac{1}{3}$, $8\frac{1}{4}$, $10\frac{1}{5}$, $12\frac{1}{6}$; $+\infty$

5. 4, $2\frac{1}{2}$, $3\frac{1}{3}$, $2\frac{3}{4}$, $3\frac{1}{5}$, $2\frac{5}{6}$; 3

7. 2, -4, 6, -8, 10, -12; no limit

9. 0, $\frac{7}{4}, \frac{26}{9}, \frac{63}{16}, \frac{124}{25}, \frac{215}{36}$; $+\infty$

11. -3, $-3\frac{1}{2}$, $-8\frac{1}{3}$, $-15\frac{3}{4}$, $-32\frac{1}{5}$, $-63\frac{5}{6}$; $-\infty$

13. -2, -2, -6, -6, -10, -10; $-\infty$

15. 1, -6, 5, -20, 27, -70; no limit

17. 2, $4\frac{1}{2}$, $7\frac{1}{3}$, $16\frac{3}{4}$, $31\frac{1}{5}$, $64\frac{5}{6}$; $+\infty$

19. $+\infty$

21. $+\infty$

23. $-\infty$

25. $-\infty$

27. $+\infty$

29. 7, 4, 19, 8, 31, 12; $+\infty$

31. 1, 1, -3, 5, -7, 9; no limit

33. $2\frac{1}{4}$, $4\frac{1}{4}$, $3\frac{5}{12}$, $4\frac{1}{16}$, $4\frac{9}{20}$, $4\frac{1}{36}$; no limit

35. $+\infty$; the value of the land increases indefinitely

Exercises 4.1

1. -2
3. -2
5. 3
7. 3
9. $\frac{3}{2}$
11. 3
13. -1
15. 1
17. -1
19. 0
21. 1
23. 3
25. 2
27. $3\frac{1}{2}$
29. $\frac{1}{5}$
31. 1
33. 2
35. $\dfrac{1}{\sqrt{7}}$
37. 1
39. If x is close to 2 but $\leqslant 2$, $f(x) = 1$; but if x is close to 2 but > 2, $f(x) = 0$. So $f(x)$ is sometimes close to 1 and sometimes close to 0 when x is close to 2, and consequently there is no limit at 2.
41. similar to 39
43. 1800
45. 300
47. similar to 39
49. 105

Exercises 4.2

1. $-\infty$
3. -9
5. $-\infty$
7. 4
9. 3
11. $3\frac{1}{2}$
13. $-\infty$

15. 0
17. -4
19. $-\infty$
21. $+\infty$
23. $+\infty$
25. 1
27.

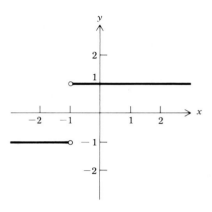

29. 0; the patient eventually forgets almost everything learned
31. $+\infty$; as the price approaches 0, virtually an unlimited amount of lawn food can be sold
33. $\lim_{x\to 0} C(x) = 120$; $\lim_{x\to +\infty} C(x) = +\infty$

Exercises 4.3

1. $(\lim_{x\to 2} 2)(\lim_{x\to 2} x) = 2.2 = 4$
3. $(\lim_{x\to 9} \frac{1}{3})(\lim_{x\to 9} x) = \frac{1}{3} \cdot 9 = 3$
9. $(\lim_{x\to -4} \frac{1}{4})(\lim_{x\to -4} x) - (\lim_{x\to -4} 2) = -3$
11. $(\lim_{x\to -3} \frac{1}{3})(\lim_{x\to -3} x) + (\lim_{x\to -3} \frac{3}{2}) = -\frac{5}{2}$
13. $(\lim_{x\to -3} x)(\lim_{x\to -3} x) + (\lim_{x\to -3} 2) = 11$
17. $(\lim_{x\to 2} \frac{-1}{2})(\lim_{x\to 2} x) - (\lim_{x\to 2} 2\frac{1}{2}) = -2\frac{1}{2}$
19. $[(\lim_{x\to 2} 2)/(\lim_{x\to 2} x)] - (\lim_{x\to 2} 1) = 0$
23. $[(\lim_{x\to 2} -3)/2(\lim_{x\to 2} x)]$
    $\quad - \lim_{x\to 2} 2\frac{1}{2} = -\frac{5}{4}$
25. $4(\lim_{x\to -2} x)(\lim_{x\to -2} x)(\lim_{x\to -2} x)$
    $\quad + 2(\lim_{x\to -2} x) - (\lim_{x\to -2} 1) = -37$
29. $[(\lim_{x\to 3} 5)/(\lim_{x\to 3} x)(\lim_{x\to 3} x)]$
    $\quad + 2(\lim_{x\to 3} x) - (\lim_{x\to 3} 1) = \frac{50}{9}$

31. $[(\lim_{x\to-2} - 4)/(\lim_{x\to-2}x)(\lim_{x\to-2}x)(\lim_{x\to-2}x)]$
    $+ [(\lim_{x\to-2}5)/(\lim_{x\to-2}x)(\lim_{x\to-2}x)]$
    $+ [(\lim_{x\to-2}1)/(\lim_{x\to-2}x)] = \frac{5}{4}$

33. $[(\lim_{x\to-1}3)/2(\lim_{x\to-1}x)(\lim_{x\to-1}x)(\lim_{x\to-1}x)]$
    $- [(\lim_{x\to-1}4)/3(\lim_{x\to-1}x)]$
    $- 7(\lim_{x\to-1}x)(\lim_{x\to-1}) = -7\frac{1}{6}$

37. $[(-1)(\lim_{x\to1}x)(\lim_{x\to1}x)(\lim_{x\to1}x) + 2(\lim_{x\to1}x)$
    $- (\lim_{x\to1}1)]/[(\lim_{x\to1}3) - (\lim_{x\to1}x)(\lim_{x\to1}x)]$
    $= 0$

41. $\lim_{x\to0}(-4x^3 + 2x^2 - x)\lim_{x\to0}(x^3 + 2x) = 0$

43. $4(\lim_{x\to1}\sqrt{x}) + (\lim_{x\to1}2) = 6$

45. $[3(\lim_{x\to1}x)(\lim_{x\to1}x)]/(\lim_{x\to1}\sqrt{x}) = 3$

51. $-4(\lim_{x\to-\infty}\frac{1}{x})(\lim_{x\to-\infty}\frac{1}{x})(\lim_{x\to-\infty}\frac{1}{x}) + (\lim_{x\to-\infty}2) = 2$

53. $-3(\lim_{x\to+\infty}\frac{1}{x})(\lim_{x\to+\infty}\frac{1}{x})(\lim_{x\to+\infty}\frac{1}{x}) + (\lim_{x\to+\infty}2) = 2$

59. $\left[2\left(\lim_{x\to4}\sqrt{x}\right) + \left(\lim_{x\to4}1/\lim_{x\to4}\sqrt{x}\right)\right]/$
    $(\lim_{x\to4}|x + 1|) = \frac{9}{10}$

61. $\lim_{x\to4}(\sqrt{x} + \frac{1}{2})\lim_{x\to4}|x + 1| = \frac{25}{2}$

Exercises 4.4

1. polynomial, -1

5. polynomial, -1

9. polynomial, $(2\sqrt{2} + 3)/6$

11. rational, 1

17. rational, $(5 + 3\sqrt{3})9$

19. polynomial, $(32\sqrt{2} + 1)/4$

21. rational, $\frac{3}{14}$

23. rational, $(128 + \sqrt{7} - 16\sqrt{5})/16\sqrt{3}$

25. rational, $-45/(32 + 8\sqrt{3})$

31. yes

33. continuous, limit exists

35. not continuous, limit exists

37. not continuous, no limit

39. continuous, limit exists

41. continuous, 1

43. continuous, 2

45. continuous, $\frac{7}{4}$

55. not continuous, $+\infty$

57. $g(x) = -3x + 2, 2$

59. $g(x) = x + 3, 6$

65. $g(x) = -\frac{x^2}{2} + \frac{3x}{4} + 1, 1$

67. 180 mph

69. 6

71. not continuous at $x \le 50$, $x = 100$, $x = 300$

73. $\lim_{x\to a}a(f \cdot g)(x) = [\lim_{x\to a}f(x)][\lim_{x\to a}g(x)]$
    $= f(a) \cdot g(a) = (f \cdot g)(a)$

Exercises 4.5

1. 20.086

3. 12.182

13. 0.26236

15. 1.4351

17. $[\exp(6)]^2 = (403.429)^2$

19. $[\exp(-7)]^2 = (0.0009)^2$

21. $\exp(-10)\exp(-3.86) = (0.0001)(0.0211)$

23. $\exp(10)\exp(0.01) = (22,026.5)(1.0101)$

27. $2\ln(6) = 2(1.7918)$

29. $\ln(10) + \ln(5.46) = 2.3026 + 1.6974$

35. $\ln(100) + \ln(1.42) = 2\ln(10) + \ln(1.42) = 2(2.3026 + 0.3507)$

37. $1/20.086$

39. $(20.086)(148.41)$

41. $1/[(20.086)(148.41)]$

49. 5

51. 20

53. -12

55. 10

61. $\exp(x - y) = \exp(x + (-y)) = \exp(x)\exp(-y)$
    $= \exp(x)(1/\exp(y)) = \exp(x)/\exp(y)$

63. $2700.00

65. 13.862 yr

67. 132,461 bacteria

69. 56 toasters

71. $9321.96

PART II   Self Test

1. 1, 3, 1, 3, 1, 3; (**3.2**)

$s(1)$ $s(2)$
$s(3)$ $s(4)$
$s(5)$ $s(6)$

2. As x approaches 0 with $x\langle 0$, $g(x) = 0$, but as x approaches 0 with $x\rangle 0$, $g(x) = 1$; thus there is no one number that $g(x)$ is close to, when x is close to 0; (**4.1**)

3. e^{11}; (**4.5**)

4. 0; (**3.4**)

5. $-\infty$; (**4.2**)

6. e^{-3}; (**4.5**)

7. -4; (**4.4**)

8. 8, from the graph; (**4.4**)

9. $-\infty$; (**4.4**)

10. 2, from the graph; (**4.4**)

11. -3; (**4.4**)

12. $\ln(7)^2 e^7$; (**4.5**)

13. 1; (**4.4**)

14. $9e$; (**4.5**)

15. 45,000; 40,500; 36,450; 32,805; (**3.3**)

16. $\lim_{t\to\infty} P(t) = +\infty$; as time goes on the population of bacteria is unlimited; (**4.2**)

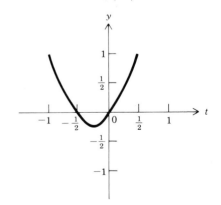

17. $33\frac{1}{3}$; (**4.4**)

18. $2,596.87$; (**4.5**)

Exercises 5.2

1. 5, 5

5. 1, 1

7. $\frac{1}{4}$, $\frac{1}{4}$

11. $-\frac{3}{5}$, $-\frac{3}{5}$

13. $4x^3$, 32

15. $3t^2$, $\frac{3}{4}$

19. $-3x^{-4}$, $-\frac{3}{16}$

21. $-4w^{-5}$, $-\frac{1}{8}$

25. $1/(2\sqrt{z}\,)$, $\frac{1}{6}$

27. $y = x/4 + 1$

29. $y = -4x - 4$

39. (2, 4)

41. (3, 27), (−3, −27)

43. (16, 4)

45. continuous, not differentiable

47. continuous, not differentiable

49. not continuous, not differentiable

51. $x_0 = 0$

53. $x_0 = 2$

Exercises 5.3

1. $\Delta x = 3$, $\Delta f = 6$

3. $\Delta x = 6$, $\Delta f = 24$

5. $\Delta x = 3$, $\Delta f = -9$

7. $\Delta x = 2$, $\Delta f = -2$

9. $\Delta x = 2$, $\Delta f = -\frac{8}{9}$

11. 4

13. 8

17. -1

19. $-\frac{5}{16}$

23. 7

29. $\frac{18}{49}$

31. 12

35. 0

37. 1

39. $-\frac{1}{8}$

43. $-\frac{3}{16}$

47. 3, 3

49. 30, 20

51. $-2, -2$

Exercises 5.4

1. 0

3. $-8x^{-5}$

5. $6x^2 + 1/(2\sqrt{x}\,)$

7. $12x^2 - 2$

11. $24x^{-5} + \frac{1}{2} - 1/(2\sqrt{x}\,)$

13. $\exp(x) - 4/x$

15. $6x^2\sqrt{x} + 6x^2 + x^3/\sqrt{x}$

21. $5x^3\sqrt{x} - x^4/(2\sqrt{x}\,)$

25. $(28x^7 - 29x^4 - 4x^3 + 6x - 1)/(2x^3 - 1)^2$

29. $-3x^2/5 + 1/(2\sqrt{x}\,) - \exp(x)\ln(x) - \exp(x)/x$

31. $-21t^3/(2\sqrt{t}\,) + 2/t$

35. $\dfrac{(-7/2)s^6 + (1/2)s^7}{\exp(s)}$

39. $\left[(r\ln(r) + r^3)\big[3\exp(r) + 1/(2\sqrt{r}\,)\big] \right.$
$\left. - \big(3\exp(r) + \sqrt{r}\,\big)(\ln(r) + 1 + 3r^2)\right] \Big/$
$(r\ln(r) + r^3)^2$

41. $D_x f(x) = 12x^2 + 4x - 3$
$D_x^2 f(x) = 24x + 4$
$D_x^3 f(x) = 24$

45. $D_x f(x) = 3x^{-1} + 4\exp(x)$
$D_x^2 f(x) = -3x^{-2} + 4\exp(x)$
$D_x^3 f(x) = 6x^{-3} + 4\exp(x)$

49. $D_w s(w) = \ln(w) + 1 - w^{-1}\exp(w) + w^{-2}\exp(w)$
$D_w^2 s(w) = w^{-1} + 2w^{-2}\exp(w) - w^{-1}\exp(w) - 2w^{-3}\exp(w)$
$D_w^3 s(w) = -w^{-2} - w^{-1}\exp(w) + 3w^{-2}\exp(w) - 6w^{-3}\exp(w) + 6w^{-4}\exp(w)$

51. $y = 8x - 17$

55. $y = x$

63. average marginal cost $= \frac{5}{2}$; marginal cost $= 3$

65. average rate of change $= \frac{5}{4}$; $1\frac{7}{8}$; 0

67. -4; $-90/(17.3026)^2 = -0.30$; P decreases as t increases

71. 150; $150 - 50\ln(2) = 115.34$

Exercises 6.1

1. $12(2x + 3)^5$

5. $3/(2\sqrt{3x + 1}\,)$

7. $-3/(x\sqrt{x}\,)$

15. $(1 - 14x + 3x^2)/(2\sqrt{x - 7x^2 + x^3}\,)$

17. $-\exp(-x)$

19. $4x\exp(2x^2 + 1)$

21. $(20x^4 + 6)/(4x^5 + 6x)$

25. $(9t^2 + 2)/(2\sqrt{3t^3 + 2t}\,)$

31. $(1/\sqrt{v} + 2v)\exp(2\sqrt{v} + v^2)$

33. $(1/s + s)/\sqrt{2\ln(s) + s^2}$

35. $2(\exp(v))^2 + 6v^2$

37. $5(3t^2 - 2)(t^3 - 2t)^4 - t/\sqrt{t^2 + 1}$

39. $-7(\sqrt{x} + 2)^{-2}/(2\sqrt{x}\,) - x/\sqrt{x^2 + 1}$

43. $(u/\sqrt{u^2 + 1}\,)\exp(2u + 3)$
$+ 2\sqrt{u^2 + 1}\,\exp(2u + 3)$

49. $4(x^6 + \exp(x))^3(6x^5 + \exp(x))/\ln(x^5)$
$- 5(x^6 + \exp(x))^4/x(\ln(x^5))^2$

51. $4t\exp(3t^4)/(2t^2 + 1) + 12t^3\ln(2t^2 + 1)\exp(3t^4)$

53. $5x(x^2 + 1)^4$

55. $t/(t^2 + 2)$

57. $2/(\sqrt{2t - 1}\,)^3$

59. $-12t/[\exp(2t^2 - 1)]^3$

61. $y = 10x - 1$

63. $y = 4x - 8$

65. $y = -8x + 2$

67. 190, 190.9

69. 218, 392

Exercises 6.2

1. relative minimum at $x = 0$; decreasing on $(-\infty, 0)$, increasing on $(0, \infty)$

3. relative minimum at $x = 0$; decreasing on $(-\infty, 0)$, increasing on $(0, \infty)$

7. relative minimum at $x = 1$; decreasing on $(-\infty, 1)$, increasing on $(1, \infty)$

11. relative minimum at $x = 1$; decreasing on $(-\infty, 1)$, increasing on $(1, \infty)$

15. relative minimum at $x = 0$, relative maximum at $x = -2$; increasing on $(-\infty, -2)$, decreasing on $(-2, 0)$, increasing on $(0, \infty)$

17. relative minima at $x = 3$ and $x = -3$, relative maximum at $x = 0$; decreasing on $(-\infty, -3)$ and $(0, 3)$, increasing on $(-3, 0)$ and $(3, \infty)$

19. relative maximum at $t = -1$, relative minimum at $t = 1$; increasing on $(-\infty, -2)$ and $(1, \infty)$, decreasing on $(-1, 1)$

21. relative maximum at $t = 0$; increasing on $(-\infty, 0)$, decreasing on $(0, \infty)$

25. relative minimum at $v = 0$; decreasing on $(-\infty, 0)$, increasing on $(0, \infty)$

29. relative minimum at $x = 1$; decreasing on $(0, 1)$, increasing on $(1, \infty)$

31. 1

33. $\pm 1/\sqrt{3}$

37. $\frac{1}{2}$

41. increasing for $a > 0$, decreasing for $a < 0$

43. $x = 20$, $y = 20$

45. 20 days; 40%

47. (a) 1800 chairs; (b) 0; (c) $12,000

49. 20 (thousand); $2000

53. length = 12, width = 12; $1488

57. $20.09; $10,055

59. 7500; $15.75

61. $P(x) = 1000(x^3 - 13.5x^2 + 60x - 10)$; relative maximum value at $x = 4$ (thousand) is $78,000; relative minimum value at $x = 5$ (thousand) is $77,500

15. relative minimum at $x = -1$; $f(-1) = -3$

17. relative maximum at $x = 2$; $g(2) = \frac{5}{3}$; relative minimum at $x = 4$; $g(4) = \frac{1}{3}$

19. relative minimum at $t = 2$; $h(2) = 3$

23. relative minimum at $t = 0$; $s(0) = 1$; relative maximum at $t = -4$; $s(-4) = 4.68$

25. increasing over $(-\infty, 0)$, decreasing over $(0, \infty)$; concave down over $(-\infty, \infty)$; relative maximum at $x = 0$; $f(0) = 7$

27. decreasing over $(-\infty, -4)$, increasing over $(-4, \infty)$; concave up over $(-\infty, \infty)$; relative minimum at $x = -4$; $f(-4) = 0$

33. decreasing over $(-\infty, -4)$, increasing over $(-4, -3)$, decreasing over $(-3, \infty)$; concave up over $(-\infty, -\frac{7}{2})$, concave down over $(-\frac{7}{2}, \infty)$; inflection point is $(-\frac{7}{2}, 14\frac{5}{12})$; relative minimum at $x = -4$; $f(-4) = \frac{43}{3}$; relative maximum at $x = -3$; $f(-3) = \frac{29}{2}$

35. decreasing over $(-\infty, -2)$, increasing over $(-2, 0)$, decreasing over $(0, 2)$, increasing over $(2, \infty)$; concave up over $(-\infty, -2/\sqrt{3})$ and $(2/\sqrt{3}, \infty)$, concave down over $(-2/\sqrt{3}, 2\sqrt{3})$; inflection points are $(-2/\sqrt{3}, -8\frac{8}{9})$ and $(2/\sqrt{3}, -8\frac{8}{9})$; relative minima at $x = -2$ and $x = 2$; $f(-2) = f(2) = -16$; relative maximum at $x = 0$; $f(0) = 0$

43. decreasing over $(-\infty, 0)$, increasing over $(0, \infty)$; concave down over $(-\infty, -1)$ and $(1, \infty)$, concave up over $(-1, 1)$; inflection points are $(-1, \ln(2))$ and $(1, \ln(2))$; relative minimum at $t = 0$; $z(0) = 0$

47. increasing over $(-\infty, 0)$, decreasing over $(0, \infty)$; concave down over $(-\infty, \infty)$; relative maximum at $x = 0$; $p(0) = -2$

51. absolute minimum at $x = 0$; $f(0) = -4$; absolute maximum at $x = 2$; $f(2) = 0$

55. absolute maximum at $x = 4$; $f(4) = 16$; absolute minimum at $x = 2$; $f(2) = -4$

57. absolute minimum at $x = -2$; $h(-2) = -20\frac{2}{3}$; absolute maximum at $x = 5$; $h(5) = \frac{8}{3}$

61. absolute minimum at $v = 0$; $p(0) = 4$; absolute maximum at $v = \sqrt{e^3 - e}$; $p(\sqrt{e^3 - e}) = 12$

65. (a) 12 (hundred)
(b) $18,000
(c) 15 (hundred)

67. 25; $75

69. width = 100 feet; length = 200 ft

Exercises 6.3

1. concave up over $(-\infty, \infty)$

5. concave down over $(-\infty, 0)$, concave up over $(0, \infty)$; inflection point is $(0, 1)$

9. concave down over $(-\infty, 1)$, concave up over $(1, \infty)$; inflection point is $(1, -1)$

11. concave up over $(-\infty, \infty)$

71. maximum profit at $x = 16$ (thousand); $p(16)$ = \$4000; minimum profit at $x = 18$ (thousand)

73. width = 3 feet, length = 6 feet, depth = 2 feet

77. minimum at $t = 10$ days; $N(10) = 70$; maximum at $t = 1$ day

79. maximum population in 1971 $(t = 8)$; $p(8) = 640$ (thousand); minimum population in 1964

83. a. $R(x) = 600x - 3x^2$
    b. $x = 100$
    c. \$30,000
    d. \$300

Exercises 6.4

5. $l(x) = 4x - 4$; $f(1.1) \approx l(1.1) = 0.4$

7. $l(x) = -x/2 + 2$; $f(2.04) \approx l(2.04) = 0.98$

11. $l(x) = -\frac{23}{36}(x + 3) - (\frac{164}{27})$; $f(-3.14) \approx l(-3.14)$ = -5.985

13. $l(x) = (5x + 87)/48$; $f(-3.02) \approx l(-3.02) = 1.498$

17. $l(x) = 4x + 1$; $f(-0.1) \approx l(-0.1) = 0.6$

21. 5.1

29. -0.86

31. 0.08

33. 0.93

37. $df(x, dx) = 2dx$; $df(3, \frac{1}{2}) = 1$; $f(x_0 + dx) \approx f(3) + df(3, \frac{1}{2}) = 10$

39. $df(x, dx) = (10x - 2)dx$; $df(2, -\frac{1}{6}) = -3$; $f(x_0 + dx) \approx f(2) + df(2, -\frac{1}{6}) = 16$

43. $df(x, dx) = (-16x^{-5} + 3/(2\sqrt{x}))\,dx$; $df(4, \frac{1}{2}) = \frac{47}{128}$; $f(x_0 + dx) \approx f(4) + df(4, \frac{1}{2}) = 6\frac{49}{128}$

49. $df(x, dx) = \left[\dfrac{\exp(x - 2)}{2\sqrt{x + 2}} + \sqrt{x + 2}\ \exp(x - 2) \right.$

    $\left. - 10x/(x^2 - 3) \right] df(2, \frac{1}{8}) = -17\frac{3}{4}$;

    $f(x_0 + dx) = f(2) + df(2, \frac{1}{8}) = -15\frac{3}{4}$

57. $P(x) = 4x^2 + 250x - 500$; $dP(x_0, dx) = 177.3$; \$5473.30

59. $dD(x_0, dx) = -11$; $D(31.65) \approx 307.91$

65. 0.18% of the syllables are forgotten; 69.82% are retained.

Exercises 6.5

1. $5x + C$

5. $-3x + C$

9. $7x^4/4 + x^2 - 3x + C$

13. $x^3/12 - x^2 - x^{-1} + C$

17. $-x^{-2} - 4x^{-1} + 5\sqrt{x} + C$

21. $x^2/2 - 3x + C$

25. $7 \exp(x) - x^{-4}/2 - x/6 + C$

27. $-\frac{1}{3}\exp(-6w) - 7 \ln(|w|) + 2w/3 + C$

29. $\frac{15}{16}\ln(|s|) - \frac{21}{20}\exp(2s/7) + 2s/3 + C$

31. $P_c(x) = -x^2/1200 + 30x + 500$

35. $V(t) = 10 - (t^2/90)$; yes; 30 minutes

39. $P(x) = 70 - 6\sqrt{x}$; $P(64) = 22\%$

PART III   Self Test

1. 3.08; (**6.4**)

2. a. $2x \exp(x) + x^2 \exp(x) - \frac{2}{x}$; (**5.4, 6.1**)

   b. $\dfrac{(6x - 10x^3 + 5)}{\exp(5x + 2)}$; (**5.4, 6.1**)

   c. $\dfrac{(4\sqrt{x} - 1)(x^3 - 7x)}{2/\sqrt{x}\,(2x - \sqrt{x}\,)} + (3x^2 - 7)[\ln(2x - \sqrt{x}\,)]$;

   (**5.4, 6.1**)

   d. $\dfrac{-3x^2 \exp(-x^3)}{2\sqrt{\exp(-x^3)}}$; (**6.1**)

3. $y = x - 1$; (**5.2**)

4. a. increasing over $(-\infty, -2)$ and $(-1, \infty)$, decreasing over $(-2, -1)$; concave up over $(-\frac{3}{2}, \infty)$, concave down over $(-\infty, -\frac{3}{2})$; point of inflection at $(-\frac{3}{2}, \frac{3}{2})$; relative maximum at $x = -2$; $f(-2) = 2$; relative minimum at $x = -1$; $f(-1)$ = 1; (**6.2, 6.3**)

   b. increasing over $(0, \infty)$, decreasing over $(-\infty, 0)$; concave up over $(-\infty, \frac{1}{3})$ and $(1, \infty)$, concave down over $(\frac{1}{3}, 1)$; points of inflection at $(\frac{1}{3}, \frac{65}{27})$ and $(1, 3)$; relative minimum at $x = 0$; $g(0) = 2$; (**6.2, 6.3**)

5. \$18; 6; \$972; (**6.3**)

6. $P(t) = -\frac{t^2}{400} + 0.5t - 1$; 8%; (**6.5**)

7. 26 dollars per unit; 32 dollars per unit; (**5.3**)

Exercises 7.2

1. 10
3. 9
7. 12
9. 12
15. $2\pi + 8$

Exercises 7.3

5. 6
7. 48
9. 9
11. $37\frac{1}{2}$
13. $2c$

Exercises 7.4

1. 10
3. -6
7. $2\frac{1}{4}, 9\frac{3}{4}, 12$
11. $8, 3\frac{1}{2}$
13. $18, 2\frac{1}{2}, 20\frac{1}{2}$
15. $2, \frac{1}{2}, 2\frac{1}{2}$

Exercises 7.5

1. 3
3. 30
7. 2
11. $\frac{15}{16}$
15. 78
17. 5
21. $\dfrac{(5 - e^6)}{3}$
23. $\frac{1}{10}$
25. $\ln(3) - \frac{3}{2}\left(\exp(2) - \exp\left(\frac{2}{3}\right)\right)$
27. 3
29. $\frac{3}{4}\left(\exp\left(\frac{8}{3}\right) - \exp\left(\frac{4}{3}\right)\right) - 255 - \frac{3}{4}\ln(2)$

33. 54
35. $2\ln(2) + 1$
39. $\frac{19}{2}$
41. $F(x) = x^2 + x + 1$
43. $F(x) = x^3 - x^2 + x + 2$
47. $F(x) = \frac{1}{2}\exp(2x) + \frac{x^3}{3} + \frac{1}{2}$
49. $F(x) = \sqrt{x} + \exp(2x) + 1$
53. 6
55. $\dfrac{(\exp(4) - 1)}{4} - \frac{4}{3}$
57. 64
59. 126 lbs
63. $1600 decrease
65. decrease of 1868.32 units
67. 4.73 miles; 4.10 miles; 14.67 miles
69. 7.22 cents; $P(x) = x + 2\ln(x) + 149$ cents; 1.66 dollars per gallon

Exercises 8.1

1. $10\frac{2}{3}$
3. $4\frac{1}{2}$
5. $\frac{38}{81}$
9. $\frac{32}{3}$
11. 8
15. 8
17. $P(t) = \frac{-0.02t^3}{3} + t^2 + 5t$; $100.59

Exercises 8.2

1. $\dfrac{(x + 2)^6}{6} + C$
3. $\dfrac{(x - 4)^{-2}}{(-2)} + C$
9. $\dfrac{e^{x^2}}{2} + C$
11. $\dfrac{(x^2 - 2x + 1)^8}{8} + C$
13. $\ln(|x^2 - 1|) + C$
17. $\dfrac{(4x^2 - 1)^2}{16} + C$

21. $\frac{1}{3}e^{x^3+6x} + C$

23. $\frac{1}{2}(\ln(3x))^2 + C$

27. $\frac{33}{5}$

29. $e - (\frac{1}{e})$

31. $323\frac{3}{4}$

33. $\ln(e^3 + 2) - \ln(3)$

39. 2542

Exercises 8.3

1. $\frac{xe^{4x}}{4} - \frac{e^{4x}}{16} + C$

3. $\frac{x(x-2)^8}{8} - \frac{(x-2)^9}{72} + C$

7. $-x^2e^{-x} - 2xe^{-x} - 2e^{-x} + C$

11. $x^3\ln(x) - \frac{x^3}{3} - x^2\ln(x) + \frac{x^2}{2} + C$

13. $\frac{2x(\sqrt{x+1})^3}{3} - \frac{4(\sqrt{x+1})^5}{15} + C$

19. $e - 2$

21. $\frac{(8e^3 - 4)}{9}$

23. $\frac{2(e^3 - 4)}{9}$

25. $6\frac{2}{3}$

Exercises 8.4

1. 1

3. 1

5. $\frac{7}{3}$

7. divergent

9. 0

11. divergent

PART IV   Self Test

1. $\frac{61}{3}$; (**8.2**)

2. $\frac{x(x+1)^{10}}{10} - \frac{(x+1)^{11}}{110} + C$; (**8.3**)

3. $36 + \frac{e^6}{2} - \frac{e^{-2}}{2}$; (**7.5**)

4. $4\sqrt{3} - \ln(3) - \frac{4}{3}$; (**7.5**)

5. $\sqrt{x^2 + 9}$; (**8.2**)

6. $2x^2\sqrt{x^2 + 9} - \frac{4}{3}(\sqrt{x^2 + 9})^3 + C$; (**8.3**)

7. $\frac{1}{3}$; (**7.5**)

8. 15 years; $1125; (**7.5, 8.1**)

9. $P(t) = -\frac{3t^2}{14} + 15t + 125$; (**7.5**)

10. $7200; (**7.5**)

Exercises 9.2

1. all pairs of real numbers (x, y); $f(1, -1) = -1$

3. all pairs of real numbers (x, y); $f(0, 1) = -1$

7. all pairs of real numbers (x, y) with $y > 0$; $f(-2, 1) = e^2$

13. all triples of real numbers (x, y, z) with $z > 1$; $f(4, 2, 2) = 4e$

15. all triples of real numbers (x, y, z) with $x \neq y$; $f(2, 6, 3) = -\frac{1}{2}\ln(2)$

27. x = number of skilled workers, y = number of unskilled workers; $W(x, y) = 11x + 5y$ dollars.

Exercises 9.3

5. $f_x(x, y) = 8x^3y - 16xy + \frac{1}{2\sqrt{x}}$; $f_x(4, 2) = 906\frac{1}{4}$; $f_y(x, y) = 2x^4 - 8x^2$; $f_y(4, 2) = 480$

7. $\frac{f_x(x, y) = y^3}{\left[\sqrt{x^2 + y^2}\,(x^2 + y^2)\right]}$; $f_x(2, 0) = 0$; $\frac{f_y(x, y) = x^3}{\left[\sqrt{x^2 + y^2}\,(x^2 + y^2)\right]}$; $f_y(2, 0) = 1$

9. $f_x(x, y) = \left[\frac{2x}{(x^2 + y)}\right] - 8y$; $f_x(-1, 2) = -16\frac{2}{3}$; $f_y(x, y) = \left[\frac{1}{(x^2 + y^2)}\right] - 8x$; $f_y(-1, 2) = 8\frac{1}{3}$

11. $f_x(x, y) = e^{y^2} - (\frac{3y}{x})$; $f_x(1, 2) = e^4 - 6$; $f_y(x, y) = xy^2e^{y^2} - \ln(x^3)$; $f_y(1, 2) = 4e^4$

13. $f_{xx}(x, y) = -8y^2$; $f_{xx}(1, -1) = -8$; $f_{yy}(x, y) = 6xy - 8x^2$; $f_{yy}(1, -1) = -14$; $f_{xy}(x, y) = f_{yx}(x, y) = 3y^2 - 16xy$; $f_{xy}(1, -1) = f_{yx}(1, -1) = 19$

19. $f_{xx}(x, y) = \left(\frac{ye^{xy}}{x^4}\right) - \left(\frac{2e^{xy}}{x^2}\right) + \left(\frac{6e^{xy}}{x^4y}\right); f_{xx}(1, 2) = -\frac{5e^2}{16};$

$f_{xy}(x, y) = \left(\frac{e^{xy}}{x}\right) - \left(\frac{2e^{xy}}{x^2y}\right) - \left(\frac{2e^{xy}}{x^3y^2}\right); f_{xy}(1, 2) = -\frac{e^2}{4};$

$f_{yy}(4, y) = x^2e^{x^2} - \left(\frac{xe^{xy}}{y}\right) - \left(\frac{e^{xy}}{y^2}\right); f_{yy}(1, 2) = \frac{e^2}{4}$

23. $f_x(x, y, z) = \dfrac{z}{\left(2\sqrt{xz + y^2}\right)} - 2xye^z; f_y(x, y, z)$

$= \left(\dfrac{y}{\sqrt{xz + y^2}}\right) - x^2e^z;$

$f_z(x, y, z) = \dfrac{x}{\left(2\sqrt{xz + y^2}\right)} - x^2ye^z$

29. $v_p(p, q, r, s) = (qse^{pqs} + rs)[\ln(p + q) - r^2s] + (e^{pqs} + rps)\left(\frac{1}{[p + q]}\right)$

$v_q(p, q, r, s) = pse^{pqs}[\ln(p + q) - r^2s] + (e^{pqs} + rps)\left(\frac{1}{[p + q]}\right)$

$v_r(p, q, r, s) = ps[\ln(p + q) - r^2s] + (e^{pqs} + rps)(-2rs)$

$v_s(p, q, r, s) = pqe^{pqs}[\ln(p + q) - r^2s] + (e^{pqs} + rps)(-r^2)$

31. $N_x(x, y) = -1000$, so $N_x(13, 1000) = -1000$; 1000 fewer dryers will be sold.

Exercises 9.4

1. relative minimum at $(0, 0)$; $f(0, 0) = -1$
3. relative minimum at $(1, 1)$; $f(1, 1) = 0$
9. relative maxima at $(0, \sqrt{2}), (0, -\sqrt{2}), f(0, \sqrt{2}) = f(0, -\sqrt{2}) = -4$
11. relative minimum at $(\frac{4}{3}, \frac{4}{3})$; no conclusion at $(0, 0)$; $f(\frac{4}{3}, \frac{4}{3}) = \frac{416}{3}$
15. relative minimum at $\left(\frac{1}{\sqrt{2}}, -1\right), \left(-\frac{1}{\sqrt{2}}, 1\right);$

saddle points at $\left(\frac{1}{\sqrt{2}}, 1, 3\sqrt{2} - 2\right),$

$\left(-\frac{1}{\sqrt{2}}, -1, -3\sqrt{2} + 2\right), f\left(\frac{1}{\sqrt{2}}, -1\right)$

$= 2 + 3\sqrt{2}, f\left(\frac{1}{\sqrt{2}}, 1\right) = 3\sqrt{2} - 2,$

$f\left(-\frac{1}{\sqrt{2}}, 1\right) = -3\sqrt{2} - 2, f\left(-\frac{1}{\sqrt{2}}, -1\right)$

$= -3\sqrt{2} + 2$

17. Minimum at $(40, 20)$; minimum cost is $20,000.
21. Maximum at $x = \$42.50$ and $y = \$35$; maximum revenue is $26,750.00.

Exercises 9.5

1. $(\frac{1}{2}, \frac{1}{2})$, minimum
3. $(5, 3)$, minimum
5. $x = y = 12$
7. $x = 1, y = 1, z = 1$

PART V   Self Test

1. all pairs of real numbers (x, y) with $x > y, f(2, 1) = 4$; (9.2)
2. all triples of real numbers (x, y, z) with $x \neq 0$ and $y > 0$; $f(3, \pi, 0) = 0$; (9.2)
3. $f_x(x, y) = 2xye^{x^2} - y^3, f_x(0, 1) = -1, f_y(x, y) = e^{x^2} - 3xy^2, f_y(0, 1) = 1$; (9.3)
4. $f_{xx}(x, y) = \left(\frac{6e^{2y}}{x^4}\right) + \left(\frac{1}{x^2}\right), f_{yy}(x, y) = \left(\frac{4e^{2y}}{x^2}\right) + \left(\frac{1}{y^2}\right), f_{xy}(x, y) = f_{yx}(x, y) = -\frac{4ye^{2y}}{x^3}$; (9.3)
5. maximum at $(0, 0, 1)$; (9.4)
6. maximum at $(\frac{1}{2}, 0, 4\frac{1}{4})$; (9.4)
7. maximum at $(-\frac{1}{2}, 4, -6)$; (9.4)
8. $x = \$33,333.33, y = \$26,666.66$ (9.4)

Index

A 6
B 7
C 8
D 9
E 0
F 1
G 2
H 3
I 4
J 5